Soil Biology

Volume 45

Series Editor

Ajit Varma, Amity Institute of Microbial Technology,
Amity University Uttar Pradesh, Noida, UP, India

More information about this series at
http://www.springer.com/series/5138

Lala Behari Sukla • Nilotpala Pradhan •
Sandeep Panda • Barada Kanta Mishra
Editors

Environmental Microbial Biotechnology

Springer

Editors
Lala Behari Sukla
Bioresources Engineering Department
CSIR - Institue of Minerals and
Materials Technology
Bhubaneswar
Odisha
India

Nilotpala Pradhan
Bioresources Engineering Department
CSIR - Institue of Minerals and
Materials Technology
Bhubaneswar
Odisha
India

Sandeep Panda
Bioresources Engineering Department
CSIR - Institue of Minerals and
Materials Technology
Bhubaneswar
Odisha
India

Barada Kanta Mishra
Mineral Processing Department
CSIR - Institute of Minerals and
Materials Technology
Bhubaneswar
Odisha
India

ISSN 1613-3382 ISSN 2196-4831 (electronic)
Soil Biology
ISBN 978-3-319-19017-4 ISBN 978-3-319-19018-1 (eBook)
DOI 10.1007/978-3-319-19018-1

Library of Congress Control Number: 2015944978

Springer Cham Heidelberg New York Dordrecht London

Printed on acid-free paper

Springer International Publishing AG Switzerland is part of Springer Science+Business Media (www.springer.com)

Preface

Rampant industrialization in the developing world is causing more and more nuisance by way of increasing global ecological devastation, and the irony is that there are not enough resources to bring the whole world up to the industrial level of the developed world. While we are increasingly becoming aware of climate change issues and the underlying causes, it is a blessing that nature has given the earth a vast variety of microorganisms which are capable of treating a variety of anthropogenic toxic chemicals, mining and industrial wastes, and other pollutants. These microscopic living beings can be deployed as such or genetically engineered for the cleaning of all the mess that man has created in the name of industrialization. The biodiversity of microorganisms is quite striking. Taking cue from the microbial metabolism, we must devise methods to clean our environment. Thanks to modern science as it evolves, we now have detailed genetic maps of hundreds of microbes that have spawned an explosion of new technologies. Looking ahead to the future, microbes will continue to dominate several areas of science and engineering and in particular it can be brought to bear on pollutants for a cleaner environment.

Understanding the biochemistry inherent to microbial biodiversity, and availability of techniques related to applied microbiology, is needed to provide a strong alternative to the conventional methodologies currently being adopted to solve environmental issues. The book "Environmental Microbial Biotechnology" explores and emphasizes on several aspects of microbial technologies to provide adequate information to researchers and industrialists to help resolve some of the major environmental issues.

The book provides updated information on applied microbiology and biochemistry of polluted ecosystem and strategies for environmental reclamation by use of bacteria, fungi, and algae. Some chapters give a detailed review on the microbial diversity and development of technology related to environmental sustainability. The book also gives detailed information on role of microorganisms in waste management, bioremediation, bioleaching, bio-hydrometallurgy, metal–microbe interaction for Fe and Mn phase transformation, biosensors, etc., related to ecological conservation. Few chapters discuss on valorization of different types of specific wastes, such as organic saline waste/wastewater, mining soil and mine area, and agriculture wastes, and the use of cotton gin waste for ethanol production for value addition and waste valorization. Microalgal technology for green fuel generation is also discussed with great hope and challenges.

We gratefully acknowledge the timely cooperation and support provided by contributing authors, without which this book would not have taken this shape. We are thankful to the series editors Prof. Ajit Varma and Dr. Jutta Lindenborn for encouragement and support.

Bhubaneswar, India

Lala Behari Sukla
Nilotpala Pradhan
Barada Kanta Mishra
Sandeep Panda

Contents

Chapter 1
The Role of Microbial Activity in Sulfide Oxidation at Dumping Sites of Sulfidic Wastes and in Abandoned Mining Areas

Irena Twardowska

1.1 Introduction

It is generally known that bacterial catalysis may accelerate process of sulfide oxidation by orders of magnitude. In the extractive industry, sulfide (mostly pyrite and pyrrhotite) oxidation is the most environmentally problematic process, being a source of acid rock drainage (ARD) generation that is toxic itself to the aquatic ecosystems due to intolerably low pH and high salinity caused by the release of sulfur in the form of sulfates. Besides, it causes mobilization of potentially toxic elements (PTEs) both from the rock matrix and from the soil during the migration of infiltrating precipitation water through the anthropogenic (extractive waste dump) and the natural vadose zone beneath the dump base. In the European perspective, the importance of this process is confirmed by the adoption of the EC Mining Directive (EC 2006), which in substantial part is focused on the environmental consequences and control of ARD. The Commission also formally adopted the BAT Reference Document (EC 2009) and gave a mandate to CEN—European Committee for Standardization—in order to develop the required sampling and analytical methods. This resulted in the publication of five standardized European documents. Three of them are focused on the susceptibility of rock leachate to acidification and on the assessment of the kinetics of this process (EN 15875 2011; CEN/TR 16363, 2012; CEN/TR 16376, 2012). In the development of this relatively recent European legislation, a rich knowledge base on ARD, by now comprising over 2,000 publications, the relevant legislation of the USA, Canada, and Australia was utilized. In other countries, where no specific regulations concerning sulfidic extractive waste exists, an interest to this kind of rock material is also high, for different reasons.

I. Twardowska (✉)
Institute of Environmental Engineering of the Polish Academy of Sciences, 34, M. Sklodowska-Curie St., 41-819 Zabrze, Poland
e-mail: irena@ipis.zabrze.pl

L.B. Sukla et al. (eds.), *Environmental Microbial Biotechnology*, Soil Biology 45,
DOI 10.1007/978-3-319-19018-1_1

In general, from the environment-related legislative standpoint, sulfide oxidation kinetics has been evaluated as a resultant of all acid generation processes in the waste rock, while the assessment of susceptibility to ARD/AMD formation and metal release considers also the neutralization potential of the material. It is well known that sulfide oxidation is a complex process involving both abiotic and biotic reactions. The activity of microorganisms may greatly influence the kinetics of this process. Since the publication of Colmer and Hinkle in *Science* (1947) indicating the significant role of microorganisms in sulfide oxidation and fundamental works of Silverman and coworkers on the mechanism of bacterial sulfide oxidation (Silverman and Lundgren 1959; Silverman et al. 1963; Silverman and Ehrlich 1964; Silverman 1967) and of Singer (1970) that also appeared in *Science* (Singer and Stumm 1970), many studies have been carried out on the biochemistry of bacterial decomposition of sulfides, and factors influencing the kinetics of this process were identified. The acceleration of sulfide (mostly pyrite FeS_2) oxidation due to bacterial activity reported by different authors (e.g., Silverman et al. 1963; Lau et al. 1970; Singer and Stumm 1970; Lundgren et al. 1972; Morth et al. 1972; Le Roux 1974) ranged from two- to threefold (Silverman et al. 1963) to some six orders of magnitude (Lundgren et al. 1972), which reflects the effect of different external geochemical factors on the microbial ecology, population, and activity. This, in turn, results in different roles and significance of microorganism activity in ARD formation and trace metal release.

The ability of microorganisms to accelerate sulfide oxidation processes has been soon recognized as a potential to utilize it for "mining with microbes" (Le Roux 1969, 1974), "biomining" (Johnson 2014), or the most widely applied "bioleaching" (e.g., Watling et al. 2014a). A large number of studies conducted since then were focused on the intensification of this process by the selection of the most efficient microbial population and on the genetic modification of microbial strains to obtain their higher activity and bigger tolerance to the external conditions at the heaps or in bioreactors, as well as on ensuring the optimal conditions for bioleaching (e.g., Schippers et al. 2010; Watling et al. 2014a, b). It is obvious that the highest attention was paid to bioleaching of the most expensive and rare elements, difficult or unsafe to extract in a conventional way (e.g., gold, uranium) (e.g., Olson 1994; Fomchenko et al. 2010; Muravyov and Bulaev 2013), or metals susceptible to biooxidation in the sulfidic ores at certain geochemical conditions that ensure substantial output and cost efficiency (mostly copper, but also cobalt, nickel, and zinc biomining from low-grade ores) (Johnson 2014). However, environmental aspects of the role of microbial mediation of sulfide oxidation under the actual conditions at waste rock dumping sites remain largely unclear. There are a large number of studies and several review articles presenting mechanisms of microbial oxidation of sulfides in sulfidic ore or mine waste under different conditions, including bioleaching of dumps and heaps, published since the article of Colmer and Hinkle (1947), which enlarged our knowledge both on the microbial recovery of metals from solids and on the microbial activity as an important factor in the environmental impact of sulfidic wastes. The most recent reviews covering a wider synthesis of microbial biooxidation processes as an integral part of waste rock hydrogeology and geochemistry were presented by Blowes et al. (2014) and Amos et al. (2015). Nevertheless, the role of iron-oxidizing bacteria in the natural systems such as dumping sites of sulfidic waste from extractive industries or abandoned mines still raises controversies, also from the regulatory point of view. Elucidation of this issue is of a crucial importance for the correct assessment of life cycle pollution potential from sulfidic waste. In this chapter, an overview of a current state of the art in this field is presented, with particular regard to the different parameters and sulfidic waste type.

1.2 Sulfide Oxidation Processes and Their Microbial Acceleration

The most abundant in the nature, and consequently the most studied sulfidic mineral, is pyrite (FeS_2) although there are a large number of other metal (metalloid)-sulfide minerals such as MoS_2 (molybdenite), WS_2 (tungstenite), As_2S_3 (orpiment), As_4S_4 (realgar), $CuFeS_2$ (chalcopyrite), FeS (troilite), Fe_7S_8 (pyrrhotite), MnS_2 (hauerite), PbS (galena), or ZnS (sphalerite). Due to its abundance, pyrite is mostly responsible for the formation of ARD/AMD (acid rock drainage/acid mine drainage) at the sulfidic waste dumps and abandoned mines, while other metal (metalloid)-sulfide minerals exert more limited spatially but similar environmental impact in the areas of their concentration in ores in large quantities and extraction. More frequently, these minerals occur in the mined seams in accessory amounts, but are mobilized in the course of sulfide oxidation processes and transported with leachate or drainage to the ground and surface waters, resulting in their pollution and posing hazard to the aquatic and terrestrial environment and to human health.

With respect to susceptibility to proton attack, pyrite FeS_2 and also MoS_2 and WS_2 may be classified as acid insoluble, while other metal/metalloid sulfides are acid soluble, which was indicated for the first time by Singer and Stumm (1970). For these two groups of metal sulfides, Shippers, Sand, and coworkers (Schippers et al. 1996, 1999; Schippers and Sand 1999; Sand et al. 2001) proposed two different sulfide oxidation mechanisms: the thiosulfate mechanism (the thiosulfate pathway) and the polysulfide mechanism (the polysulfide pathway).

1.2.1 Insoluble Sulfide Oxidation (The Thiosulfate Pathway)

The process of insoluble sulfide oxidation can be exemplified in pyrite FeS_2 oxidation that was studied for decades by many researchers. The overall reaction and consecutive oxidation steps were also presented with some modifications in many publications. Here, a synthesis of the process is given after Appelo and Postma (2007) and Vera et al. (2013).

The overall reaction of pyrite oxidation involves both the oxidation of S_2^{2-} and Fe^{2+} and illustrates its strong generation of acid:

$$FeS_2 + 15/4\,O_2 + 7/2\,H_2O \rightarrow Fe(OH)_3 + 2\,SO_4^{2-} + 4\,H^+ \quad (1.1)$$

The different steps of pyrite oxidation are summarized in Fig. 1.1.

The initial step of FeS_2 reaction with O_2 occurs either directly (**a**) or through dissolution (**a′**):

$$FeS_2 + 7/2\,O_2 + H_2O \rightarrow Fe^{2+} + 2\,SO_4^{2-} + 2\,H^+ \quad (1.2)$$

It is followed by the oxidation of Fe(II) to Fe(III) (**b**):

$$Fe^{2+} + 1/4O_2 + H^+ \rightarrow Fe^{3+} + ½\,H_2O \quad (1.3)$$

Subsequently, Fe(II) precipitates as hydroxide (**d**) generating most of the acidity:

$$Fe^{3+} + 3\,H_2O \rightarrow Fe(OH)_3 + 3H^+ \quad (1.4)$$

Oxidation of FeS_2 by oxygen is a slow process. Another (thiosulfate) pathway of pyrite oxidation is by Fe(III) ion and can be described by the reactions (Vera et al. 2013):

$$FeS_2 + 6\,Fe^{3+} + 3\,H_2O \rightarrow S_2O_3^{2-} + 7\,Fe^{2+} + 6\,H^+ \quad (1.5)$$

$$S_2O_3^{2-} + 8\,Fe^{3+} + 5\,H_2O \rightarrow 2\,SO_4^{2-} + 8\,Fe^{2+} + 10\,H^+ \quad (1.6)$$

These reactions are fast, produce low pH, and can be described by the overall equation that reflects the step (**c**) in Fig. 1.1:

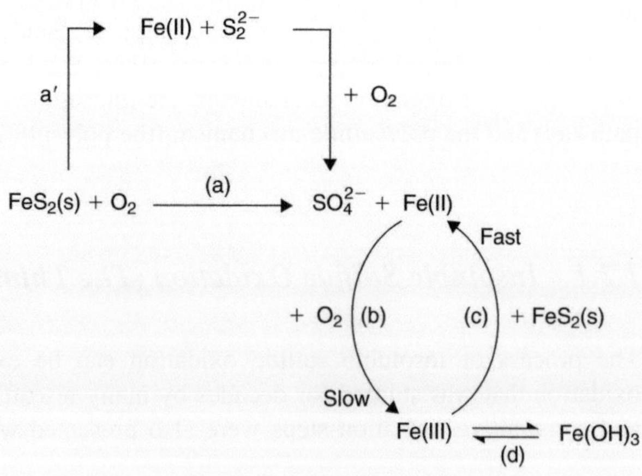

Fig. 1.1 Process of pyrite oxidation (after Stumm and Morgan 1996)

$$FeS_2 + 14\,Fe^{3+} + 8\,H_2O \rightarrow 2\,SO_4{}^{2-} + 15\,Fe^{2+} + 16\,H^+ \qquad (1.7)$$

Generated intermediate sulfur compounds may be oxidized to sulfate and Fe(II) to Fe(III) (**b**) in chemical and/or biological reactions.

1.2.2 Acid-Soluble Sulfide Oxidation (The Polysulfide Pathway)

While acid-insoluble sulfide (mainly pyrite) oxidation plays a major role in the environmental issues related to ARD generation at the sulfidic waste dumping sites and abandoned mines due to pyrite abundance in the environment, a large variety of metal/metalloid sulfides, as denoted above, belong to another large group of acid-soluble sulfides. Despite a high variety, they mostly occur in spatially limited areas and are of high interest predominantly to mining companies. For this reason, in contrast to pyrite, improving of metal extraction and not ARD and its control in leachate is of a primary focus. However, also in the case of acid-soluble sulfides, issues of ARD generation and metal enrichment in the aquatic and terrestrial environment must be controlled to fulfill the environmental requirements. This means that both facilitation and mitigation of ARD and metal mobilization in such sites must be equally considered.

The mechanism of acid-soluble metal sulfide oxidation involves oxidation of sulfur in metal sulfides mostly to elemental sulfur, with formation of polysulfides as intermediate sulfur compounds.

The polysulfide mechanism of sulfide oxidation can be described by the following reactions, where MS means "acid-soluble metal sulfides," M metal/metalloid, and H_2S_n various polysulfides specific for the reactions (Schippers and Sand 1999; Vera et al. 2013):

$$MS + Fe^{3+} + H^+ \rightarrow M^{2+} + 0.5\,H_2S_n + Fe^{2+} (n \geq 2) \qquad (1.8)$$

$$0.5\,H_2S_n + Fe^{3+} \rightarrow 0.125\,S_8 + Fe^{2+} + H^+ \qquad (1.9)$$

Besides, it was observed that elemental sulfur, which is chemically inert in the environment, can be oxidized to sulfuric acid under the conditions of bioleaching (Schippers and Sand 1999; Vera et al. 2013):

$$0.125\,S_8 + 1.5\,O_2 + H_2O \rightarrow SO_4{}^{2-} + 2\,H^+ \qquad (1.10)$$

This reaction is also a source of protons and, consequently, of ARD and metal mobilization. However, there is a basic question, how much optimized conditions of bioleaching differ from the conditions occurring at the dumping sites of wastes not intended for bioleaching. In contrast to that purpose, the major aim at such sites is ARD prevention and control; therefore, microbial activity that might accelerate ARD formation is highly undesirable there.

1.2.3 Role of Microorganisms

Comparison of both thiosulfate and polysulfate mechanisms of metal sulfide oxidation and the role of microbial species are demonstrated in Fig. 1.2.

In both pathways of sulfide oxidation, aerobic mesophilic or moderately thermophilic Fe(II)-oxidizing acidophilic microorganisms, mostly *Acidithiobacillus ferrooxidans/thiooxidans* and *Leptospirillum ferrooxidans*, take part in the oxidation of intermediate sulfur compounds that are products of chemical decomposition of the either acid-insoluble (A) or acid-soluble metal sulfides (B). Microorganisms oxidize Fe(II) to Fe(III) ions under acidic and oxic conditions. In turn, Fe(III) ions act as oxidants of metal sulfides and subsequently of intermediate sulfur compounds, i.e., thiosulfates and polysulfides. Besides, microbial catalysis may support the oxidation of intermediate sulfur compounds to sulfuric acid, which occurs mostly abiotically.

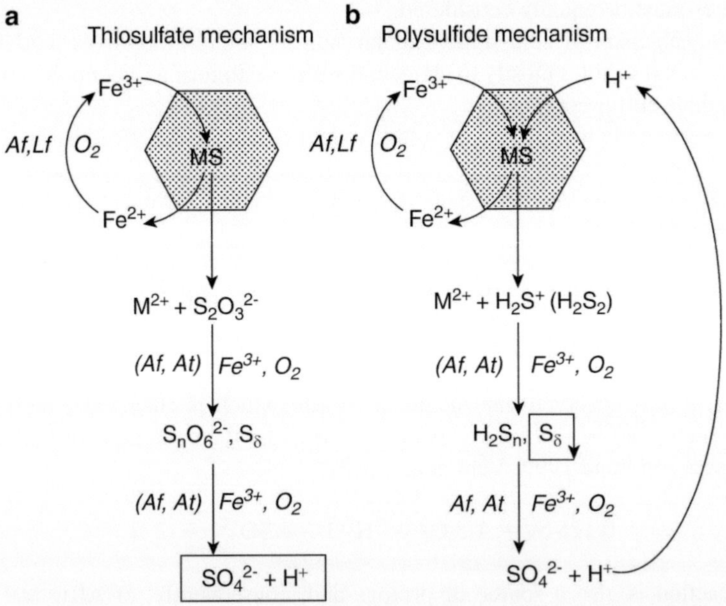

Fig. 1.2 Thiosulfate (**a**) and polysulfide (**b**) oxidation mechanisms of acid-insoluble and acid-soluble metal sulfides (by Schippers and Sand 1999 and Sand et al. 2001, modified, after Vera et al. 2013). *Af*, *Acidithiobacillus ferrooxidans*; *At*, *Acidithiobacillus thiooxidans*; *Lf*, *Leptospirillum ferrooxidans*; *In boxes, final reaction products are indicated*

The thiosulfate and polysulfate mechanisms are basically similar with respect to processes of Fe(II) recycling in the abiotic or bacteria-mediated way. However, in the case of the polysulfide pathway of acid-soluble metal sulfide oxidation (Fig.1.2b), dissociated sulfuric acid is not a final reaction product, but additionally, metal sulfides are attacked by protons.

As far as the natural, not mediated anthropogenically, systems are concerned, such as dumping sites of extractive sulfidic waste or abandoned mines, the mechanisms of Fe(II)/Fe(III) cycling are the same. This can be derived from the comparison of schematic pathways presented in Figs. 1.1 and 1.2, although worse conditions and higher limitations for the bacterial activity in such systems can be anticipated. Bacterial catalysis accelerates oxidation of acid-insoluble metal sulfides (pyrite) at low pH (<4) according to reactions (**c**) and (**b**) (Fig. 1.1). At higher pH, precipitation of $Fe(OH)_3$ causes reduction of Fe(III) in the system that suppresses microbial activity; hence, the rates and share of sulfide biooxidation in such systems, along with the variety and activity of bacterial strains, might be considerably lower.

1.3 Microbiology of Sulfidic Waste and ARD

1.3.1 Growth Environment of Microorganisms

Microorganisms taking part in metal sulfide decomposition are known to grow predominantly attached to the surface of sulfides. Attachment is mediated mostly by extracellular polymeric substances (EPS) surrounding the cells (Fig. 1.3), primarily through electrostatic forces between positively charged cells and negatively charged sulfide (pyrite) surface and through hydrophobic interaction (Sand et al. 1995; Vera et al. 2013).

It was found that microorganisms are not attached to metal sulfide surface randomly, but concentrate preferentially (>80 %) to the sites with surface imperfections and lower degree of crystallization. This leads to an assumption that "attachment to specific sites on the mineral surface is related to different attractants, most likely caused by charge imbalances on the surface as caused by e.g. oxidation processes" (Vera et al. 2013).

Although the majority of the cells are growing attached to the sulfide surfaces, thus forming a biofilm, some minor number of cells (<5 %) were observed to remain in a liquid in a planktonic state. Hallberg (2010) attributes the planktonic phase to the newly described heterotrophic acidophilic species *A. ferrivorans*.

A comprehensive review of a surface science related to the microbial cell attachment mechanisms, along with the biofilm formation and its role in sulfide oxidation processes, is presented by Vera et al. (2013).

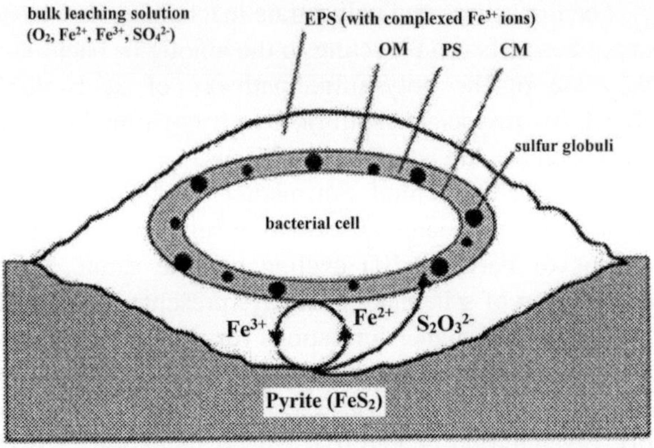

Fig. 1.3 Bacterial cell attachment to pyrite surface (Vera et al. 2013, after Sand et al. 1995, modified)

1.3.2 Microorganisms Detected in Mine Dumps and Heaps

Considering different climatic conditions and variable mineralogy, physical properties, geochemistry, and hydrogeology of sulfidic waste and the indubitable impact of these factors on the dump microbiology and microbial activity, an identification of microorganisms whose activity is essential for the rate of ARD generation is of particular importance. On the basis of a review of over 70 publications from the period of 1969–2009 conducted predominantly on mine tailings, but also on waste rock, mining-impacted sediments, leaching heaps and dumps, effluents and leaching solutions, and mining waste dumps, including own studies, Schippers et al. (2010) provided a list of identified microorganisms belonging to three domains, Bacteria, Archaea, and Eukarya. The data originated from different continents and several countries of diverse climatic zones (from semiarid to arctic), while the rock material represented mostly tailings from metal ores, predominantly Cu, Cu/Zn, Pb/Zn, Au, and U, and polymetallic ores, e.g., Zn/Ag/Au/Pb or Ni/Cu/Zn/Co, and rarely from coal mining. Heaps and dumps represented mainly metal biomining facilities and occasionally disposal sites.

The major metal sulfide-oxidizing microorganisms belong to extremely acidophilic ($pH < 3$) bacteria and archaea. The pyrite/Fe(II) and other metal sulfide- and inorganic sulfur-oxidizing microorganisms that belong to the domain Bacteria are represented by mesophilic or moderately thermophilic aerobes such as *Acidithiobacillus*, *Alicyclobacillus*, and *Sulfobacillus* genera. The Archaea of the same oxidation properties comprise extremely thermophilic *Acidianus*, *Metallosphaera*, and *Sulfolobus*. Some microorganisms oxidize pyrite/Fe(II) and other metal sulfides but do not oxidize sulfur, among them bacteria of *Leptospirillum* genus and archaea of *Ferroplasma* genus. Other bacterial genera have even more limited oxidation ability and oxidize either pyrite/Fe(II) (*Acidimicrobium*, *Gallionella*) or other metal sulfides and sulfur (*Thiobacillus*, *Thiomonas*) or sulfur only (*Acidiphilium*, *Pseudomonas*). Among Archaea, *Sulfurisphaera* genus is also a sulfur oxidizer only. In turn, Fe(III) and oxidized sulfur compounds under appropriate conditions are being reduced by numerous anaerobic bacterial genera that results in the release of metals and metal sulfides.

Eukarya present in the studied material did not show either oxidation or reduction properties. In general, more universal microorganisms of a wider adaptation properties and motility show better growth and higher activity under the changeable conditions of waste rock disposal sites. Also, dominance of a certain mining waste (or a group of wastes) in the environment determines predominance of the microbial community specific for this waste. These features appear to display Bacteria domain, in particular *Acidithiobacillus* bacterial genus. Within the *Acidithiobacillus* spp., the most widespread species appear to be Fe(II)- and sulfur-oxidizing *Proteobacteria Acidithiobacillus ferrooxidans* (*Af*, Fig. 1.2) and *Acidithiobacillus thiooxidans* (*At*, Fig. 1.2). These bacteria show high diversity and currently are classified in 23 strains and 7 subgroups (Vera et al. 2013), while a high number of indigenous strains are probably not yet classified.

Other bacteria that seem to play a significant, however lesser, role as metal sulfide oxidizers are *Leptospirillum* spp. (*Leptospirillum ferrooxidans—Lf*, Fig. 1.2) and to the least extent *Sulfobacillus* spp. In contrast to Bacteria, Archaea was detected in sulfidic waste in the accessory quantities (Kock and Schippers 2008; Schippers et al. 2010; Vera et al. 2013).

There is not much data on the abundance of microorganisms in mining waste dumps. Available data for waste rock and tailings are presented as maximum numbers that do not give a clear idea about the range and mean values that better characterize the actual activity of the microorganisms. However, these data indicate a very high growth of sulfide oxidizers and hence their high activity in both kinds of sulfidic waste dumps in seven countries, at the total level of 10^6–10^8 cells/g (dw). It is remarkable that the similar, comparable cell number of acidophilic chemolithoautotrophic S-oxidizers and Fe(II)-oxidizers and of neutrophilic S-oxidizers was detected. Cell number of heterotrophic acidophilic microorganisms was lower by 1–4 orders of magnitude, distinctly lower in tailing dumps (10^2–10^3 cells/g, dw) than in waste rock dumps (10^3–10^6 cells/g, dw) (Schippers et al. 2010, after Bosecker et al. 2004).

Apart from the indication of the similar high growth of acidophilic and neutrophilic sulfur-oxidizing bacteria in the heterogeneous environment within sulfidic waste dumps (which suggests occurrence within the dump both acidic and near-neutral conditions), in freshly deposited material, pH is always alkaline or near neutral and undergoes acidification in the course of time. Over some time, alkaline or near-neutral conditions prevail within the dump that influences the microbial composition. It should be underlined that in the studies on the microbiology of sulfidic mine waste, most experiments have been carried out in the strongly acidic ($pH < 3$) environment. However, the changes of microbial community composition over time during progressive acidification of sulfidic waste were reported already in the last two decades (e.g., Blowes et al. 1995; Elberling et al. 2000). In the recent researches, the growing attention has been paid to the biogeochemistry of weathering transformations of heterogeneous environments within sulfidic waste (e.g., Chen et al. 2013; Korehi et al. 2014; Liu et al. 2014; Watling et al. 2014a, b).

These studies are focused mostly on pH as the major parameter governing the development and composition of the microbial communities in different sulfidic wastes, such as pyrrhotite-containing waste rock from processing of polymetallic (Ni/Cu/Zn/Co) pyrrhotite-containing ore in Botswana (Schippers et al. 2007; Korehi et al. 2014); pyrite-containing Pb/Zn (Huang et al. 2011) and Cu mine tailings in China (Liu et al. 2014); pyritic mine tailing dumps in Sweden and Germany, the latter containing pyrite, arsenopyrite, sphalerite, and galena (Korehi et al. 2014); or polymetallic ores (Co, Ni, Cu, Zn) in Australia (Watling et al. 2014a, b). In some cases, such factors as temperature (Watling et al. 2014a, b), Fe(II) and Fe(III) contents, and Eh were also considered (Liu et al. 2014). Due to heterogeneity and differences in geoenvironments, substantial differences in the composition of microbial communities were observed between the studied objects. However, significant common features also could be noticed. The microbial communities were found to change dramatically in the course of progressive acidification. The highest microbial diversity was observed in the uppermost, moderately acidic waste layers (pH 6.4–6.5). *Proteobacteria*, *Firmicutes*, and *Actinobacteria* were found to be most abundant in these wastes. The microbial diversity decreased with decreasing pH. The microbial community in the deeper more acidic layers (at pH 3.2–5.3) was represented mostly by *Firmicutes* phylum (Korehi et al. 2014). The similar pattern of changes in microbial community during progressive sulfide oxidation and pH decrease was observed also in copper mine tailing, at simultaneous increasing abundance of *Euryarchaeota* and *Firmicutes* and decreasing occurrence of *Actinobacteria* and *Proteobacteria* phyla (Liu et al. 2014). In turn, at pH <3, a significant decrease of *Actinobacteria*, *Proteobacteria*, and *Firmicutes* and domination of *Euryarchaeota* and *Nitrospirae* occurred (Chen et al. 2014). It appears that at the early stages of pyrite weathering (pH $>$ 5), spore-forming and sulfur-oxidizing genera such as *Tumebacillus* and *Thiobacillus* prevail in a large variety of other microorganisms. With the progress of acidification ($5 > \text{pH} > 3$), these genera become replaced at the dominating position by *Alicyclobacillus* that shows relatively high pH tolerance (in the range pH 2–6) and is able to oxidize, besides sulfur, also pyrite and other metal sulfides. At the final stage (pH $<$ 3), strongly acidophilic Fe(II)-oxidizing genera become prevalent: *Ferroplasma*, *Leptospirillum*, *Sulfobacillus*, *Alicyclobacillus*, and *Acidithiobacillus*; thus a deep change of the microbial community occurs (Chen et al. 2014; Watling et al. 2014a, b). Hence, following the transformation of the composition of microbial communities at the different stages of sulfidic waste oxidation, also different oxidation rates and oxidation kinetics should be anticipated, along with the acceleration of the process at the final, strongly acidic stage. On the one hand, this trend is favorable for biomining; on the other hand, it results in ARD/AMD generation and metal release that poses hazard to the environment.

1.3.3 Microbiology of Acid Mine Drainage

Many studies are related to the acid rock/acid mine drainage (ARD or AMD) that (as it has been already pointed out) is a worldwide environmental problem associated with mining (e.g., Sima et al. 2011; Sahoo et al. 2014; Hindar and Nordstrom 2014). As ARD/AMD is generated in sulfidic mine waste dumps and in the abandoned mines where seams are exposed to air and water, the processes of sulfide oxidation continuously proceed. The geochemistry of AMD has been discussed in many publications (e.g., Hindar and Nordstrom 2014; Pinto et al. 2014; Sahoo et al. 2014; Simate and Ndlovu 2014); of these, the chapter by Blowes et al. (2014) summarizes the state of the art in this field.

It is obvious that along with dissolved substances, which are mostly products of iron sulfide oxidation, i.e., iron Fe(II), sulfates SO_4^{2-}, and protons H^+, determining ARD/AMD acidity and metal loads from leached rock materials, also microbial cells are being leached. Therefore, the more abundant is the microflora within the waste rock or sulfidic rock seam, the higher number of cells can be leached out. Apart from a relatively small number of microorganisms (<5 % of total, according to Vera et al. 2013), a substantially higher cell number washed out with ARD/AMD from the biofilm attached to the metal sulfide surface is a good indicator of the total number of cells in the rock material. From this standpoint, ARD/AMD is a good source of indirect information about the microbial activity within a waste dump or abandoned mine. A comparison of AMD from six different mining sites located in the UK, Norway, Spain, and the USA showed that concentrations of Fe_t, Fe(II), and sulfate significantly grow in parallel with decreasing pH (in the range from pH 6.2 to pH 0.5–1) (Hallberg 2010). In the percolation stage, free acidity in ARD was found to play an important role in the release of acid-soluble metals from ore (Watling et al. 2014a, b). Concentrations of metals other than iron (Zn, Cu) in AMD also increased, although in this case, the metal enrichment in a primary rock apparently played no less important role than pH. Therefore, attributing by the author (Hallberg 2010) of metal enrichment in AMD entirely to pH is wrong. AMD from coal mine is depleted of Zn and Cu not due to the lower acidity, but primarily due to the low content of these metals in coal seams, which was confirmed by own experience of long-term (over decades) monitoring of drainage from coal mining waste dumps of several mines in the Upper Silesia coal basin (USCB), the largest hard coal deposits in Europe (Szczepanska and Twardowska 1999, 2004). Similarly, as can be derived from characteristics of AMD presented by Hallberg (2010), AMD from Zn-rich waste of pH 3.4 from Sn mining contained from over two- to sixfold more Zn than more acidic AMD (pH 2.5–2.7) from copper mines. In contrast, Cr concentrations in highly acidic AMD (pH 2.5–0.5), from all Cu mines and Zn in extremely acidic AMD from RM mine (pH 0.5–1) showed distinct pH dependence, occurring in the range from 10 to 10^3 mg Cu/L and from 10 to 10^4 mg Zn/L.

It should be emphasized that physicochemical characteristics of AMD are formed within the rock material, similarly as microbial communities. Although a wide range of moderately and strongly acidophilic iron- and sulfur-oxidizing and iron-reducing microorganisms were isolated from AMD (Baker and Barnfeld 2003; Hallberg 2010; Auld et al. 2013), they belonged to the genera growing in the sulfidic rock material and represented the succession corresponding to the environment actually occurring in the leached matrix (e.g., Chen et al. 2014; Korehi et al. 2014; Liu et al. 2014). However, distinctly lower (2–4 orders of magnitude) cell numbers of these microorganisms in AMD (Hallberg 2010) than in sulfidic tailing dumps (Schippers et al. 2010), along with their known predominant attachment to the surfaces of mineral sulfides (Vera et al. 2013), confirm their secondary character. Reported by Hallberg (2010), cell numbers of moderate Fe-oxidizers in AMD ranged from $<10^2$ to 10^4 cell/ml, cell numbers of extreme Fe-oxidizers from $<10^2$ to 10^5 cell/ml, and these of S-oxidizers from <50 to 10^2 cell/ml. In the recent study of Bozecki (2013, unpublished), the isolated cell number of *Acidithiobacillus* spp. in 11 AMDs from abandoned lignite mines of average pH 2.75–5.40 was at the mean level of 10^2 to $<10^3$ cell/ml, while significant seasonal variations were also noticed. The reported cell numbers in AMD are mostly too low to play a significant role in iron oxidation processes. In addition, the AMD environment is neither specific nor appropriate for biooxidation, which is in line with the conclusion of Kirby et al. (1999) that oxidation of iron in AMD at acidic pH under the natural conditions is not due to microbial activity.

1.4 Dumping Sites of Sulfidic Waste Rock and Abandoned Mines

1.4.1 The Need of Coupling Factors Determining the Role and Activity of Microorganisms

The environmental issues related to mining, including problem of pollution potential from extractive activities related to sulfidic waste disposal and generation of acid rock drainage (ARD) and acid mine drainage (AMD) affecting mainly abandoned mines, emerged in advance to the idea of biomining and developed in parallel. The environmental problems related to sulfidic mining wastes and the role and significance of microorganisms in ARD/AMD generation have its specificity and differ substantially from those arising in biomining field. However, as can be derived from the discussion presented above, many findings related to microbiology of biomining may substantially facilitate the measures focused on ARD/AMD prevention and control. At the same time, the information regarding microbial communities, their activity, growth, sulfide oxidation rates, and overall contribution to pollution potential of the specific object should not be directly applied to sulfidic waste rock deposited at dumps or to AMD from the abandoned mines, as this may lead to overestimation (or in some cases underestimation) of the role and share of bacteria in pollution load formation and, consequently, to wrong and costly prevention/attenuation/control measures.

There are several reasons and aspects that play a crucial role in the assessment of the pollution potential of a sulfidic waste dump. Of them, data on the source and amounts of waste rock from mining operations and its characteristics (mineralogical, textural, physical, hydrogeochemical, biochemical), along with the characteristics of the dump construction, water balance, water, gas and heat transport, hydrology/hydrogeology of the site, and climatic conditions affecting sulfide oxidation processes in the dump, are required. These data would allow performing site- and object-specific environmental impact assessment related to pollution loads resulted from ARD generation and required for elaborating effective control strategies. The variability of waste rock properties and conditions within the dump, as well as of climatic conditions, requires site-specific evaluation of the sulfide oxidation rate and makes the role of the microbial mediation in the process kinetics uncertain. In particular, there are substantial differences between construction of a bioleaching heap and a sulfidic waste dump (Fig. 1.4). While the bioleaching heap is evenly aerated from the bottom and inoculated/irrigated with inoculum/acid solution from the top, sulfidic waste dump has uneven access and diffusion of air, unstable water balance, and different temperature conditions. These specific conditions are dependent on the fluctuations of ambient temperature, heat generation and transport within the heap due to sulfide (pyrite) oxidation, availability and transport of atmospheric oxygen and CO_2, pH values dependent on ANC (acid neutralization capacity) determined by NP/AP (neutralization potential/acid potential ratio), Fe(III) precipitation at higher pH, and occurrence of only indigenous microbial strains.

Many of these issues have been presented in the comprehensive review of Amos et al. (2015) summarizing studies on the sulfidic waste rock hydrogeology and geochemistry conducted in the last decade, along with waste rock piles and waste rock heterogeneity: physical characteristics (internal structure, particle size, permeability to air, thermal conductivity), as well as mineralogy with particular regard to sulfides occurring in waste rock and to the presence of acid-neutralizing/acid-buffering minerals at several sulfidic dumps located in different countries. A review of geochemical processes and microbial population in the dumps related to sulfide oxidation showed substantial diversity of microbial communities within different dumps. The cases discussed by Amos et al. (2015) comprised piles of mining waste rock representing sulfide (pyrrhotite/pyrite) content from 0.035 wt% S_S to as high as 7.0 wt % S_S. Also, a different mineralogy of sulfidic rock material has been overviewed, from coarse-grained marble hornfels and calcium-bearing exoskarn of pyritic sulfur content within the range 2 to >3 wt% S, through aluminosilicates with sulfide minerals occurring as pyrrhotite, pyrite, and secondary marcasite at sulfur content from 0.035 to 0.35 wt% S_S, up to sericite schist and diorite (porphyry copper deposit) rocks containing up to 7 wt% S_S and having low NP (neutralization potential). High diversity of sulfide reactivity and neutralization potential of matrix, along with the diverse access of atmospheric oxygen, resulted in different trends to

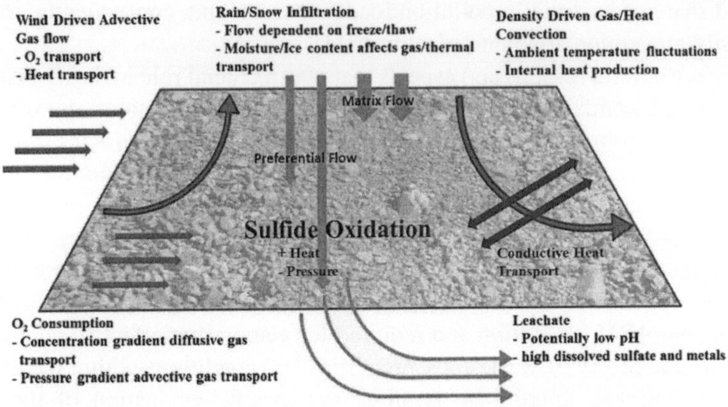

Fig. 1.4 Processes and conditions occurring within the sulfidic waste dump (Lefebre et al. 2001, modified after Amos et al. 2015)

the rock/leachate acidification and in the growth and activity of species responsible for microbiological oxidation.

In particular, in piles of calcite-/dolomite-rich waste rock, despite high acid potential, the pH was invariably neutral or slightly alkaline (pH 7–8.5), while a low number or absence of microorganisms was detected. At the same time, elevated metal contents in leachate, especially Zn (up to 45 mg/L), were detected. This, along with other observations conducted at these waste piles, when up to 10^8 cells/g of S-oxidizing neutrophilic microbial species was locally found, leads to the conclusion that under heterogeneous conditions occurring within the high-sulfide dump, the formation of local acidic microenvironments appropriate for neutrophilic or acidophilic bacteria growth may occur. This may explain the aforementioned elevated concentrations of metals at near-neutral pH, although an overall microbial activity at such objects is low.

In the piles of waste with a low acid potential (AP), but also with a low acid neutralization capacity (NP/AP < 1), a gradual acidification of the rock occurred from the near-neutral pH to as low values as pH < 5, accompanied with an increase of SO_4, Fe, and other metals (Zn, Cd, and Cu) in leachate. Simultaneously, a pH-dependent succession of microbial communities was observed, as it was discussed previously. The microbial population changed from the domination in effluent of neutrophilic S-oxidizers (max 10^6 cells/ml), mostly *Thiobacillus* spp., and a minor number of acidophilic S- and Fe-oxidizers ($<10^3$ cells/ml) at near-neutral pH to the prevalence after pH decrease to ~ 3.5, of acidophilic Fe- and S-oxidizers, mainly *Acidithiobacillus* spp., in the similar proportion. The contact with atmosphere was found to enhance growth of these species.

In the piles of high acid potential (AP), but low acid neutralization capacity (NP/AP < 1), fast acidification to pH 2–3 of the material extending from top to bottom of the pile was observed, along with the formation of abundant and strongly acidophilic microbial community.

The summarized overview of the state of the art presented by Amos et al. (2015) shows that the actual role of sulfur-oxidizing bacteria at the different sulfidic waste dumps depends on the number of site- and waste-specific parameters, such as climatic conditions, dump construction that determines water balance and flow, gas and heat transport, as well as physical characteristics, chemistry, and mineralogy of sulfidic waste material along with its temporal and spatial geochemical, hydrogeochemical, and biochemical transformations, which may differ substantially within particular objects. Both overestimation and underestimation of these factors lead to the wrong assessment and management practice aimed either at the enhancement (for biomining purposes) or at the attenuation of microbial activity at sulfidic waste dumping sites within the environment protection strategy. The understanding of the complex analysis of these factors and their effect on the formation of microbial communities and on the growth and activity of microorganisms on the extent of acceleration of sulfide oxidation are increasing. Many questions are though still remaining unanswered. One of them is the issue of the actual effect of microbial processes on ARD/AMD formation and metal mobilization in mining wastes other than extensively studied low-buffered metamorphic rocks of the "porphyry copper deposit" type, predominantly used for biomining of copper, gold, uranium, more rarely zinc, lead, or polymetallic ores that render the process economically and technically viable.

However, in the mineral extraction practice worldwide, traditional mining and mineral processing prevail, at a large variety of sulfide-bearing rock types and their textural, petrological, mineralogical, and physicochemical characteristics. Pyrite (FeS_2), the most abundant sulfide mineral in the world, occurs in many forms of different crystallinity and development of specific surface. These forms show a big variety—from metacolloidal and framboidal to highly crystalline forms of different reactivities in sulfidic sediments and sedimentary rocks (Rickard 2012). Such variety of sulfide forms is common in hard coal and lignite seams, formed from coal, silicate, clay minerals, and carbonates. It accompanies also other sulfide minerals in sulfide ore deposits that are an important source of production of Cu, Ni, Zn, Pb, Co, Mo, Bi, Sb, and Hg. Some formations such as porphyry deposits are predominant source of Cu. Massive sulfide deposits contain large concentrations of mixed sulfide minerals, such as Cu, Ni, Pb, or Zn. In many sulfide ores, admixtures of highly valuable metals such as Pt, Au, Ag, In, Se, and Te are also present. Sulfide ores contain also other minerals, such as silicate, aluminosilicate, and calcite, and often occur as veins and massive replacements in limestone. Besides massive ores, in which sulfides are dominating minerals, in disseminated ores, sulfides occur as inclusions in other minerals. Mineralogy of sulfides and co-occurring minerals highly influences the composition of sulfidic wastes, which are a potential source of ARD/AMD, and the processes that determine the rate of sulfide oxidation, acid and neutralization potentials, ARD generation, and its composition (Lottermoser 2007). Due to the sensitivity of microbial processes to external conditions, their participation in ARD generation and the possible extent of acceleration of this process should be better investigated and clarified. This statement may be exemplified in researches on the desulfurization of coal and on the role of microbial oxidation of pyrite at the coal mining waste dumps.

Coal and coal mining waste appear to be very good objects for elucidating the role of microbiological processes of pyrite oxidation (which is the most abundant sulfide mineral in the lithosphere, mostly occurring as inclusions in rock matrix) under the natural conditions of exposure to the atmosphere. Due to the insignificant presence of other sulfidic minerals, the effect of textural forms of pyrite and lithological/mineralogical characteristics of pyrite-bearing rocks on oxidation processes can be better followed, unhindered by other factors. In brief, in the older stratigraphic rock formations, the proportion of crystalline forms of pyrite increases, while metacolloidal and framboidal forms are the most abundant in younger seams. Besides, it was found that the crystallization of sulfides appears to be the most advanced in sandstone, weaker in coal, and the least advanced in shale, where the finest microcrystalline aggregates were dominant (Szczepanska and Twardowska 2004; Sahoo et al. 2014). It determines different reactivities of pyrite from different formations. It should be added that another mineralogical form of iron sulfide, pyrrhotite, which is usually absent in coal seams, but besides pyrite is a frequent component of metal ores, has been observed to be the most reactive form of iron disulfide (Amos et al. 2015). However, a secondary reactive form of iron sulfide, marcasite, often can be found in the younger coal seams (Szczepanska and Twardowska 1999, 2004). More details on pyrite forms are presented in recent publications focused on sulfide minerals, e.g., in an excellent monograph by Rickard (2012).

Pyrite in coal seams and coal is an undesirable mineral responsible for ARD/AMD formation and SO_2 emission at coal combustion causing acid rains; thus, there are a substantial number of studies on coal desulfurization, assessment of coal mining waste pollution potential, and their environmentally safe disposal directed to ARD/AMD attenuation. All these researches provide mutually complementary information that allows for evaluation of different factors playing a crucial role in the correct assessment and attenuation of coal mining waste pollution potential, including the actual role of microbial oxidation.

1.4.2 Biodesulfurization of Coal: Successes and Failures

Studies on biodesulfurization of coal have been carried out since the 1960s of the twentieth century (Silverman et al. 1963) and are continued with different intensities up to now. However, despite good results (e.g., Dugan and Apel 1978; Tillet and Meyerson 1987; Olson and Kelly 1988; Klein et al. 1988; Uhl et al. 1989; Dugan 1991; Andrews et al. 1991, 1992; Tripathy et al. 1998; Acharya et al. 1997, 1999; Malik et al. 2000, 2001a, b; Aller et al. 2001), they have never been applied even in pilot installations that indicates the scale of this problem.

Biodesulfurization have been considered to be performed in two major types of bioreactors, heap bioleaching and airlift slurry bioreactors, while acidophilic autotrophic *Acidithiobacillus* has been the genus most frequently found and investigated in coal and coal mining dumps and most widely used for biodesulfurization. As autotrophs, Fe- and S-oxidizing strains *A. ferrooxidans* and *A. thiooxidans* need O_2 (at a saturation level >10–15 %) and continuous CO_2 supply as a source of carbon for their growth and activity. The inoculation ensuring maximum cell growth and activity, availability of fresh pyrite surface, optimum pH (~2.0) temperature (~28–35 °C), Fe(III)/Fe(II) ratio, and concentrations of both ions in the microbial environment are essential for biodesulfurization process at high rates.

The first published results of small-scale experiments on batch biodesulfurization by *A. ferrooxidans* of Indian (Assam) coal containing 4–6 % S_t, in this 1.3–2.0 % S_p (pyritic sulfur), resulted in about 25 % S_p removal at the selected parameters (pulp density 2 % (w/v), initial pH 2.5) that was considered a promising result (Acharya et al. 1997). Further experiments on chemical pretreatment of this coal with NH_4OH and Na_2CO_3 solutions before bacterial leaching with a mixed culture of *Acidithiobacillus* spp. showed an improvement of sulfur bioleaching to 35–38 wt% compared to 5–27 wt% for the raw coal (Tripathy et al. 1998). Also, due to bacterial preconditioning with adapted strains of *Acidithiobacillus* spp. followed with froth flotation, over 23 % of total sulfur was removed, mostly of pyritic sulfur, whereas organic sulfur S_{org} remained at the same level (Acharya et al. 1999).

Studies on biodesulfurization of the same fine-grained high-sulfur coal conducted in a 20 L laboratory batch bioreactor inoculated with *A. ferrooxidans* cells at 10 % (w/v) pulp density, temperature 30 °C, and initial pH 1.8 showed that the process occurs in a three-step mode: direct oxidation of pyrite by bacteria (0–4 days, exponential growth phase) resulted in 28 wt% S_p removal (I step) followed by both direct and indirect oxidation (4–10 days, stationary growth phase) when 51 % S_p was removed (II step), and final cease of the process due to shielding pyrite surface with jarosite and ferric sulfate precipitates, and reduction of Fe(III) concentration in solution (III step) (Dastidar et al. 2000; Malik et al. 2001a). High concentrations of Fe(II) and Fe(III) in solution also caused retardation of the pyrite oxidation process (Dastidar et al. 2000). Therefore, the full cycle of biodesulfurization at the reported parameters, according to the authors, would last some 8–14 days, resulting in roughly 80 % S_p removal.

The sulfur removal was found to be substantially enhanced by using pulse feeding and leachate recycle in 30-day biodesulfurization cycle—from 72 % in the conventional batch process to 90–97 % (Malik et al. 2001b). Also an addition to the conventional biodesulfurization batch system of *A. ferrooxidans* grown on sulfur was reported to increase S_p removal to 93 % in 15 days of cycle compared to 77 % in the control (Malik et al. 2000).

These results show high capability of biodesulfurization in removing pyritic sulfur from coal. However, the process parameters were very far from those occurring under the natural conditions and were adjusted to the optimum conditions for *Acidithiobacillus* growth. The measures undertaken to maintain optimum parameters for *Acidithiobacillus* growth explain the lack of practical applications, as the costs of microbial desulfurization of coal in batch reactors and the size of installations required are much higher than that of the treated coal, which makes this technique unviable.

A much more cost-efficient process would be heap leaching. Such experiments were performed by the author's team (Twardowska and Galimska 1993) on high pyrite coal (~2 % S_p) with the use of a mixed culture of two selected and adopted the most efficient native *Acidithiobacillus* strains of *A. ferrooxidans* and *A. thiooxidans*. In the laboratory conditions, the strains grown on 9 K medium with the addition of unprocessed pulverized coal showed high efficiency, by removing 90–95 wt% S_p at cell number 10^7–10^8 per 1 ml of medium. The next stage of process performance optimization was carried out using column bioleaching with pulverized coal layer 2 m thick and inoculation from the top of a column, with adjustment of pH leaching medium to 3.5, at the ambient temperature from 10.5 to −5 °C. Both parameters were close to the natural conditions of pulverized coal storage.

In two series of bioleaching that comprised application of inoculum on 6 columns filled with pulverized coal, followed by 20 days of retention (no aeration was applied), two sets of results were obtained. In series I, where the number of microorganisms in the inoculum was only 10^4 cells/ml, unsatisfactory results were obtained, as the maximum number of microorganisms in inoculated coal was 10^5 cells/ml, while in control non-inoculated columns, the number of native cells was 10^3 cells/ml. With pH of coal at the moderately acidic level (pH 4.5–5.5), S_p removal was low and only by 24–30 % higher than in non-inoculated control columns. In contrast, the increase of cell number to 10^7 cells/ml in inoculum applied to columns resulted in *Acidithiobacillus* growth in inoculated coal to 10^7–10^8 cells/g, dw, and the S_p removal by roughly 47 wt% from the finer pulverized coal and by 23 % from the coarser material under the unfavorable conditions of low winter ambient temperatures. The highest decomposition of dispersed and framboidal forms of pyrite was observed.

These experiments showed also that the native microorganisms isolated from coal are highly adaptable and capable to ensure relatively high pyrite removal, at a moderate, but carefully selected, improvement of the cell growing conditions.

Similar conclusions were derived by other authors (Aller et al. 2001) that obtained the best extent of pyrite removal (76 to >92 wt% S_p) from six different coals in laboratory batch experiments by using enriched native cultures isolated from coal as inoculum, at working parameters: pH 2.0, temperature 32 °C, pulp density of 5 %, in 10 days leaching cycle. Another observation based on the presented results is a variable but generally relatively high extent of a purely chemical pyrite removal from sterilized and non-inoculated coals (from 0.6 to 41.5 wt%), which means that microbial acceleration of the process under the optimum conditions ranges from 2.2- to 123-fold, but most frequently in two- to ~tenfold range.

Recent publications (Gonsalvesh et al. 2008, 2012) report the case of using for desulfurization of subbituminous and lignite high-sulfur coals containing pyritic and organic sulfur, also fungi. Inorganic sulfide sulfur was removed by 79 wt% with mixed culture of bacteria and subsequent removal of ~13 wt% of organic sulfur (S_{org})

with fungi, at maximum removal of total sulfur by 26 wt% S_t (Gonsalvesh et al. 2008). In another experiment, chemical depyritization was applied prior to biodesulfurization, resulting in the removal of 25–54 wt% S_t, in this of 60–77 wt% S_p. By the subsequent biodesulfurization using either white rot fungi or thermophilic and acidophilic archaea *Sulfolobus*, further reduction of S_t and S_{org} by 24 wt% with fungi, or S_t by 17 wt% and S_{org} by 18 wt% with archaea was attained.

These examples show, on the one hand, high potential efficiency of microorganisms, in particular of indigenous wild-growing strains for sulfide, but also organic sulfur removal from sulfidic rock material. On the other hand, it becomes clear that high extent of sulfur, in particular sulfide sulfur removal by these organisms, requires assuring stable optimum conditions for their growth and activity in a narrow range of parameters normally not occurring in the real heterogeneous environment. This explains technical problems with biodesulfurization of coal caused by failure in changing scale from thoroughly successful laboratory into even a semi-pilot one. Taking into consideration the relative efficiency of more robust chemical processes of sulfide oxidation, attributing the responsibility for ARD/AMD generation in sulfidic waste dumps or abandoned mines predominantly, or at least in significant part, to the microbial activity without site- and object-specific evaluation of its actual role may result in applying wrong, costly, and inefficient control measures.

1.4.3 The Role of Microbiological Processes of Sulfide Oxidation at Coal Mining Waste Dumps

The role of microbiological decomposition of sulfides under the natural conditions of sulfidic mining waste, and specifically of coal mining waste disposal and use, is a subject of considerable controversy that was particularly strong in the 1970s of the twentieth century and remains unresolved since then (Twardowska 1986, 1987; Szczepanska and Twardowska 2004). While many authors (e.g., Singer 1970; Harrison 1978; Olson et al. 1979, 1981) attributed ARD/AMD formation to the growth and activity of *Acidithiobacillus* bacteria (formerly known as *T. ferrooxidans*), Lau et al.(1970) or Palmer (1978) questioned the dominant role of these bacteria in ARD/AMD generation. Belly and Brock (1974) tried to define the ecology of iron-oxidizing bacteria in sulfidic rocks associated with coal and evaluate an effect of such parameters as depth below the dump surface, waste age, and pH. Some researchers assumed that the microbial growth may occur in the microenvironments providing appropriate conditions, different from the prevalent ones (e.g., Olson et al. 1979; Szczepanska and Twardowska 2004; Dockrey 2010). Nevertheless, not the specific microenvironments but the overall growth and number of active cells in all parts of a dump indicate the level of microbial activity and its participation in sulfide decomposition and ARD formation.

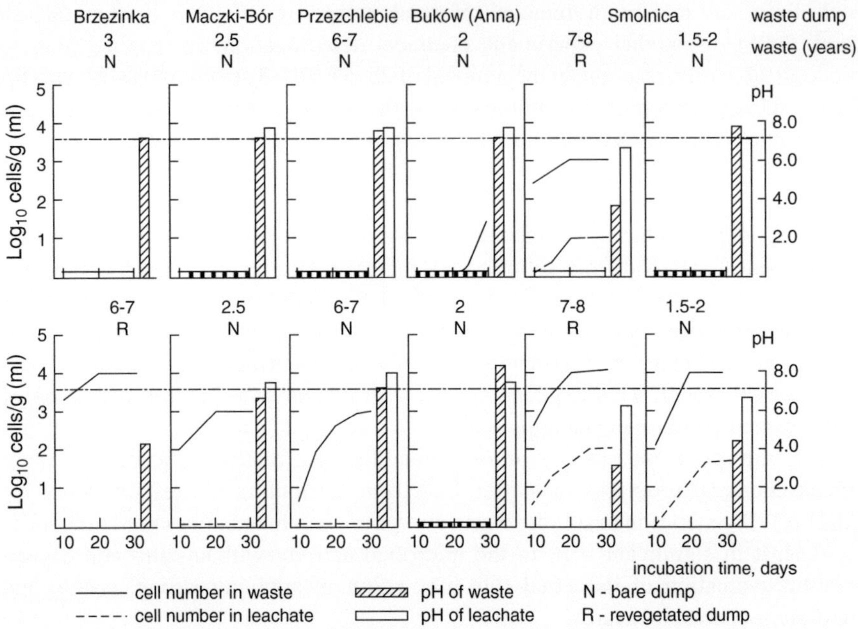

Fig. 1.5 Seasonal variations of pH and cell number of *A. ferrooxidans* in the uppermost 0–5 cm layer of coal mining dumps of different ages and buffering capacities of waste, USCB, Poland (after Twardowska 1987). (**a**) Samples taken in June; (**b**) samples taken in September

Own studies on the occurrence of the *A. ferrooxidans* species in the hard coal mining dumps (located in the largest in Europe Upper Silesia coal basin USCB, Poland) show high variability and instability of their growth. Seasonal variations of *A. ferrooxidans* in the surface 0–5 cm layer of five coal mining dumps of different ages and acid neutralization capacities showed high dependence on the pH of waste, moisture content, and ambient temperatures (Fig. 1.5).

No active cells were found in samples collected in June either in relatively freshly disposed waste or in a highly buffered material at pH alkaline or close to neutral (e.g., Bukow—Anna mine). The seasonal variations of pH and cell number in the uppermost waste layer reflected strong influence of ambient atmospheric conditions such as temperature and moisture content. The decrease of waste pH below 7 in the waste rock sampled in September was accompanied by the bacterial growth in the range of 10^3–10^4 cells/g, dw. The highest number of 10^4 cells/g, dw, occurred at pH $\leq$ 4.4. This was considerably lower than in the laboratory bench investigations, where up to 10^8 cells/g, dw, in the same waste rock was detected (Twardowska 1986). This confirms the strong inhibitory effect of changeable climatic conditions on microbial population in the uppermost waste dump layer.

Much lower cell numbers of *A. ferrooxidans* detected in the rock drainage than in the rock material, or even their absence (Fig. 1.5), confirm the conclusion that the rock material is the major environment of species activity, while water is mostly a transportation means for the cells.

The growth of *A. ferrooxidans* along vertical profiles of a surface layer 1 m thick at the Smolnica dump of unbuffered sulfidic waste (NPR = NP/AP <1) generating ARD (Figs. 1.6 and 1.7) showed the absence or low cell number of bacteria in the uppermost surface layer (0–1 cm), apparently due to the direct exposure to the variable weather conditions. An increase of cell number and activity of bacteria and decrease of pH with depth were observed, up to the occurrence of the highest population (up to 10^5 cells/g, dw) and the lowest pH values (pH 3.0–4.4) in the layers at different depths, from 2–20 cm to 35–65 cm below the surface (Fig. 1.6), followed by the decrease of cell numbers to 10^2–10^3 cells/g, dw, and the increase of pH (to pH 5–6) in the deeper layers (Fig. 1.7). In one profile (P-104-P), the formation of a hard subsurface pan in the layer 1–3 cm resulted in the occurrence of a peak bacteria growth and a sharp drop of cell number in the layer below due to insulation from O_2 and CO_2 access. Detected highly variable spatial and vertical distribution of *A. ferrooxidans* in the dump generating ARD and overall low microbial growth despite relatively favorable pH in the coal mining waste of the low buffering capacity susceptible to acidification lead to the assumption of a marginal role of bacterial activity in pyrite decomposition at this and the majority of other coal mining dumps.

Besides the detected low growth of *A. ferrooxidans* on the sulfidic waste rock deposited at the Smolnica dump, this assumption was strongly supported by the kinetics of sulfide decomposition evaluated in 23-year lysimetric studies (1984–2007) on 1.5 m-thick layers of this rock containing sulfides (pyrite) of a moderate reactivity (lysimeters SM1 and SM2) versus the buffered rock from another dump containing sulfide of a high reactivity occurring in the form of framboidal pyrite and marcasite (lysimeter TS1) (Fig. 1.8). Despite proven overall low buffering capacity, due to heterogeneity of the waste rock, the material with a higher carbonate content and thus of a higher buffering capacity and neutralization potential was sampled and packed into the SM1 lysimeter, while another one (SM2) was filled with rock of an average low buffering capacity. Basic characteristics of the waste rock and leachate from lysimeters for the years 1984–1996 are presented in Table 1.1 and Figs. 1.8 and 1.9.

Kinetics of sulfide decomposition that displayed a very regular character (Fig. 1.8) and the time of total sulfide decomposition were assessed on the basis of sulfate loads released to leachate, from the first-order equation (Szczepanska and Twardowska 2004):

$$(G_s)_t = (G_s)_0 \exp(-kt) \quad (1.11)$$

where

$(G_s)_t$—content of residual sulfidic sulfur (wt%) in waste rock after t days since the experiment beginning

$(G_s)_0$—initial content of sulfidic sulfur (wt%) in waste rock at the experiment beginning ($t = 0$)

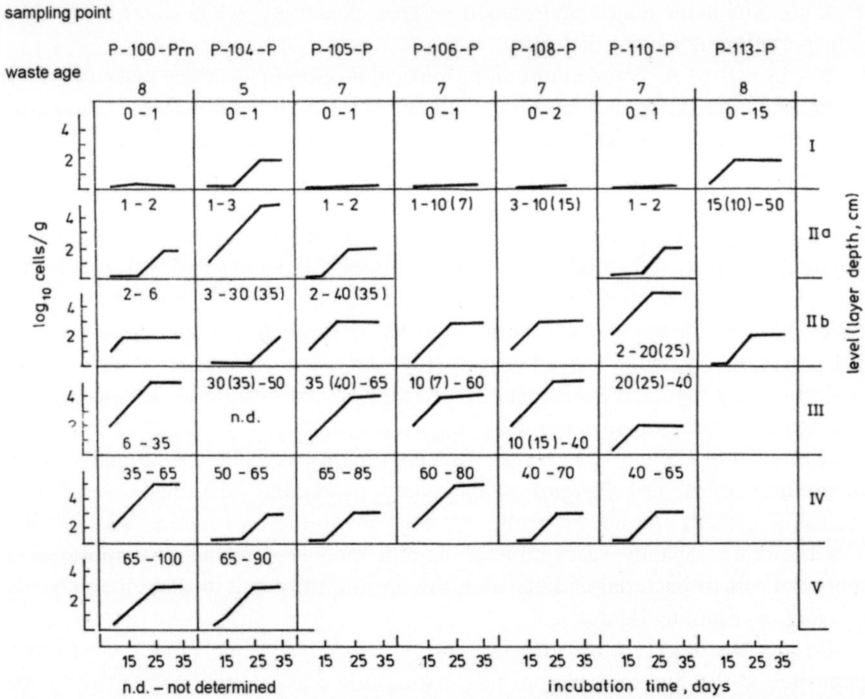

Fig. 1.6 Growth of *A. ferrooxidans* within the surface layer 0–100 cm of unbuffered mining waste disposed at the ARD-generating Smolnica coal mining dump, USCB, Poland (after Twardowska 1987)

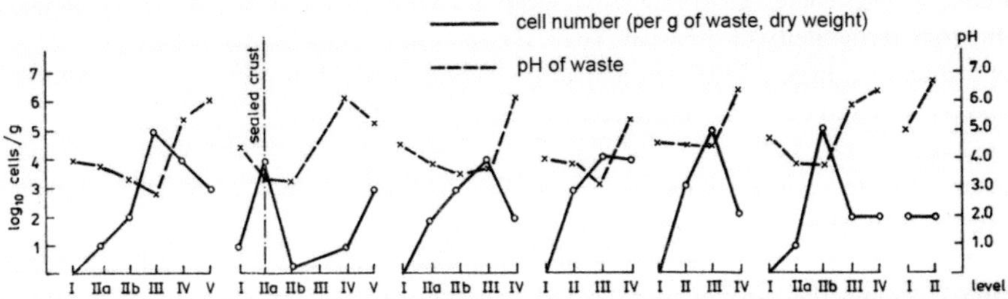

Fig. 1.7 Occurrence of active *A. ferrooxidans* cells and pH of the coal mining waste (Smolnica dump, USCB, Poland) along the vertical profiles of the surface layer 0–100 cm thick (after Twardowska 1987)

k—kinetics constant

t—time (days) from the beginning of the experiment

The time of total sulfide decomposition computed for the buffered (SM1) rock at pH constantly within alkaline to near-neutral range (pH 9.1–6.4) was 21 years and for unbuffered rock (SM2) at acidic pH 6.3–2.6 was 19 years. Both decomposition kinetic patterns and sulfide sulfur depletion periods are very similar and proven by

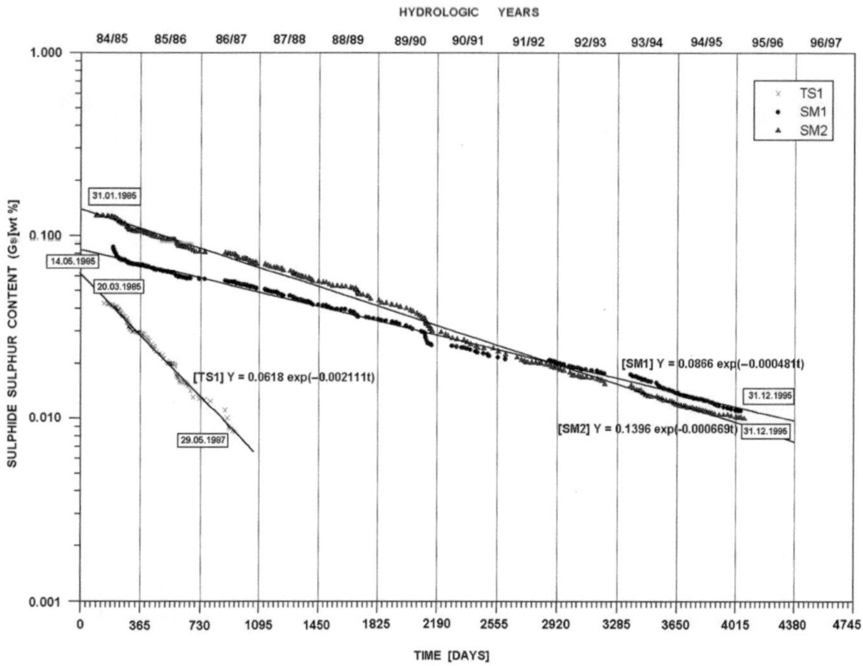

Fig. 1.8 Kinetics of sulfide decomposition in the surface layer 1.5 m thick of coal mining waste. Lysimetric studies in the 12-year natural hydrologic cycle (1985–1997) (after Szczepanska and Twardowska 2004). TS1—highly reactive sulfidic waste from Westphalian C coal seams (Poland, USCB, Siersza mine, 1.48 wt% S_p); SM1 - moderately reactive near-neutral/alkaline sulfidic waste, Namurian C coal seams (Poland, USCB, Smolnica mine, 0.86 wt% S_p); SM2 - moderately reactive acidic sulfidic waste, Namurian C coal seams (Poland, USCB, Smolnica mine, 0.81 wt% S_p)

Table 1.1 Basic parameters of coal mining waste, pH, and sulfate range in leachate in the natural 12-year hydrologic cycle (1984–1996) during long-term lysimetric studies (after Szczepanska and Twardowska 2004)

Lysimeter	Rock origin (coal seams)	Buffering capacity	S_S content in mining waste, %	Parameter range in leachate		Sulfide 99 % decomposition time (years)
				pH	SO_4, mg/L	
TS 1	Westphal C	NP/AP > 1	1.48	8.5–6.8	3.46- 372	6
SM1	Namur C	NP/AP > 1	0.86	9.1–6.4	7.97- 332	21
SM2	Namur C	NP/AP < 1 (0.6)	0.81	6.3–2.6	6.15- 298	19

AP acid potential, *NP* neutralization potential, NP/AP > 1 (buffered waste), NP/AP < 1 (0.6) (unbuffered waste)

S_p sulfide sulfur content

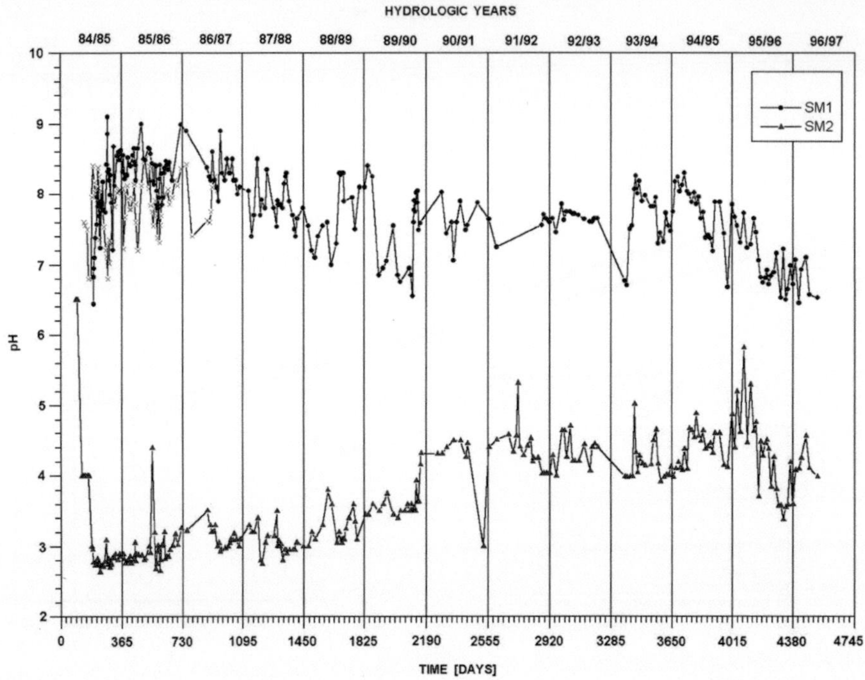

Fig. 1.9 pH values of leachate from the surface layer of coal mining waste 1.5 m thick

the evaluation at the end of oxidation process when thorough pyrite depletion occurred (Szczepanska and Twardowska 2004). Along with the observed rather poor growth of *A. ferrooxidans* in acidic waste and their absence or much lesser occurrence in alkaline material (Figs. 1.5 and 1.7), this similarity and a long duration of the process suggest predominantly chemical oxidation and a marginal effect of the biooxidation by the native microbial strains.

In turn, relatively fast sulfide depletion in coal mining waste in TS1 lysimeter at alkaline to near-neutral pH range, evaluated for 6 years (Table 1.1, Fig 1.8), was mostly due to the occurrence of sulfides in the material as marcasite and as highly reactive framboidal forms, but rather not due to the microbial activity, limited at this pH range.

To conclude, the role of indigenous microorganisms in the sulfide oxidation and ARD formation in coal mining waste dumps seems to be limited for many reasons; the major of them are "matrix impact"; instable, often inappropriate climatic conditions; low flexibility and adaptability of microorganisms to the ambient conditions different from the parameters required to be maintained for growth and activity of microbial communities. Coal mining waste is a heterogeneous material, in which sulfides occur as inclusions, while properties of a matrix decide about specific conditions occurring within the material. Besides chemical properties, also waste rock particle size is of high importance with respect to the access of organisms to sulfide surfaces, circulation of O_2, CO_2, water balance (humidity/dryness), and temperature conditions. For this reason, dominating or significant effect of microbiological processes on sulfide decomposition in heterogeneous wastes and at variable climatic conditions is rather doubtful. A remarkably high stability of the kinetics of pyrite decomposition for over two decades reported above confirms that processes of variable kinetics to which natural microbiological processes under changeable ambient environmental conditions belong do not exert strong effect on the slow, but stable, chemical processes.

1.4.4 The Role of Microbiological Processes of Sulfide Oxidation in Other Sulfidic Wastes

Coal mining wastes, as have been mentioned, are a representative of heterogeneous sulfidic wastes, where sulfides (mostly pyrite Fe_2S) are unwanted inclusions in sedimentary rocks and occur in amounts from <1 wt% to several wt%. Also many metal ores, where metals occur in sulfide forms and sulfides are extracted components, occur in the heterogeneous matrix together with pyrite. Although tailings from metal ore processing are more enriched with sulfide, they represent similar types of heterogeneous waste of different acid and neutralization potentials and buffering capacities. In this case, the role of microbiological processes in ARD/AMD formation may be marginal, similar to that in coal mining waste.

Overall, even in other more sulfide-enriched waste rocks, microbiological processes may be suppressed due to some unfavorable parameter. An essential role of microorganisms in the ARD/AMD generation in sulfidic ore processing waste disposed at dumps is not at all proven in many cases, although generally considered explicit.

From 15-week- to 9-month (36 weeks)-long column studies on the effect of weathering processes on the microbial activity in tailings with 63.4 wt% pyrite and minor amounts of Cu, Pb, and Zn sulfides (around 1 wt%) disposed in a tailing pond, showed diversity of microorganisms, their different abundances and activity depending on the oxygen and moisture availability under vadose/saturation state of tailings during dry/humid periods (Garcia et al. 2007). While no or very poor growth of Fe- and S-oxidizing bacteria in waterlogged/saturated tailings at unchanged initial pH 6–7 was detected, in the material under vadose zone conditions, the highest growth and bacterial activity (up to 10^4 cells/g, dw, of both Fe- and S-oxidants) at a general prevalence of S-oxidizers and pH decrease to pH 2.5–3.5 were observed in the surface layers having the direct contact with atmospheric oxygen. The highest total cell number of chemolithotrophic aerobic bacteria in tailings at the full availability of oxygen (surface layer) accounted for 10^6–10^8 cells/g, dw, whereas in the deeper parts of the tailing layer 1.5 m thick, the total cell number dropped to 10^2 cells/g, dw, proving high dependence from availability of O_2. Similarly to the studies on bacterial growth in coal mining wastes, the cell number in drainage was by 1–2 orders of magnitude lower than that in the rock material, at predominance of S-oxidizers. This confirms strong association of bacteria with the solid phase and a weaker attachment of S-oxidizers to the solid surface.

In the reported bench weathering experiments, a constant optimum temperature 30 °C was maintained that excluded temperature effect on the microorganism population. In the real world, temperature variations particularly strongly affected the uppermost layer of the dump that resulted in the poor growth of microorganisms at the surface of coal mining dumps despite the best availability of oxygen. There is a strong probability that the same effect of temperature variability will occur at every dump of sulfidic waste, suppressing the microbial growth and activity. Taking into consideration detected variability of microbial growth and activity within a layer of pyritic waste of a very high acid potential and otherwise appropriate chemical parameters reported in the weathering studies (Garcia et al. 2007), there is no certainty that even in the case of high-pyritic waste disposed at the dumping sites, microbial oxidation processes dominate over chemical ones.

Bioleaching experiments and practice show that acidophilic chemolithotrophic microorganisms are able to substantially accelerate the process of sulfide decomposition in the sulfidic material if the active cell number is at the level of 10^6–10^8 cell/mg, dw, while the most frequently detected microbial growth at the dumps is 10^1–10^4 cells/mg, dw, particularly in the subsurface and deeper parts of a dump. This however cannot change much the kinetics of sulfide oxidation. Therefore, the role and significance of microbial activity at the sulfidic waste dumps in the generation of ARD/AMD remain an open question.

1.5 Concluding Remarks

Although extensive research on microbiological processes of sulfide oxidation that have been carried out for over five decades significantly extended our knowledge and allowed for successfully utilizing it in ore biomining, there are still large gaps in the correct assessment of the role and significance of microbial activity in sulfide decomposition at the dumps of sulfidic mining wastes and in abandoned mines, resulting in the environmentally problematic ARD/AMD formation. High variability of mining waste rock properties and climatic conditions, along with a dump construction, causes instability and variability of hydrogeochemical and thermal conditions in the dump as a whole and in specific parts/layers within the dump, thus affecting microbial activity and consequently differentiating it within the dump. The overall pollution load and ARD generation and release are a resultant of hydrogeochemical and microbiological processes. Some measures undertaken for prevention, attenuation, and control of ARD (which is a transporting means for pollutants generated in a waste dump/abandoned mine), such as sealing against oxygen and water penetration, are common for both abiotic and biotic processes and are a part of the best management practice (EC 2009). However, the potentially decisive role of microbial activity in the kinetics of sulfide decomposition at some dumps that might cause significant acceleration of pollutant generation and release requires a differentiated assessment of a joint effect and eventual development and application of efficient measures specific for suppression of microbial processes (e.g., Schippers et al. 2001; Sand et al. 2007; Hallberg 2010; Kazadi Mbamba et al. 2012). A correct waste- and site-specific evaluation of the role and share of microbiological processes in the generation of pollution load in ARD would allow for elaborating adequate pollution control strategies. This needs designing and conducting complex studies that link host rock and sulfide mineralogy, climatic conditions, hydrogeochemistry, water balance, water, and gas and heat transport within the dump with pure chemical and biochemical processes.

References

Acharya C, Kar RN, Mishra SK, Twardowska I, Sukla LB (1997) Microbial depyritization of Assam coal. Indian J Microbiol 37:21–24

Acharya C, Rao GV, Twardowska I, Kar RN, Sukla LB (1999) Removal of sulfur from Assam coal by bacterial conditioning and froth flotation. J Chem Technol Biotechnol 74:945–948

Aller A, Martinez O, de Linaje J, Méndez R, Morán A (2001) Biodesulphurization of coal by microorganisms isolated from the coal itself. Fuel Process Tech 69:45–57

Amos RT, Blowes DW, Bailey BL, Sego DC, Smith L, Ritchie AIM (2015) Waste-rock hydrogeology and geochemistry. Appl Geochem 57:140–156

Andrews GF, Dugan PR, Stevens CJ (1991) Combining physical and bacterial treatment for removing pyritic sulfur from coal. In: Dugan PR, Quigley DR, Attia YA (eds) Processing and utilization of high-sulfur coals IV. Elsevier, Amsterdam

Andrews GF, Dugan PR, McIlwain ME, Stevens CJ (1992) Microbial desulfurization of coal. Idaho National Engineering Laboratory EG&G Idaho, Inc., Idaho Falls, EGG-2669, Prepared for the U.S. Department of Energy

Appelo CAJ, Postma D (2007) Geochemistry, groundwater and pollution, 2nd edn. Balkema, Leiden, The Netherlands

Auld RR, Myre M, Mykytczuk NCS, Leduc LG, Merritt TJS (2013) Characterization of the microbial acid mine drainage microbial community using culturing and direct sequencing techniques. J Microbiol Methods 93(2):108–115

Baker BJ, Barnfeld JF (2003) Microbial communities in acid mine drainage. FEMS Microbiol Ecol 44(2):139–152

Belly RT, Brock TD (1974) Ecology of iron-oxidizing bacteria in pyritic materials associated with coal. J Bacteriol 117:726–732

Blowes DW, Lortie L, Gould WD, Jambor JL (1995) Microbiological, chemical and mineralogical characterization of the Kidd Creek mine tailings impoundment, Timmins area, Ontario. Geomicrobiol J 13:13–31

Blowes DW, Ptacek CJ, Jambor JL, Weisener CG, Paktunc D, Gould WD, Johnson DB (2014) The geochemistry of acid mine drainage. In: Holland H, Turekian K (eds) Earth systems and environmental sciences. Environmental geochemistry. Treatise in geochemistry, vol 11, 2nd edn. Elsevier, Amsterdam, pp 131–190

Bosecker K, Mengel-Jung G, Schippers A (2004) Geomicrobiological risk assessment of abandoned mining sites. In: Thezos M, Hatzikioseyian A, Remoundaki E (eds) Biohydrometallurgy: a sustainable technology in evolution, Part 1. University of Athens, Greece, pp 585–593

Bozecki P (2013) Study on sediments formation in the area of brown coal extraction in Leknica region (Muzhakow belt). Ph.D. Dissertation, AGH University of Science and Technology Krakow, Poland (in Polish)
CEN/TR 16363 (2012) Characterization of waste – kinetic testing for assessing acid generation potential of sulfidic waste from extractive industries. CEN – European Committee for Standardization, Brussels
CEN/TR 16376 (2012) Characterization of waste - overall guidance document for characterization of waste from extractive industries. CEN – European Committee for Standardization, Brussels
Chen LX, Li JT, Chen YT, Huang LN, Hua ZS, Hu M et al (2013) Shifts in microbial community composition and function in the acidification of a lead/zinc mine tailings. Environ Microbiol 15:2431–2444
Chen Y, Li J, Chen L, Hua J, Huang L, Liu J et al (2014) Biogeochemical processes governing natural pyrite oxidation and release of acid metalliferous drainage. Environ Sci Technol 48:5537–5545
Colmer AR, Hinkle ME (1947) The role of microorganisms in acid mine drainage. A preliminary report. Science 106(2751):253–256
Dastidar MG, Malik A, Roychoudhury PK (2000) Biodesulphurization of Indian (Assam) coal using *Thiobacillus ferrooxidans* (ATCC 13984). Energy Convers Manag 41:375–388
Dockrey JW (2010) Microbiology and geochemistry of neutral pH waste rock from the Antamina Mine, Peru. M.Sc Dissertation, University of British Columbia, Vancouver, Canada
Dugan PR (1991) Bioprocessing of coals program at the Idaho National Engineering Laboratory. In: 3rd Symposium on biotechnology of coal and coal-derived substances, Saalzbau, Essen, Germany, 23–24 September 1991
Dugan PR, Apel WA (1978) Microbial desulfurization of coal. In: Murr LE, Thorma EA, Brierley JA (eds) Metallurgical applications of bacterial leaching and related microbiological phenomena. Academic, New York
EC (2006) Directive 2006/21/EC of the European Parliament and of the Council of 15 March 2006 on the management of waste from the extractive industries. OJ L 102 of 11.4.2006
EC (2009) BAT Reference document on Best Available Techniques for the management of tailings and waste-rock in mining activities. http://eipph.jrc.ec.europa.eu/reference/BREF/mmr-_adopted_0109.pdf
Elberling B, Schippers A, Sand W (2000) Bacterial and chemical oxidation of pyritic mine tailings at low temperatures. J Contam Hydrol 41:225–238
EN 15875 (2011) Characterization of waste – static test for determination of acid potential and neutralisation potential of sulfidic waste. CEN – European Committee for Standardization, Brussels
Fomchenko N, Muravyov MI, Kondrat'eva TF (2010) Two-stage bacterial-chemical oxidation of refractory gold-bearing sulfidic concentrates. Hydrometallurgy 101:28–34
Garcia C, Ballester A, González F, Blázquez ML (2007) Microbial activity in weathering columns. J Hazard Mater 141:565–574
Gonsalvesh L, Marinov SP, Stefanova M, Yürüm Y, Dumanli AG, Dinler-Doganay G, Kolankaya N, Sam M, Carleer R, Reggers G, Thijssen E, Yperman J (2008) Biodesulphurized subbituminous coal by different fungi and bacteria studied by reductive pyrolysis. Part 1: Initial coal. Fuel 87:2533–2543
Gonsalvesh L, Marinov SP, Stefanova M, Carleer R, Yperman J (2012) Organic sulphur alterations in biodesulphurized low rank coals. Fuel 97:489–503
Hallberg KB (2010) New perspectives in acid mine drainage microbiology. Hydrometallurgy 104:448–453
Harrison AP (1978) Microbial succession and mineral leaching in an artificial coal spoil. Appl Environ Microbiol 36(6):861–869
Hindar A, Nordstrom K (2014) Effects and quantification of acid runoff from sulfide-bearing rock deposited during construction of Highway E18, Norway. Appl Geochem. doi:10.1016/j.apgeochem.2014.06.016

Huang LN, Zhou WH, Hallberg KB, Wan CY, Li J, Shu W-S (2011) Spatial and temporal analysis of the microbial community in the tailings of a Pb-Zn mine generating acidic drainage. Appl Environ Microbiol 77:5540–5544
Johnson DB (2014) Biomining – biotechnologies for extracting and recovering metals from ores and waste materials. Curr Opin Biotechnol 30:24–31
Kazadi Mbamba C, Harrison STL, Franzidis J-P, Broadhurst JD (2012) Mitigating acid rock drainage risks while recovering low-sulfur coal from ultrafine colliery wastes using froth flotation. Miner Eng 29:13–21
Kirby CS, Thomas HM, Southam G, Donald R (1999) Relative contributions of abiotic and biological factors in Fe(II) oxidation in mine drainage. Appl Geochem 14:511–530
Klein J, Beyer M, Afferden M, Hodek W, Pfeifer F, Seewald H, Wolff-Fishe E, Jungten H (1988) Coal in biotechnology. In: Rehm HJ, Reed G (eds) Biotechnology, vol 6(16). VCH, Weinheim
Kock D, Schippers A (2008) Quantitative microbial community analysis of three different sulfidic mine tailings dumps generating acid mine drainage. Hydrometallurgy 83:167–175
Korehi H, Blöthe M, Schippers A (2014) Microbial diversity at the moderate acidic stage in three different sulfidic mine tailings dumps generating acid mine drainage. Res Microbiol 165:713–718
Lau CM, Shumate KS, Smith EE (1970) The role of bacteria in pyrite oxidation kinetics, 114–119, in: Proc. 3rd Symposium on coal mine drainage research, Mellon Institute, Pittsburgh, Pennsylvania, 19–20 May 1970
Le Roux NW (1969) Mining with microbes. New Sci 43(668):12–16
Le Roux NW (1974) Mineral attack by microbiological processes. In: Miller JDA (ed) Microbial aspects of metallurgy. Medical and Technical Publishing, Aylesbury
Lefebre R, Hockley D, Smolensky J, Gélinas P (2001) Multiphase transfer processes in waste rock piles producing acid mine drainage 1: Conceptual model and system characterization. J Contam Hydrol 52:137–164
Liu J, Hua ZS, Chen ZN, Kuang JL, Li SL, Shu WS et al (2014) Correlating microbial diversity patterns with geochemistry in an extreme and heterogeneous mine tailings environment. Appl Environ Microbiol 80:3677–3686
Lottermoser BG (2007) Mine wastes. Characterization, treatment, environmental impacts, 2nd edn. Springer, Berlin
Lundgren DG, Vestal JR, Tabita FR (1972) The microbiology of mine drainage pollution. In: Mitchel K (ed) Water pollution microbiology. Wiley, New York
Malik A, Dastidar MG, Roychoudhury PK (2000) Biodesulphurization of coal: rate enhancement by sulphur-grown cells. Biotechnol Lett 22:273–276
Malik A, Dastidar MG, Roychoudhury PK (2001a) Biodesulphurization of coal; mechanism and rate limiting factors. J Environ Sci Health A Tox Hazard Subst Environ Eng 36(6):1113–1128
Malik A, Dastidar MG, Roychoudhury PK (2001b) Biodesulphurization of coal: effect of pulse feeding and leachate recycle. Enzyme Microb Technol 28(1):49–56
Morth AH, Smith EE, Shumate KS (1972). Pyritic systems. A mathematical model. Environmental Protection Agency. Report Number EPA-R2-72-002. Proj. Design 14019 EAH, Contract No. 14-12-589. Environmental Protection Technology Series
Muravyov MI, Bulaev AG (2013) Two-step oxidation of a refractory gold-bearing sulfidic concentrate and the effect of organic nutrients on its biooxidation. Miner Eng 45:108–114
Olson GJ (1994) Microbial oxidation of gold ores and gold bioleaching. FEMS Microbiol Lett 119 (1–2):1–6
Olson GJ, Kelly RM (1988) Microbiological processing of coal. In: Bioprocessing of fossil fuels. CRC Press, Boca Raton, FL
Olson GJ, Turbak SC, McFeters GA (1979) Impact of western coal mining. II. Microbiological studies. Water Res 13:1033–1041
Olson GJ, McFeters GA, Temple KL (1981) Occurrence and activity of iron- and sulfur-oxidizing microorganisms in alkaline coal strip mine spoils. Microb Ecol 7:39–50

Palmer ME (1978) Acidity and nutrient availability in colliery spoil. In: Goodman GT, Chadwick MJ (eds) Environmental management of mineral wastes. Sijthoff & Noordhoff, Alphen aan den Rijn, pp 85–126

Pinto PX, Al-Abed SR, Holder C, Reisman DJ (2014) Evaluation of metal partitioning and mobility in a sulfidic mine tailing pile under oxic and anoxic conditions. J Environ Manage 140:135–144

Rickard D (2012) Sulfidic sediments and sedimentary rocks. Developments in Sedimentology, vol 65. Elsevier, Amsterdam

Sahoo PK, Tripathy S, Panigrahi MK, Equeenuddin SkMd (2014) Geochemical characterization of coal and waste rocks from a high sulfur bearing coalfield, India; Implication for acid and metal generation. J Geochem Explor 145:135–147

Sand W, Giehrke T, Hallman R, Schippers A (1995) Sulfur chemistry, biofilm, and the (in)direct attack mechanism – a critical evaluation of bacterial leaching. Appl Microbiol Biotechnol 43:961–966

Sand W, Giehrke T, Jozsa PG, Schippers A (2001) (Bio)chemistry of bacterial leaching – direct vs. indirect (bio)leaching. Hydrometallurgy 59:159–175

Sand W, Jozsa P-G, Kovacs Z-M, Săsăran N, Schippers A (2007) Long-term evaluation of acid rock drainage mitigation measures in large lysimeters. J Geochem Explor 92(2–3):205–211

Schippers A, Sand W (1999) Bacterial leaching of metal sulfides proceeds by two indirect mechanisms via thiosulfate or via polysulfides and sulfur. Appl Environ Microbiol 65:319–321

Schippers A, Jozsa P, Sand W (1996) Sulfur chemistry in bacterial leaching of pyrite. Appl Environ Microbiol 62:3424–3431

Schippers A, Rohwerder T, Sand W (1999) Intermediary sulfur compounds in pyrite oxidation: implications for bioleaching and biodepyritization of coal. Appl Microbiol Biotechnol 52:104–110

Schippers A, Jozsa P-G, Kovacs ZM, Jelea M, Sand W (2001) Large scale experiments for microbiological evaluation of measures for safeguarding sulfidic mine waste. Waste Manag 21:139–146

Schippers A, Kock D, Schwartz M, Böttcher ME, Vogel H, Hagger M (2007) Geomicrobiological and geochemical investigation of a pyrrhotite-containing mine waste tailings dam near Selebi-Phikwe in Botswana. J Geochem Explor 92:151–158

Schippers A, Breuker A, Blazejak A, Bosecker K, Kock D, Wright TL (2010) The biogeochemistry and microbiology of sulfidic mine waste and bioleaching dumps and heaps, and novel Fe (II)-oxidizing bacteria. Hydrometallurgy 104:342–350

Silverman MP (1967) Mechanism of bacterial pyrite oxidation. J Bacteriol 94:1046–1051

Silverman MP, Ehrlich HL (1964) Microbial formation and degradation of minerals. Adv Appl Microbiol 6:153–206

Silverman MP, Lundgren DC (1959) Studies on the chemoautotrophic iron bacterium *Ferrobacillus ferrooxidans*. J Bacteriol 77:642–647

Silverman MP, Rogoff MH, Wender I (1963) Removal of pyritic sulfur from coal by bacterial action. Fuel 15:113–124

Sima M, Dold B, Frei L, Senila M, Balteanu D, Zobrist J (2011) Sulfide oxidation and acid mine drainage formation within two active tailings impoundments in the Golden Quadrangle of the Apuseni Mountains, Romania. J Hazard Mater 189:624–639

Simate GS, Ndlovu S (2014) Acid mine drainage. Challenges and opportunities. J Hazard Mater 2 (3):1785–1803

Singer PC (1970) Oxygenation of ferrous iron. Contract PH 36-66-107 for Federal Quality Administration. Harvard University, Cambridge, MA

Singer PC, Stumm W (1970) Acidic mine drainage: the rate-determining step. Science 167:1121–1124

Stumm W, Morgan JJ (1996) Aquatic chemistry, 3rd edn. Wiley, New York

Szczepanska J, Twardowska I (1999) Distribution and environmental impact of coal mining wastes in Upper Silesia, Poland. Environ Geol 38(3):249–258

Szczepanska J, Twardowska I (2004) Mining waste. In: Twardowska I, Allen HE, Kettrup AAF, Lacy WJ (eds) Solid waste: assessment, monitoring and remediation, Waste Management Series 4(C). Elsevier, Amsterdam, pp 319–385

Tillet DM, Meyerson AS (1987) The removal of pyritic sulfur from coal employing *Thiobacillus ferrooxidans* in a packed column reactor. Biotechnol Bioeng 29(1):146–150

Tripathy SS, Kar RN, Mishra SK, Twardowska I, Sukla LB (1998) Effect of chemical pretreatment on bacterial desulphurisation of Assam coal. Fuel 77:859–864

Twardowska I (1986) The role of *Thiobacillus ferrooxidans* in pyrite oxidation in colliery spoil tips. I Model investigations. Acta Microbiol Pol 35(3/4):291–304

Twardowska I (1987) The role of *Thiobacillus ferrooxidans* in pyrite oxidation in colliery spoil tips. II Investigation of samples taken from spoil tips. Acta Microbiol Pol 36(1/2):101–107

Twardowska I, Galimska R (1993) Microbiological desulfurization of pulverized coal from the Janina mine, USCB, Poland. Paper presented at the International symposium on the extraction and processing of hard coal of Poland and Czech Republic in the environmentally safe way. Ostrava, 17–18 May 1993

Uhl W, Hone HJ, Beyer M, Klein J (1989) Continuous microbial desulfurization of coal. Biotechnol Bioeng 34:1341

Vera M, Schippers A, Sand W (2013) Progress in bioleaching: fundamentals and mechanisms of bacterial metal sulfide oxidation – part A. Appl Microbiol Biotechnol 97:7529–7541

Watling HR, Collinson DM, Li J, Mutch LA, Perrot FA, Rea SM, Reith F, Watkin ELJ (2014a) Bioleaching of a low-grade copper ore: linking leach chemistry and microbiology. Miner Eng 56:35–44

Watling HR, Collinson DM, Fjastad S, Kaksonen AH, Li J, Morris C, Perrot FA, Rea SM, Shiers DW (2014b) Column bioleaching of a polymetallic ore: effects of pH and temperature on metal extraction and microbial community structure. Miner Eng 58:90–99

Chapter 2
Microbe–Mineral Interactions: Exploring Avenues Towards Development of a Sustainable Microbial Technology for Coal Beneficiation

Srabani Mishra, Sandeep Panda, Nilotpala Pradhan, Surendra Kumar Biswal, Lala Behari Sukla, and Barada Kanta Mishra

2.1 Introduction

Microbes are ubiquitous in nature, and their specific interactions with the surrounding environment play a key role in the functioning of several processes occurring in our day-to-day life. The numerous microbe–mineral interactions act as a link to understand the innate mechanisms of certain phenomena taking place in nature and assist to study the mineral surfaces and changes occurring in minerals with respect to time and action of microbial species (Mapelli et al. 2012). Over the past few years, the interactions of selected microorganisms for beneficiation and bioleaching of the low-grade minerals/mineral wastes have been reported (Baba et al. 2011; Erust et al. 2013; Esther et al. 2013; Panda et al. 2012a, b, 2013a, 2015) and gradually gaining lot of importance for providing eco-friendly methods of environmental reclamation.

Coal is one such fossilized mineral comprising mainly of carbon and is formed due to the burial and conversion of dead plant matter under high pressure and temperature conditions. For several years, coal has been used worldwide as a principal source of energy in the generation of heat, electricity, and in the refining

S. Mishra (✉) • N. Pradhan
Bioresources Engineering Department, CSIR-Institute of Minerals and Materials Technology, Bhubaneswar, 751013 Odisha, India

Academy of Scientific and Innovative Research, CSIR-Institute of Minerals & Materials Technology, Bhubaneswar, 751013 India
e-mail: srabani_seema@yahoo.co.in

S. Panda • L.B. Sukla
Bioresources Engineering Department, CSIR-Institute of Minerals and Materials Technology, Bhubaneswar, 751013 Odisha, India

S.K. Biswal • B.K. Mishra
Mineral Processing Department, CSIR-Institute of Minerals and Materials Technology, Bhubaneswar, 751013 Odisha, India

L.B. Sukla et al. (eds.), *Environmental Microbial Biotechnology*, Soil Biology 45,
DOI 10.1007/978-3-319-19018-1_2

of metals. The multiple advantages of coal and its applications in diverse fields have made it a suitable area of research in the current era. For efficient and judicious use of coal, it is essential to enhance the quality of coal and make it a better fuel for several industries. However, the usage of coal also has some environmental issues. By-products generated by combustion of coal have resulted in greenhouse effect and acid rain. Environmental pollutants such as sulfur dioxide, sulfuric acid, and hydrogen sulfide have their origin from high-sulfur coal combustion. The present law framed by the governments all over the world has deals to lower level of sulfur emissions from coal industries associated with coal burning.

Thus to enhance the quality of coal, one of the integral components like sulfur needs to be removed (Prayuenyong 2002). Many strategies have been adopted to enhance the quality of coal by removing its sulfur component. Some of the conventional chemical and physical methods adopted for removal of sulfur from coal are shown in Fig. 2.1. Although such methods have responded effectually toward removal of sulfur, the use of hazardous chemical and elevated operating temperatures make such processes unattractive and less eco-friendly. The present scenario of research is particularly focused on the development of eco-friendly techniques.

To achieve this objective, it is necessary to understand the complex structure of coal and the various stages of coal formation which determine the rank of coal. Microbe–mineral interactions are very pervasive in nature. Since coal is a chief source of energy for wide variety of industries, the importance of microbe–mineral interaction is indispensible for developing a sustainable microbial coal biotechnology. The underlying necessity of microbe–mineral interaction is also linked with acid mine drainage that is a universal environmental problem in iron- and sulfur-rich environments. In the view of the fact that microbes act as a storehouse of several novel biomolecules or enzymes, they can be used for bioprocessing on an industrial scale incorporating innovative ideas and advanced technologies.

This review highlights some of the important aspects related to coal and its types along with possible interactions occurring between microbes and the mineral phases associated with coal. In addition, the available biological methods for industrial application ranging from lab-scale experiments to commercial-scale ventures are summarized and discussed. Moreover, understanding the inherent

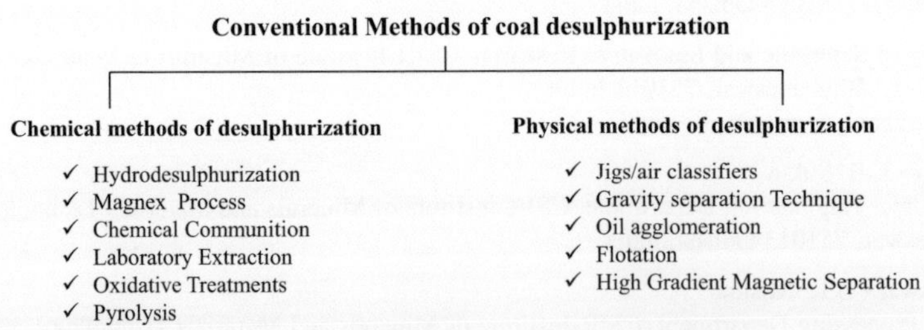

Fig. 2.1 A generalized flowchart depicting several conventional physicochemical methods adopted for desulfurization of coal (summarized from Soleimani et al. 2007; Kawatra and Eisele 2001, IPRO-346 report 2006)

mechanisms responsible in the microbe–mineral interactions would lead to the development of an efficient and eco-friendly technology toward the judicious use of natural resources and pave way toward a sustainable microbial coal biotechnology.

2.2 Types of Coal and Its Sulfur Content

Coal is a complex macromolecular structure that is formed by the combination of various biological and geological factors (Bhutto et al. 2013). Various organic and inorganic components constitute the complex structure. According to their stages of formation, different ranks of coal arise by the continuous burial of dead organic plant matter over time. The metamorphosis or the degree of alteration occurring during coal formation subsequently decides the so-called rank of a coal (Fakoussa and Hofrichter 1999). With increase in time, coal undergoes changes and passes through different ranks based on increase in burial pressure and temperature. High-rank coals have higher energy, lower volatile content, and shiny appearance than the low-rank coals. In general, most of the coals have a sulfur content varying from 0 to 4 %, but it can also go up to 10 % depending upon the region from which it has been extracted. In India, the sulfur content is also seen to vary in different regions of the state of Assam (Barooah and Baruah 1996). According to Soleimani et al. (2007), the average amount of sulfur in the USA coals ranges around 1.04–5.25 %. The organic sulfur of many coals may sometimes vary from 50 to 70 % of total sulfur content in coal (Soleimani et al. 2007). Since coal is a cheap and profuse source of energy, understanding the framework and composition of coal can assist in improving the energy value of coal. Some of the characteristic features of different ranks of coal are discussed below.

2.2.1 Peat

The process of coal formation begins with peat. Peat is considered to be the prerequisite for coal formation and is generally formed in water-logged regions like peat bogs, swamps, etc. (see http://en.wikipedia.org/wiki/Coal). It consists of various plant debris, decomposing plants, and plant products which undergo bacterial decay. Such materials experience continuous burial, supplemented with heat, time, and pressure leading to the formation of different ranks of coal. Basically, it is a decomposed fibrous material with typical content of volatile matter/weight percentage which is greater than 50 %. The typical weight percentage of carbon is 50–55 %, hydrogen 5–10 %, and oxygen 40 %. It acts as a useful absorbent for oil spills and has the ability to hold water and gradually release it which makes it a suitable conditioner (http://en.wikipedia.org/wiki/Coal). Seawater, freshwater, vegetation, and extraneous mineral matter can serve as an originating point for primary sulfur (S). Although freshwater having 0–10 ppm S is continuously circulated through peat, it cannot donate much sulfur to coal (Ryan and Ledda 1997).

2.2.2 *Lignite (Brown Coal: Immature)*

Lignite is the lowest rank of coal formed during coalification of plant matter concomitant with loss of some water and volatile matter by the action of increasing pressure for a prolonged period of time. In other words, with increase in burial pressure and temperature, peat is metamorphosed into lignite. Lignite has a lower heat content, high moisture, and high volatile matter content. Basically, it contains woody material with hard blocky appearance. Typical content of volatile matter/weight percentage is about 50 %. The typical element analysis/weight percentage for carbon is 60 %, hydrogen 6 %, and oxygen 30 %. Lignite is mostly used as a fuel in the generation of electricity. But the transportation issues associated with lignite due to its high moisture content and spontaneous combustion property make it a less preferable fuel worldwide as compared to other high-rank coals (see http://en.wikipedia.org/wiki/Coal). Some lignite contains a sulfur percentage of less than 1 % (see http://energy.about.com/od/Coal/a/Lignite.htm; Fisher 1974). However, typical Indian lignite contains a high sulfur percentage of 1–7 % (Selvakumaran et al. 2013).

2.2.3 *Subbituminous Coal*

With increase in time, the increase in temperature and pressure converts the light weight and soft lignite into a more advanced form of coal, i.e., the dark subbituminous type. Subbituminous coal has high water content with properties varying from lignite to that of bituminous coal. Basically, it is a dark material with slight developing luster. Typical content of volatile matter/weight percentage is 45 %. The typical element analysis/weight percentage for carbon is 75 %, hydrogen 5 %, and oxygen 20 %. They are mainly employed for steam-electric power generation (see http://en.wikipedia.org/wiki/Coal). Most of the subbituminous coal found in different regions have low sulfur content than bituminous coal (discussed in the next section and may have less than 1–2 % of sulfur content (see http://energy.about.com/od/Coal/a/Sub-Bituminous-Coal.htm; http://www.britannica.com/EBchecked/topic/570576/subbituminous-coal).

2.2.4 Bituminous Coal

As the process of coal maturation continues, the subbituminous coal experiences physical and chemical changes which thereby convert it into a more hard coal—the bituminous type. Bituminous coal has a black and shiny appearance. It is of improved quality when compared with lignite and subbituminous coal but has lesser carbon content as compared to anthracite (discussed below). The major constituents of bituminous coal are macerals such as liptinite and vitrinite. Typical content of volatile matter/weight percentage is 35 %. The typical element analysis/weight percentage for carbon is 85 %, hydrogen 5 %, and oxygen 10 %. The application of bituminous coal in various areas is attributed to its distinctive grade of plasticity, volatility, and lower ash content specifically carbonate, phosphorus, and sulfur. Bituminous coals are mostly used for production of coking coal, in steel making, and for power generation (see http://en.wikipedia.org/wiki/Bituminous_coal). The sulfur content may vary from 0.7 to 4 % in this rank of coal.

2.2.5 Anthracite Coal (Mature Coal)

It is the highest rank of coal. Following a strong maturation of the bituminous variety, the anthracite coal forms the ultimate classification of coal. Anthracite coal or black coal is a hard mineral with metallic luster, high relative density, low volatile matter, and high fixed carbon content. Typical content of volatile matter/weight percentage is <10 %. The typical element analysis/weight percentage for carbon is 90 %, hydrogen 4 %, and oxygen 5 %. It is mainly used for metallurgical purpose as a replacement for coke in practices such as sintering and pelletizing. In many countries, anthracite is being used as a fuel for smelting of iron. In the present era, anthracite finds its major use as a domestic fuel in stoves and furnaces. This hard coal has a sulfur content of less than 1 % with high heating value and higher density. The latter interferes with its ignition properties (see http://energy.about.com/od/Coal/a/Anthracite-Coal.htm).

The summary of the formation of different ranks of coal is shown as a flowchart in Fig. 2.2.

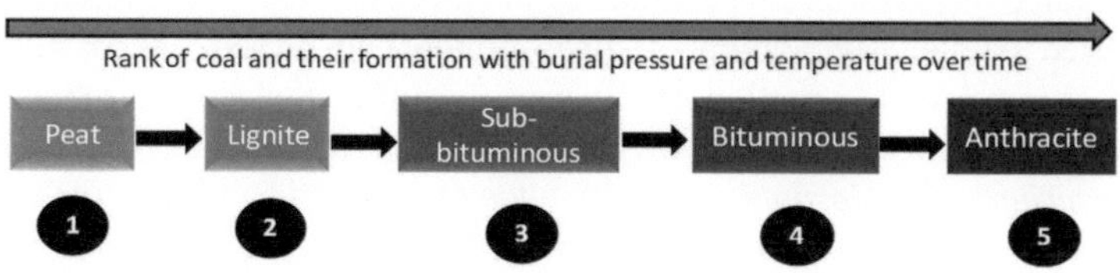

Fig. 2.2 A generalized flowchart for the formation of different ranks of coal with time

2.3 Microbes Reported for Beneficiation of Coal Through Desulfurization

Bacterial species belonging to the group α-, β-, γ-, and δ-proteobacteria have been found to predominate in the coal mine areas, while certain other bacterial lines of descent fall under the division *Nitrospira*, *Firmicutes*, and *Acidobacteria*. Similarly, archeal families comprising of *Thermoplasmatales* and *Sulfolobales* have been reported from the acid mine drainages (Siezen and Wilson 2009). Specific groups of microbes associate symbiotically with each other in order to fulfill their nutritional requirements and endure their growth and survival. However, few reports have additionally focused on the existence of eukaryotes in these sites (Baker and Banfield 2003). Majority of these microbial species employ essential enzymes required for various key transformations occurring in an acid mine drainage. Certain autotrophic groups are capable of oxidizing or reducing iron or sulfur to obtain their sole source of energy, whereas certain other heterotrophic groups are adept in utilizing organic compounds for their sustenance. Thus, most of the subsurface interactions occurring in the environment proceed through iron or sulfur oxidation, organic carbon oxidation, fixation of carbon or nitrogen, extracellular polymeric slime production, as well as iron and sulfur reduction.

Certain microorganisms are capable of modifying the structure of coal and degrading it. Both bacterial and fungal species have been found to disintegrate the coal structure and utilize the compounds present in coal as a source of carbon and sulfur to meet their nutrient requirements (Catcheside and Ralph 2002). Some bacterial species like *Pseudomonas fluorescens* release a very effective surfactant into the medium which lowers the surface tension of the medium, converts some coal into a more hydrophilic substance so that they become water soluble and can be easily taken up by the bacterial species to meet their nutrient requirement (Fakoussa 1981).

2.3.1 Organic S Removal

Initial efforts to remove organic sulfur from coal have been considered unsuccessful in view of the fact that the sulfur was not specifically removed from coal, and the calorific value of coal was also lost during the process. Most of the research works are focused on the degradation of dibenzothiophene present in coals, since it occupies a considerable portion of the organic sulfur. Several microbial species including *Rhodococcus erythropolis*, *Arthrobacter* spp., *Flavobacterium* spp., *Mycobacterium* spp., *Gordonia* spp., *Pseudomonas alcaligenes*, *Corynebacterium* spp., *Bacillus* spp., *Nocardia* spp., *Paenibacillus* spp., etc. are reported of having the capability to degrade various sulfur bearing organic compounds. Among these, certain species like *Microbacterium* ZD-M2 (Li et al. 2005; Soleimani et al. 2007),

Rhodococcus erythropolis IGTS8, *Rhodococcus erythropolis* D-1, *Mycobacterium* spp. G3, *Gordonia desulfuricans*, *Nocardia globerula* R-9 (Mingfang et al. 2003; Soleimani et al. 2007), and *Arthrobacter* species (Lee et al. 1995; Soleimani et al. 2007) could degrade dibenzothiophene via the 4S pathway or sulfur-specific pathway (discussed in Sect. 2.5.2.2). Other than these, certain thermophilic strains like *Bacillus subtilis* WU-S2B and *M. phlei* WU-F1 (Kirimura et al. 2004; Soleimani et al. 2007) and *Paenibacillus* spp. A11-2 (Ishii et al. 2000; Soleimani et al. 2007) also possessed the ability to remove sulfur from DBT via the sulfur-specific pathway. However, some other species such as *Pseudomonas alcaligenes*, *Pseudomonas putida* (Hartdegen et al. 1984; Monticello et al. 1985; Soleimani et al. 2007), *Brevibacterium* sp. DO (Nojiri et al. 2001; Dahlberg et al. 1993; Soleimani et al. 2007), and *Arthrobacter* K3b (Soleimani et al. 2007) were found to follow the carbon destructive pathway, i.e., the Kodama pathway of dibenzothiophene degradation (discussed in Sect. 2.5.2.1). Anaerobic strain like *Desulfovibrio desulfuricans* M6 have been efficient in degrading 96 and 42 % sulfur from benzothiophene and dibenzothiophene, respectively (Kim et al. 1990; Soleimani et al. 2007), by converting it to biphenyl and H_2S, while certain other species like *Desulfovibrio longreachii* and *Desulfomicrobium scambium* were found capable of degrading only 10 % dibenzothiophene dissolved in kerosene (Yamada et al. 2001; Soleimani et al. 2007). Strains of *Pseudomonas putida* were genetically modified to obtain better yield, since these strains are solvent tolerant and have a higher growth rate (Soleimani et al. 2007). Resting cells of genetically modified strain *Pseudomonas putida* A4 having solvent-tolerant capacity along with desulfurizing properties of *R. erythropolis* was able to degrade 86 % DBT in 10 % p-xylene in 6 h (Tao et al. 2006; Soleimani et al. 2007).

2.3.2 Inorganic S Removal

Inorganic sulfur mainly occurs in the form of sulfates and sulfides in coal. Since the sulfide particles exist as fine discrete crystals and are not bonded to the coal matrix, hence, they can be easily removed by physicochemical methods as well as biological methods, while sulfate sulfur easily gets dissolved and comes out when washed with water. Many sulfur-oxidizing microbes, namely *Acidithiobacillus ferrooxidans* (*A. ferrooxidans*), *Leptospirillum ferrooxidans* (*L. ferrooxidans*), *Beggiatoa*, *Sulfolobus*, *Acidithiobacillus thiooxidans* (*A. thiooxidans*) (Acharya et al. 2001; Masau et al. 2001), *Sulfolobus acidocaldarius* (*S. acidocaldarius*) (Kargi 1982), are reported of degrading sulfur compounds to form elemental sulfur or metal sulfides (Ryu et al. 1993). Among these, mesophiles such as *A. ferrooxidans*, *L. ferrooxidans*, and *A. thioxidans* are mostly involved in pyritic sulfur removal from coal as discussed by Prayuenyong (2002). Both *A. ferrooxidans* and *L. ferrooxidans* are capable of carrying out pyrite oxidation, whereas *A. thioxidans* is unable to utilize pyrite but can grow on sulfur released due to iron oxidation (Rawlings et al. 1999; Prayuenyong 2002). *A. ferrooxidans* is also reported of removing a small amount of organic sulfur from coal along with pyritic sulfur removal (Chandra et al. 1980). *L. ferrooxidans* are more adapted to lower pH conditions and posses higher affinity for ferrous iron when compared to *A. ferrooxidans*. They are also more resistant toward inhibition by higher ferric iron concentration, which assists them in overriding *A. ferrooxidans* (Prayuenyong 2002). Certain thermophiles such as *Thiobacillus caldus*, *Metallosphaera sedula* (*M. sedula*), *Acidianus brierleyi* (*A. brierleyi*), and *S. acidocaldarius* are also reported of pyrite oxidation (Schippers et al. 1999). A thermophilic and acidophilic facultative autotroph, *S. acidocaldarius*, is also reported of removing pyritic as well as organic sulfur from bituminous coal (Kargi and Robinson 1986). Reports have shown that it was capable of removing 90 % pyritic and approximately 19% of organic sulfur from coal (Kargi and Robinson 1986). However, while going for a commercialization, mesophiles are more preferred over thermophiles, since they operate at a moderate temperature and provide higher output with lower residence time (Rossi 1993; Schippers et al. 1999).

2.4 Susceptibility of Coal to Microbial Action

In general, the coal is a highly complex structured material. These complex structures have been worked out, proposed, discussed, and reviewed by many researchers (Fakoussa and Hofrichter 1999; Levine et al. 1982). Sulfur accounts for one of the major impurities in coal and mainly occurs in the form of organic sulfur and inorganic sulfur. The inorganic sulfur is predominantly present as sulfides and sulfates. For example, pyrite (FeS_2), sphalerite (ZnS), galena (PbS), arsenopyrite (FeAsS), etc. constitute the sulfide minerals, and barite ($BaSO_4$, gypsum ($CaSO_4$. $2H_2O$), anhydrite ($CaSO_4$), and a number of iron sulfates constitute the sulfate minerals. On the other hand, the organic sulfur in coal exists in four different groups as reviewed by Prayuenyong (2002). They are the following:

(a) Aliphatic or aromatic thiols (mercaptans, thiophenols)
(b) Aliphatic, aromatic, or mixed sulfides (thioesthers)
(c) Aliphatic, aromatic, or mixed disulfides (dithioesthers)
(d) Heterocyclic compounds or thiophene type (dibenzothiophenes)

In case of the inorganic sulfur in coal, pyrite forms the major portion. It is very interesting to know that pyrite is randomly distributed throughout the coal. The percentage of inorganic sulfur in coal changes from 0.5 to 5 % depending on the type of coal. The inorganic sulfur usually exists in the form of discrete crystals and gets easily separated from the coal structure by the help of physicochemical techniques of sulfur removal (Kawatra and Eisele 2001). On the other hand, the organic sulfur is covalently bonded to the complex carbon matrix of coal which makes it very difficult to be removed by any physicochemical means.

Some coals are composed of a matrix which cannot be degraded by microbial action and are termed as hard coal, whereas a mobile phase in coal is composed of nonoxidized compounds or volatile components like alkylated benzenes and alkanes and polyaromatic hydrocarbons which can be easily attacked by microbe. The mobile phase mostly includes phenols, benzoic acids, biphenyls, biphenyl ethers, n-alkenes, n-alkanols, and wax like aliphatics. Hard coals contain higher proportion of condensed aromatic rings and have a lower oxygen content and higher hydrophobicity due to which they are resistant to microbial attack (Hodek 1994). Fathoming the underlying relationship between the different aliphatic and aromatic hydrocarbons including other heteroatomic forms provides detail information about the coal chemistry and the possible chemical reactions it can undergo. The aromatic and naphthenic rings are interlinked to each other by aliphatic chains or heteroatomic bridges. Other polar groups like the hydroxyl group are also involved in the formation of a stable coal structure through electrostatic binding. The percentage of oxygen in coal plays an essential role in determining the coal structure and reactivity (Levine et al. 1982).

2.5 Microbial Coal Desulfurization

The following subsections discuss in details the microbe interactions with the inorganic and organic part of coal for sulfur removal. In addition, the details of the microbial systems for organic sulfur removal are also discussed.

2.5.1 Interaction Mechanisms for Inorganic Sulfur Removal

As already discussed in Sect. 2.4, pyrite forms the major portion of inorganic sulfur in coal. Certain group of chemolithoautotrophic meso-acidophilic bacteria has proved to be the correct answer to biodegradation of pyrite (Acharya et al. 2001). Such microorganisms utilize the iron source of pyrite for their growth and metabolism and are frequently used in bioleaching (Sand et al. 2001; Schippers and Sand 1999; Ciftci and Akcil 2010). The iron oxidizer *Acidithiobacillus ferrooxidans* (*A. ferrooxidans*) is considered to be the industrially important microorganism (Panda et al. 2012a; Pradhan et al. 2008). It is very important to note that *A. ferrooxidans* act as both an iron and sulfur oxidizer (under aerobic condition) and reducer under anaerobic conditions (Osorio et al. 2013). Different types of mechanisms have been proposed regarding the biological oxidation of pyrite by *A. ferrooxidans*, namely, the "*direct* (*contact*)" and "*indirect (noncontact*)" and "*cooperative*" bioleaching (Sand et al. 2001; Schippers and Sand 1999). Further, with the advance in the area bioleaching, two most coherent explanations are presently available for the indirect mechanisms, namely, the "*thiosulfate*" and "*polysulfide*" mechanisms (Schippers and Sand 1999; Schippers et al. 1996; Rohwerder et al. 2003). As shown in Fig. 2.3, the thiosulfate mechanism involves a ferric iron (Fe III) attack on the acid-insoluble metal sulfides, with thiosulfate

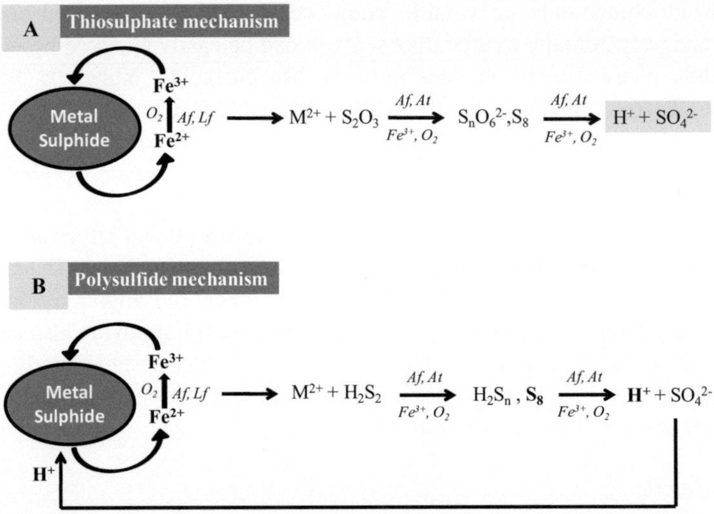

Fig. 2.3 Schematic comparison of two indirect mechanisms for metal sulfide (e.g., pyrite) degradation. (**a**) Thiosulphate mechanism and (**b**) polysulfide mechanism. M^{2+} - metal cations, *Af - Acidithiobacillus ferrooxidans, At - Acidithiobacillus thiooxidans*, and *Lf - Leptospirillum ferrooxidans* (Adapted from Schippers and Sand 1999)

being the main intermediate and sulfate being the main end product (Eqs. 2.1 and 2.2).

$$FeS_2 + 6Fe^{3+} + 3\,H_2O \rightarrow S_2O_3{}^{2-} + 7Fe^{2+} + 6H+ \quad (2.1)$$

$$S_2O_3{}^{2-} + 8Fe^{3+} + 5H_2O \rightarrow 2SO_4{}^{2-} + 8Fe^{2+} + 10H^+ \quad (2.2)$$

Similarly, in the polysulfide mechanism, the acid-soluble metal sulfide gets combined attack by ferric iron and protons, with elemental sulfur as the main intermediate. This elemental sulfur may be oxidized to sulfate by sulfur-oxidizing microbes (Fig. 2.3).

$$MS + Fe^{3+} + H^+ \rightarrow M^{2+} + 0.5H_2S_n + Fe^{2+}(n \geq 2) \quad (2.3)$$

$$0.5H_2S_n + Fe^{3+} \rightarrow 0.125S_8 + Fe^{2+} + H^+ \quad (2.4)$$

$$0.125S_8 + 1.5O_2 + H_2O \rightarrow SO_4{}^{2-} + 2H^+ \quad (2.5)$$

The ferrous iron may be reoxidized by iron-oxidizing organisms to ferric iron.

$$2Fe^{2+} + 0.5O_2 + 2H^+ \rightarrow 2Fe^{3+} + H_2O \quad (2.6)$$

The role of the microorganisms in the solubilization of metal sulfides as described through these mechanisms is to provide sulfuric acid for a proton attack and to keep the iron in the oxidized ferric state for an oxidative attack on the mineral.

These microorganisms are able to grow on pyrite and derive their energy. It is important to note that the natural geomicrobiological processes and bioleaching operations enable conversion of insoluble metal sulfides into water-soluble metal sulfates. At neutral pH and aerobic conditions, ferrous Fe (II) is rapidly oxidized to ferric iron [Fe (III)]. The accessibility of Fe (III) in biological systems becomes low as ferric iron is poorly soluble due to formation of iron hydroxides (as low as 10–18 M at pH 7.0) (Bonnefoy and Holmes 2012). However, it is interesting to note that under acidic conditions in the presence of atmospheric oxygen, Fe (II) is stable. This feature provides an excellent opportunity for the acidophilic microorganisms to use Fe (II) as an electron donor and as an energy source which is not a common feature for microorganisms inhabiting circumneutral environments (Austin and Dopson 2007; Bonnefoy and Holmes 2012).

2.5.2 Interaction Mechanisms for Organic Sulfur Removal

Sulfur is an essential requirement for microbes to sustain their biological activities and growth. It is involved in the formation of disulfide bonds in proteins and is also an integral part of amino acids such as cysteine and methionine. Certain enzyme cofactors such as biotin, thiamine, and coenzyme A incorporate sulfur in order to complete their structure. Since dibenzothiophene (DBT) constitutes a major organic portion of coal, microbes follow one of the different proposed pathways to meet their sulfur requirement from dibenzothiophene. The proposed pathways have been discussed below.

2.5.2.1 Aerobic Bio-desulfurization: The Oxidative C–C Cleavage (Kodama Pathway)

The aerobic bio-desulfurization via the Kodama pathway is followed by microbial species like *Pseudomonas alcaligenes*, *Pseudomonas putida*, *Burkholderia fungorum*, *Brevibacterium* spp., etc. (Kodama et al. 1970; Monticello et al. 1985; Yamada et al. 1968). Several microorganisms that follow the 4S pathway have been discussed in Sect. 2.3.1. These microorganisms use DBT as their carbon source through a series of enzymatic oxidation steps leading to the breakage of carbon–carbon (C–C) bond present in the ring structure. They form water-soluble end products, which inhibit further microbial growth and oxidation of DBT. The pathway is initiated by the attack on one of the phenyl rings and proceeds through three major steps, namely, hydroxylation, ring cleavage, and hydrolysis (Gupta and Roychoudhury 2005; Soleimani et al. 2007). This pathway is reported to be based on microbial plasmids that are associated in certain *Pseudomonas* spp. like *Pseudomonas alcaligenes* and *Pseudomonas putida* (Monticello et al. 1985). These plasmids are believed to mediate biodegradation of some other aromatic compounds such as naphthalene and salicylate in addition to DBT. It has been reported that a DNA fragment of length 9.8 kb has been responsible in encoding of degrading enzymes in the Kodama pathway (Soleimani et al. 2007). Since sulfur is not specifically attacked in this pathway, it is not considered as an efficient approach for DBT biodegradation. Therefore, this process is often referred to as destructive bio-desulfurization (Soleimani et al. 2007).

2.5.2.2 Aerobic Bio-desulfurization: The Oxidative C–S Cleavage (4S Pathway)

Owing to the nonspecific process of sulfur removal from DBT thiophene ring through the Kodama pathway, several attempts were made to isolate a variety of aerobic and anaerobic microbes that can specifically remove sulfur nondestructively. Several microorganisms that follow the 4S pathway have been discussed in Sect. 2.3.1.

With the isolation and characterization of the strain *Rhodococcus erythropolis* IGTS8 (formerly referred to as *Rhodococcus rhodochrous* IGTS8), a major advancement in DBT biodegradation was achieved, namely, the 4S pathway (Kilbane 1990). The 4S pathway represents a sequential degradation of DBT to DBT-sulfoxide followed by DBT-sulfone, DBT-sulfinate, HBP (hydroxybiphenyl) and finally to sulfite. Through a selective oxidation approach, the bacteria oxidize the sulfur of DBT without affecting the C–C bond. This process is often referred to as the specific oxidative desulfurization (Soleimani et al. 2007). Though bacteria such as *Brevibacterium* sp. DO and *Arthrobacter* sp. K3b partly followed the 4S pathway, they were reported to be carbon destructive because of the degradation of aromatic compounds during bio-desulfurization.

2.5.2.3 Anaerobic Bio-desulfurization: The Reductive C–S Cleavage

Apart from the role of aerobic microbes, several anaerobes have also shown promising results toward biodegradation of DBT. Certain anaerobic bacteria such as *Desulfovibrio desulfuricans*, *Desulfomicrobium scambium*, and *Desulfovibrio longreachii* have shown positive response toward DBT biodegradation. *Desulfovibrio desulfuricans* have been reported to degrade 96 % of BT and 42 % of DBT which can be related to the conversion of DBT to biphenyl and H_2S (Kim et al. 1990). On the other hand, *Desulfomicrobium scambium* and *Desulfovibrio longreachii* have been reported to desulfurize 10 % of DBT dissolved in kerosene. Interesting information through GC analysis was obtained which indicated that these microbes follow a somewhat different pathway from other anaerobes for bio-desulfurization (Yamada et al. 2001). The most fascinating part in anaerobic bio-desulfurization is the minimal formation of undesirable compounds such as colored and gum-forming products upon oxidation of hydrocarbons which calls for an exciting research area on reductive bio-desulfurization (McFarland 1999). However, maintenance of proper anaerobic conditions and precise attack of isolated microbial strains on DBT are some of the major challenges faced by the investigators for the establishment of a successful anaerobic desulfurization technique (Armstrong et al. 1995).

2.6 Role of Few Microbial Enzymes in Biodegradation of Sulfur Compounds

Various lignolytic enzymes like peroxidases, phenol oxidases, and other supporting enzymes are involved in the depolymerization of coal. Some oxidative enzymes such as monooxygenases are also responsible for distortion of the complex structure of coal.

2.6.1 *Oxygenase*

Oxygenases are one of the most vital enzymes involved in the degradation of aromatic compounds. They are responsible for the ring cleavage of aromatic ring structure, and based on the number of oxygen atoms they incorporate into, the aromatic ring structure is classified into monooxygenase and dioxygenase (Arora et al. 2009, 2010). Based on their cofactors monooxygenases are classified into two types, namely, flavin-dependent monooxygenase, which contains flavin as prosthetic group and require NADP or NADPH as coenzyme, and P450 monooxygenases, which are heme-containing oxygenases (Pazmiño et al. 2010). Some other bacterial monooxygenases such as pterin-dependent monooxygenases and metal ion-dependent monooxygenases do not have flavin or heme as cofactor (Pazmiño et al. 2010). Monooxygenases, which do not include a cofactor, require only oxygen for their activities and use the substrate as a reducing agent. Similarly, dioxygenases are also grouped into two different classes according to their mode of action, namely, the aromatic ring hydroxylation dioxygenases and aromatic ring cleavage dioxygenases (Arora et al. 2009, 2010).

2.6.2 *Lignin Peroxidases*

Since coal is composed of dead plant matter, lignin is the major constituent which is present in coal. The coalification process is accompanied by alterations in the physical properties and chemical structures of the basic plant material. But traces of the typical structure of lignin can be found in certain coals. Lignin peroxidase is one of the lignolytic enzymes that was first found in the white rot fungus *Phanerochaete chrysosporium* (Glenn et al. 1983; Tien and Kirk 1983) and later also isolated from other basidiomycete fungi like *Phlebia radiata* (Hatakka et al. 1987), *Trametes versicolor* (Dodson et al. 1987), *Bjerkandera adusta* (Kimura et al. 1991), and *Nematoloma frowardii* (Hofrichter and Fritsche 1997) and an ascomycete *Chrysonilia sitophila* (Duran et al. 1987). Lignin peroxidase has broad substrate specificity and can act on both phenolic and nonphenolic compounds (Odier et al. 1988; Schoemaker et al. 1985). The catalytic activity of lignin peroxidase is similar to that of horseradish peroxidase.

2.6.3 Manganese Peroxidases

Manganese peroxidase was identified from *P. chrysosporium* for the first time (Kuwahara et al. 1984). Its activity is similar to that of lignin peroxidase but has MnI and MnII as the specific substrate and mediators, respectively. Manganese peroxidase is also found in some other white rot and litter-decaying basidiomycetes like *T. versicolor* (Johansson and Nyman 1987), *P. radiata* (Nikku-Pavola et al. 1990), *Panus tigrinus* (Maltseva et al. 1991), *Lentinula edodes* (Forrester et al. 1990), *Ceriporiopsis subvermispora* (Ruettimann et al. 1992), *Pleurotus ostreatus* (Becker and Sinitsyn 1993, *Agaricus bisporus* (Bonnen et al. 1994), *N. frowardii* (Schneega et al. 1997), *Stropharia rugosoannulata* (Scheibner and Hofrichter 1998), *Clitocybula dusenii* (Ziegenhagen and Hofrichter 1998).

2.6.4 Laccase

It is a copper-containing polyphenol oxidase (Reinhammar and MalstroÈm 1981) produced by most of the lignolytic fungi like *Trametes versicolor* (Fahreus and Reinhammar 1967), *Phlebia radiata* (Kantelinen et al. 1989), *Pycnoporus cinnabarinus* (Eggert et al. 1996), and *Nematoloma frowardii* (Hofrichter and Fritsche 1997). The enzyme oxidizes various compounds simultaneously reducing O_2 to H_2O with a total reduction of four electrons. Laccase activity on phenolic aromatics results in the formation of phenoxy radicals, which can easily participate in radical–radical coupling, disproportionation, deprotonation, and nucleophilic attack by H_2O. Several reactions ultimately result in alkyl–aryl cleavage, C-α-oxidation, or demethoxylation of phenolic reductants (Kirk and Shimada 1985).

2.7 Future Prospects and Potential for a Microbial Technology

Microbial processing of several industrial wastes and/or low-grade ores has been gaining a lot of importance over the past few years (Baba et al. 2011; Panda et al. 2012c, 2013b, 2014). As a result, much focus is being generated for the application of microorganisms towards development of an industrial technology. Coal being a cheap source of renewable energy with numerous industrial applications is a potential material for future development of a microbial biotechnology. Furthermore, research in this field is particularly aimed at removal of sulfur from coal for environmental benefits and enhancement in the calorific value of coal, which would ultimately result in the development of a proficient source of energy with higher output. As already discussed in the above sections, biological application has gained importance to remove sulfur from coal using an eco-friendly approach. Therefore, it is a necessity to improve biological performance to bring about an efficient bio-based technology for coal beneficiation.

Many researchers have proposed numerous methods to improve bio-desulfurization of coal. Improving strain performance is an important aspect in coal biotechnology. Identification of novel strain with higher desulfurization capabilities can be expected to improve overall performance of the biological system employed for coal bio-desulfurization. Very recently, identification of novel microorganisms (iron- and/or sulfur-oxidizing chemolithoautotrophic acidophiles) through a web-based tool has been proposed to identify novel consortiums for bioleaching process (Parida et al. 2014). With the application of such web-based tool, a novel isolated consortium of iron- and/or sulfur-oxidizing chemolithoautotrophic acidophiles can be proposed and tested for bio-desulfurization (inorganic sulfur) efficiency for iron-rich coal. With such knowledge, more and more bioinformatic applications can also be proposed to identify many novel isolates for removal of organic sulfur from coal.

In addition to the microbial aspects of coal bio-desulfurization, nature and mineralogy of coal needs better understanding prior to microbial applications. Since coal varies in rank, composition and occurrence from region to region and its in-depth characterization would immensely help to enable the selection of microbes/microbial consortia, which can be employed to achieve maximum desulfurization for developing an efficient process with the application of novel technologies. For example, very recently, a two-step bio-desulfurization method has been proposed which sequentially removes organic sulfur followed by inorganic sulfur from coal (Mishra et al. 2014). The study presents a two-step sequential method using a microbe *Sinomonas flava* 1C to efficiently remove organic sulfur followed by removal of inorganic sulfur by *A. ferrooxidans*. In the coming years, more and more research in such interesting aspect of coal bio-desulfurization would improve its characteristic features and energy value, arousing the hope to cultivate an industrial-scale bioengineering system for an efficient eco-friendly coal biotechnology.

2.8 Conclusion

Microbe–mineral interactions are the fundamental basis of various geochemical and biogeochemical processes occurring in nature. A detail study into these interactions can open up novel areas of research, which are still waiting to be explored. Since coal is a major source of energy in the present era, understanding the fundamental mechanism behind coal formation, maturation, weathering, and degradation can provide more structural information on coal which would further help to enhance the calorific value, increase the energy output, and decrease negative aspects related to the use of this fossil fuel. In the end, understanding the microbe–mineral interactions in coal would facilitate in achieving the ultimate goal of efficiently engineered microbial processes, which would produce cleaner coal at ambient temperature and pressure in eco-friendly and economic way.

Acknowledgments The first author Mrs. Srabani Mishra and the second author Mr. Sandeep Panda are thankful to the Council of Scientific and Industrial Research (CSIR), New Delhi for the award of Senior Research Fellowship. NP and SKB are thankful to Ministry of Steel for financial support. All the authors would like to thank the Director of CSIR-IMMT for his kind permission to publish this paper.

References

Acharaya C, Kar RN, Sukla LB (2001) Bacterial removal of sulphur from three different coals. Fuel 80:2207–2216

Armstrong SM, Sankey BM, Voordouw G (1995) Conversion of dibenzothiophene to biphenyl by sulfate reducing bacteria isolated from oil field production facilities. Biotechnol Lett 17:1133–1136

Arora PK, Kumar M, Chauhan A, Raghava GP, Jain RK (2009) OxDBase: a database of oxygenases involved in biodegradation. BMC Res Notes 2:67

Arora PK, Srivastava A, Singh VP (2010) Application of monooxygenases in dehalogenation, desulphurization, denitrification and hydroxylation of aromatic compounds. J Bioremediat Biodegrad 1:1–8

Austin CB, Dopson M (2007) Life in acid: pH homeostasis in acidophiles. Trends Microbiol 15:165–171

Baba AA, Adekola FA, Atata RF, Ahmed RN, Panda S (2011) Bioleaching of Zn (II) and Pb (II) from Nigerian sphalerite and galena ores by a mixed culture of acidophilic bacteria. Trans Nonferrous Met Soc China 21:2535–2541

Baker BJ, Banfield JF (2003) Microbial communities in acid mine drainage. FEMS Microbiol Ecol 44:139–152

Barooah PK, Baruah MK (1996) Sulphur in Assam coal. Fuel Process Technol 46:83–97

Becker HG, Sinitsyn AP (1993) Mn-peroxidase from Pleurotus ostreatus: the action on the lignin. Biotechnol Lett 15:289–294

Bhutto AW, Bazmi AA, Zahedi G (2013) Underground coal gasification: from fundamentals to applications. Prog Energy Combust Sci 39:189–214

Bonnefoy V, Holmes D (2012) Genomic insights into microbial iron oxidation and iron uptake strategies in extremely acidic environments. Environ Microbiol 14:1597–1611

Bonnen AM, Anton LH, Orth AB (1994) Lignin-degrading enzymes of commercial button mushroom, Agaricus bisporus. Appl Environ Microbiol 60:960–965

Catcheside DEA, Ralph JP (2002) Bioconversion of coal by Fungi. Mycota 10:343–354

Chandra D, Roy P, Mishra AK, Chakrabarti JN, Prasad NK, Chaudhuri SG (1980) Removal of sulphur from coal by *Thiobacillus ferrooxidans* and by mixed acidophilic bacteria present in coal. Fuel 59:249–252

Ciftci H, Akcil A (2010) Effect of biooxidation conditions on cyanide consumption and gold recovery from a refractory flotation gold concentrate. Hydrometallurgy 104:142–149

Dahlberg MD, Rohrer RL, Fauth DJ, Sprecher R, Olson GJ (1993) Biodesulfurization of dibenzothiophene sulfone by *Arthrobacter* sp. and studies with Illinois no. 6 coal. Fuel 72:1645–1649

Dodson PJ, Evans CS, Harvey PH, Palmer JM (1987) Production and properties of an extracellular peroxidase from *Coriolus versicolor* which catalyses Ca-Cb cleavage in a lignin model compound. FEMS Microbiol Lett 42:17–22

Duran N, Ferrer I, Rodriguez J (1987) Ligninases from Chrysonilia sitophila (TFB-27441). Appl Biochem Biotechnol 16:157–167

Eggert C, Temp U, Eriksson KEL (1996) Laccase from the ligninolytic fungus Pycnoporus cinnabarinus. Appl Environ Microbiol 62:1151–1158

Erust C, Akcil A, Gahan CS, Tuncuk A, Deveci H (2013) Biohydrometallurgy of secondary metal resources: a potential alternative approach for metal recovery. J Chem Tech Biotechnol 88:2115–2132

Esther J, Panda S, Behera SK, Sukla LB, Pradhan N, Mishra BK (2013) Effect of dissimilatory Fe (III) reducers on bio-reduction and nickel–cobalt recovery from Sukinda chromite-overburden. Bioresour Technol 146:762–766

Fahreus G, Reinhammar B (1967) Large scale production and purification of laccase from cultures of the fungus *Polyporus versicolor* and some properties of laccase A. Acta Chem Scand 21:2367–2378

Fakoussa RM, Hofrichter M (1999) Biotechnology and microbiology of coal degradation. Appl Microbiol Biotechnol 52:25–40

Fakoussa RM (1981) Coal as a substrate for microorganisms: investigation of the microbial decomposition of (untreated) bituminous coals. Doctoral Thesis, Rhein Friedrich-Wilhelms University, Bonn

Fisher WL (1974) Texas lignite: near-surface and deep-basin resources. Bureau of economic geology The university of Texas at Austin, Texas 78712: report of investigations—79

Forrester IT, Grabski AC, Mishra C, Kelley BD, Strickland WN, Leatham GF, Bugess RR (1990) Characteristics and N-terminal amino acid sequence of a manganese peroxidase purified from Lentinula edodes culture grown on a commercial wood substrate. Appl Microbiol Biotechnol 33:359–365

Glenn JK, Morgan MA, Mayfield MB, Kuwahara M, Gold MH (1983) An extracellular H_2O_2-requiring enzyme preparation involved in lignin biodegradation by the white rot basidiomycete Phanerochaete chrysosporium. Arch Biochem Biophys 242:329–341

Gupta N, Roychoudhury PK (2005) Biotechnology of desulphurization of diesel: prospects and challenges. Appl Microbiol Biotechnol 66:356–366

Hartdegen FJ, Coburn JM, Roberts RL (1984) Microbial desulfurization of petroleum. Chem Eng Prog 80:63–67

Hatakka A, Kantelinen A, TervilaÈ -Wilo A, Viikari L (1987) Production of ligninases by Phlebia radiata in agitated conditions. In: Odier E (ed) Lignin-enzymic and microbial degradation. INRA, Paris, pp 335–340

Hodek W (1994) The chemical structure of coal in regard of microbiological degradation. Fuel Process Technol 40:369–378

Hofrichter M, Fritsche W (1997) Depolymerization of low-rank coal by extracellular fungal enzyme systems. II. The ligninolytic enzymes of the coal-humic-acid-degrading fungus Nematoloma frowardii b19. Appl Microbiol Biotechnol 47:419–424

IPRO 346 – fall (2006) Design of coal desulfurization processes to improve the environment midterm report, Illinois Institute of Technology, pp. 25

Ishii Y, Konoshi J, Okada H, Hirasawa K, Onaka T, Suzuki M (2000) Operon structure and functional analysis of the genes encoding thermophilic desulfurizing enzymes of Paenibacillus sp. A11-2. Biochem Biophys Res Commun 270:81–88

Johansson T, Nyman PO (1987) A manganese (II)-dependent extracellular peroxidase from the white-rot fungus Trametes versicolor. Acta Chem Scand B 41:762–765

Kantelinen A, Hatakka A, Viikari A (1989) Production of lignin peroxidase and laccase by Phlebia radiata. Appl Microbiol Biotechnol 31:234–239

Kargi F (1982) Microbial coal desulphurization. Enzyme Microb Technol 4:13–19

Kargi F, Robinson JM (1986) Removal of organic sulphur from bituminous coal: use of the thermophilic microorganism Sulfolobus acidocaldarius. Fuel 65:397–399

Kawatra SK, Eisele TC (2001) Coal desulphurization, high-efficiency preparation methods. Taylor and Franscis, New York

Kilbane JJ (1990) Sulfur-specific metabolism of organic compounds. Resour Conservat Recycl 69:590–596

Kim HY, Kim TS, Kim BH (1990) Degradation of organic compounds and the reduction of dibenzothiophene to biphenyl and hydrogen sulfide by Desulfovibrio desulfuricans M6. Biotechnol Lett 12:761–764

Kimura Y, Asada Y, Kuwahara M (1991) Molecular analysis of a Bjerkandera adusta lignin peroxidase gene. Appl Microbiol Biotechnol 35:510–514

Kirimura K, Harada K, Iwasawa H, Tanaka T, Iwasaki Y, Furuya T, Ishii Y, Kino K (2004) Identification and functional analysis of the genes encoding dibenzothiophene-desulfurizing enzymes from thermophilic bacteria. Appl Microbiol Biotechnol 65:703–713

Kirk TK, Shimada M (1985) Lignin biodegradation: the microorganisms involved and the physiology and biochemistry of degradation by white-rot fungi. In: Higuchi T (ed) Biosynthesis and biodegradation of wood components. Academic, Orlando, pp 579–605

Kodama K, Nakatani S, Umehara K, Shimizu K, Minoda Y, Yamada K (1970) Stoichiometry. Part III. Isolation and identification of products from dibenzothiophene. Agric Biol Chem 34:1320–1324

Kuwahara M, Glenn JK, Morgan MA, Gold MH (1984) Separation and characterization of two extracellular H_2O_2-dependent oxidases from ligninolytic cultures of Phanerochaete chrysosporium. FEBS Lett 169:247–250

Lee MK, Senius JD, Grossman MJ (1995) Sulfur-specific microbial desulfurization of sterically hindered analogs of dibenzothiophene. Appl Environ Microbiol 61:4362–4366

Levine DG, Schlosberg RH, Silbernagel BG (1982) Understanding the chemistry and physics of coal structure. Proc Natl Acad Sci 79:3365–3370

Li W, Zhang Y, Wang MD, Shi Y (2005) Biodesulfurization of dibenzothiophene and other organic sulfur compounds by a newly isolated Microbacterium strain ZD-M2. FEMS Microbiol Lett 247:45–50

Maltseva OV, Niku-Paavola MZ, Leontievsky AA, Myasoedova NM, Golovieva LA (1991) Ligninolytic enzymes of the white-rot fungus Panus tigrinus. Biotechnol Appl Biochem 13:291–302

Mapelli F, Marasco R, Balloi A, Rolli E, Cappitelli F, Daffonchio D, Borin S (2012) Mineral-microbe interactions: biotechnological potential of bioweathering. J Biotechnol 157:473–481

Masau RJ, Oh JK, Suzuki I (2001) Mechanism of oxidation of inorganic sulphur compounds by thiosulfate-grown *Thiobacillus thiooxidans*. Can J Microbiol 47:348–358

McFarland BL (1999) Biodesulfurization. Curr Opin Microbiol 2:257–264

Mingfang L, Zhongxuan G, Jianmin Z, Huizhou L, Jiayong C (2003) Microbial desulfurization of model and straight-run diesel oils. J Chem Technol Biotechnol 78:873–876

Mishra S, Panda PP, Pradhan N, Satapathy D, Subudhi U, Biswal SK, Mishra BK (2014) Effect of native bacteria *Sinomonas flava* 1C and *Acidithiobacillus ferrooxidans* on desulphurization of Meghalaya coal and its combustion properties. Fuel 117:415–421

Monticello DJ, Bakker D, Finnerty WR (1985) Plasmid-mediated degradation of dibenzothiophene by Pseudomonas species. Appl Environ Microbiol 49:756–760

Nikku-Pavola ML, Karhunen E, Kantelinen A, Viikari L, Lundell T, Hatakka A (1990) The effect of culture conditions on the production of lignin-modifying enzymes by white-rot fungus Phlebia radiata. J Biotechnol 13:211–221

Nojiri H, Habe H, Omori T (2001) Bacterial degradation of aromatic compounds via angular dioxygenation. J Gen Appl Microbiol 47:279–305

Odier E, Mozuch MD, KalyanaramanB KTK (1988) Ligninase-mediated phenoxy radical formation and polymerization unaffected by cellobiose: quinone oxidoreductase. Biochemie 70:847–852

Osorio H, Mangold S, Denis Y, Ñancucheo I, Esparza M, Johnson DB, Bonnefoy V, Dopson M, Holmes DS (2013) Anaerobic Sulfur metabolism coupled to dissimilatory iron reduction in the extremophile *Acidithiobacillus ferrooxidans*. Appl Environ Microbiol 79:2172–2181

Panda S, Parhi PK, Pradhan N, Mohapatra UB, Sukla LB, Park KH (2012a) Extraction of copper from bacterial leach liquor of a low grade chalcopyrite test heap using LIX 984N-C. Hydrometallurgy 121–124:116–119

Panda S, Sanjay K, Sukla LB, Pradhan N, Subbaiah T, Mishra BK, Prasad MSR, Ray SK (2012b) Insights into heap bioleaching of low grade chalcopyrite ores: a pilot scale study. Hydrometallurgy 125–126:157–165

Panda S, Sarangi CK, Pradhan N, Subbaiah T, Sukla LB, Mishra BK, Bhatoa GL, Prasad MSR, Ray SK (2012c) Bio-hydrometallurgical processing of low grade chalcopyrite for recovery of copper metal. Korean J Chem Eng 26:781–785

Panda S, Parhi PK, Nayak BD, Pradhan N, Mohapatra UB, Sukla LB (2013a) Two step meso-acidophilic bioleaching of chalcopyrite containing ball mill spillage and removal of the surface passivation layer. Bioresour Technol 130:332–338

Panda S, Pradhan N, Mohapatra UB, Panda SK, Rath SS, Nayak BD, Sukla LB, Mishra BK (2013b) Bioleaching of copper from pre and post thermally activated low grade chalcopyrite contained ball mill spillage. Front Environ Sci Eng 7:281–293

Panda S, Rout PC, Sarangi CK, Mishra S, Pradhan N, Mohapatra UB, Subbaiah T, Sukla LB, Mishra BK (2014) Recovery of copper from a surface altered chalcopyrite contained ball mill spillage through bio-hydrometallurgical route. Korean J Chem Eng 31:452–460

Panda S, Mishra S, Rao DS, Pradhan N, Mohapatra UB, Angadi S, Mishra BK (2015) Extraction of copper from copper slag: mineralogical insights, physical beneficiation and bioleaching studies. Korean J Chem Eng 32((4):667–676. doi:10.1007/s11814-014-0298-6

Parida BK, Panda S, Misra N, Panda PK, Mishra BK (2014) BBProF: an asynchronous application server for identification of potential bioleaching bacteria. Geomicrobiol J 31:299–314

Pazmiño DET, Winkler M, Glieder A, Fraaije MW (2010) Monooxygenases as biocatalysts: classification, mechanistic aspects and biotechnological applications. J Biotechnol 146:9–24

Pradhan N, Nathsarma KC, Rao KS, Sukla LB, Mishra BK (2008) Heap bioleaching of chalcopyrite. Miner Eng 21:355–365

Prayuenyong P (2002) Coal biodesulphurization process. J Sci Technol 24:493–507

Rawlings DE, Tributsch H, Hansford GS (1999) Reasons why 'Leptospirillum'-like species rather than *Thiobacillus ferrooxidans* are the dominant iron-oxidizing bacteria in many commercial processes for the biooxidation of pyrite and related ores. Microbiology 145:5–13

Reinhammar B, MalstroÈm BG (1981) Blue copper-containing oxidases. In: Spiro TG (ed) Copper proteins (Metal ions in biology, Vol 3). Wiley, New York, pp 109–149

Rohwerder T, Gehrke T, Kinzler K, Sand W (2003) Bioleaching review part A: progress in bioleaching: fundamentals and mechanisms of bacterial metal sulfide oxidation. Appl Microbiol Biotechnol 63:239–248

Rossi G (1993) Biodepyritization of coal: achievements and problems. Fuel 72:1581–1592

Ruettimann C, Schweber E, Salas L, Cullen D, Vicuna R (1992) Ligninolytic enzymes of the white-rot basidiomycete Phlebia brevispora and Ceriporiopsis subvermispora. Biotechnol Appl Biochem 16:64–76

Ryan B, Ledda A (1997) A Review of Sulphur in Coal: with Specific Reference to the Telkwa Deposit, North-Western British Columbia. Geological Fieldwork, 22 p.

Ryu HW, Chang YK, Kim SD (1993) Microbial coal desulphurization in airlift bioreactor by sulphur-oxidizing bacterium *Thiobacillus ferrooxidans*. Fuel Process Technol 36:267–275

Sand W, Gehrke T, Jozsa PG, Schippers A (2001) (Bio) chemistry of bacterial leaching—direct vs. indirect bioleaching. Hydrometallurgy 59:159–175

Scheibner K, Hofrichter M (1998) Conversion of aminonitrotoluenes by fungal manganese peroxidases. J Basic Microbiol 38:51–59

Schippers A, Sand W (1999) Bacterial leaching of metal sulfides proceeds by two indirect mechanisms via thiosulfate or via polysulfides and sulfur. Appl Microbiol Biotechnol 65:319–321

Schippers A, Jozsa P, Sand W (1996) Sulfur chemistry in bacterial leaching of pyrite. Appl Microbiol Biotechnol 62:3424–3431

Schippers A, Rohwerder T, Sand W (1999) Intermediary sulphur compounds in pyrite oxidation: implications for bioleaching and biodepyritization of coal. Appl Microbiol Biotechnol 52:104–110

Schneega I, Hofrichter M, Scheibner K, Fritsche W (1997) Purification of the main manganese peroxidase isoenzyme MnP2 from the white-rot fungus Nematoloma frowardii. Appl Microbiol Biotechnol 48:602–605

Schoemaker HE, Harvey PJ, Bowen RM, Palmer JM (1985) On the mechanism of enzymatic lignin breakdown. FEBS Lett 183:7–12

Selvakumaran P, Lawrence A, Lakshminarasimhan M, Bakthavatsalam AK (2013) Mineralogical influence of mining intrusions in CFB combustion of Indian lignite. Int J Energy Environ Eng 34:1–11

Siezen RJ, Wilson G (2009) Genomics update: bioleaching genomics. J Microb Biotechnol 2:297–303

Soleimani M, Bassi A, Margaritis A (2007) Biodesulphurization of refractory organic sulphur compounds in fossil fuels. Biotechnol Adv 25:570–596

Tao F, Yu B, Xu P, Ma CQ (2006) Biodesulfurization in biophasic systems containing organic solvents. Appl Environ Microbiol 72:4604–4609

Tien M, Kirk TK (1983) Lignin-degrading enzyme from the hymenocete Phanerochaete chrysosporium Burds. Science 221:661–663

Yamada K, Minoda Y, Kodama K, Nakatani S, Aasaki T (1968) Microbial conversion of petro-sulfur compounds. Part I. Isolation and identification of dibenzothiophene-utilizing bacteria. Agric Biol Chem 32:840–845

Yamada KO, Morimoto M, Tani Y (2001) Degradation of dibenzothiophene by sulfate-reducing bacteria cultured in the presence of only nitrogen gas. J Biosci Bioeng 91:91–93

Ziegenhagen D, Hofrichter M (1998) Degradation of humic acids by manganese peroxidase from the white-rot fungus Clitocybula dusenii. J Basic Microbiol 38:289–299

Chapter 3
Effect of Pollution on Aquatic Microbial Diversity

Anirban Chakraborty and Punyasloke Bhadury

3.1 Introduction

Microbes are ubiquitous across various environments ranging from soil to aquatic ecosystems. Based on salinity, aquatic environment can be categorized into two major division—freshwater environment and marine environment. Prokaryotic communities such as archaea and bacteria are predominant in both these types of environment and play key role in biogeochemical cycling and thereby actively contributing to ecosystem level processes.

A wide variety of synthetic compounds contaminate aquatic environment routinely as a result of mainly anthropogenic discharges such as from chemical and industrial processes. For example, organic loads can enter riverine or estuarine water thereby altering organic pool and affect degradation processes undertaken by *in situ* aquatic microbial communities. Many chemical compounds used for industrial enterprises, which are usually toxic and persistent, can enter aquatic domains, thereby affecting the biota, although some of the *in situ* microbial communities have the potential to degrade these chemical compounds. However, the ability of

A. Chakraborty
Integrative Taxonomy and Microbial Ecology Research Group, Department of Biological Sciences, Indian Institute of Science Education and Research Kolkata, Mohanpur, Nadia 741246, West Bengal, India

Department of Biological Sciences, The University of Calgary, Calgary, AB, Canada T2N 1N4

P. Bhadury (✉)
Integrative Taxonomy and Microbial Ecology Research Group, Department of Biological Sciences, Indian Institute of Science Education and Research Kolkata, Mohanpur, Nadia 741246, West Bengal, India
e-mail: pbhadury@iiserkol.ac.in

L.B. Sukla et al. (eds.), *Environmental Microbial Biotechnology*, Soil Biology 45,
DOI 10.1007/978-3-319-19018-1_3

microbial communities to degrade chemical compounds can be also controlled by physical and chemical factors that may persist in aquatic domains. For example, it is well known that aquatic contaminants which are hydrophobic can interact with dissolved organic matter and particulate organic matter present in the water column, and this can lead to physical partitioning (Lin et al. 2003; Akkanen et al. 2004). As a result, susceptible substrate can get into closer association with microbial communities capable of degrading such compounds. However, partitioning can also result in concentration of the contaminant to toxic levels, thus leading to suppression of microbially mediated biodegradation processes, and this can ultimately affect overall biological communities in an aquatic ecosystem. It is also well known that biological compounds such as lipids, proteins, nucleic acids, and amino acids can increase the solubility of certain chemical compounds such as chlorinated biphenyls and polycyclic aromatic hydrocarbons.

In this chapter, pollutants that routinely enter freshwater and marine environments have been discussed, and the ability of microbial communities to effectively biodegrade some of these pollutants has been also detailed. For marine environment, discussion has been mainly focused on pollutants such as oil and polycyclic biphenyl compounds that affect marine biota (planktonic and benthic) on a global scale.

3.2 Effect of Pollution on Freshwater Microbial Diversity

Freshwater generally has a salt content (salinity) of less than 1 g/L and makes up only 2.5 % of Earth's hydrosphere, of which 68.7 % is perennially frozen and 29.8 % is groundwater (Carpenter et al. 2011). Nevertheless, freshwater ecosystems are often referred to as the "blood of society" because of their overall significance in sustaining life and establishing civilizations throughout history (Wetzel 2000). Human settlements worldwide are concentrated near freshwater ecosystems, with over half of the world's population living within 20 km of a permanent river (Small and Cohen 2004). Understandably, a majority of Earth's accessible and renewable freshwater bodies are consistently being subjected to increasing pressure resulting from a variety of anthropogenic events, including but not limited to agricultural, industrial, and domestic activities. This in turn leads to water contamination through release of numerous anthropogenic and geogenic compounds. Consequently, it is no surprise that chemical pollution of natural waters has, over the course of time, become a major public concern (Schwarzenbach et al. 2006).

Microorganisms, especially prokaryotes, constitute a significant proportion of biomass in freshwater ecosystems and play crucial roles in elemental biogeochemical cycling. Freshwater aquatic environment is essential to many ecosystem services which are primarily mediated by microbial communities. Their significant position in aquatic food webs does not merely stem from their principal role in mineralizing organic matter to carbon dioxide, but also results from their active participation in catalysis and synthesis of organic matter (Pernthaler 2013). Surface freshwater systems such as lakes and rivers are generally more productive than open oceans and also receive higher loads of inorganic and organic carbon from a suite of terrestrial sources. In these systems, microbial respiration represents a large sink of organic carbon, and microbial activities are central to different pollutant degradation pathways. Diversity and richness of microbial populations in contaminated freshwater systems are generally regulated by the type and concentration of contaminants as well as adaptability of certain microorganisms to these environments by way of tolerating or utilizing the contaminants. The interplay between microbial communities and pollutants in freshwater systems will be explored in this section. We will limit our discussion to major freshwater habitats such as surface water (rivers, streams, and lakes), groundwater, and wetlands.

3.2.1 Surface Water Habitats

There is a great ecological diversity in surface waters. Lotic systems aside, even stagnant surface waters (lakes and ponds) represent rich diversity of microbial habitats. In comparison with marine ecosystems, surface freshwaters vary much more in their hydrological characteristics, productivity, water chemistry, and influences from surrounding terrestrial habitats (Hahn 2006). Microorganisms are introduced in surface waters from a multitude of sources, such as terrestrial habitats, biofilms, and sediments. Nevertheless, microbial community composition in lentic and lotic systems has been observed to be roughly comparable (Pernthaler 2013). Till the 1980s, there was a generalized assumption that freshwater bacterial species do not differ taxonomically from bacteria inhabiting surrounding terrestrial habitats (Hiorns et al. 1997). This assumption was based only on cultivation-dependent studies that probably resulted in successful cultivation of similar opportunistic taxa, while overlooking the potentially vast diversity of uncultivable microbes. However, this perception radically changed after the publication of the first investigations on the microbial diversity of freshwater lakes on the basis of culture-independent, molecular phylogenetic analysis in the 1990s (Hiorns et al. 1997; Zwart et al. 1998). Zwart and co-workers performed a meta-analysis of previously published studies that showed that the majority of bacterial sequences in different freshwater habitats were phylogenetically closely related (Zwart et al. 2002). Additionally, using habitat-specific clustering, the authors demonstrated that most bacteria inhabiting freshwater environments are indigenous and there is little compositional overlap with communities of terrestrial and marine habitats. Relative to the large number of different surface water habitats, however, the number of investigations studying microbial diversity of surface waters remains insufficient to this date. Therefore, it can be assumed that certain members of microbial communities in surface waters are so far undiscovered.

Rivers and streams are extremely heterogeneous ecosystems characterized by fluctuating input of inorganic and organic materials and various arrangements of habitat patchiness (Proia et al. 2012). Most pollutants enter river networks from direct application or from terrestrial runoff, e.g., wastewater treatment plants and industrial sewage that remain responsible for introducing some well-known toxicants as well as new emerging pollutants into these systems. Microorganisms in contaminated riparian systems possess certain characteristics such as different environmental responses within a functional group and divergent adaptive capabilities, thereby enhancing the stability of microbial communities. Branco and co-workers investigated the effects of continuous inputs of chromium-contaminated wastewater on the microbial assemblages of a river and its two tributaries (Branco et al. 2005). *Actinobacteria*, *Firmicutes*, and *Proteobacteria* were observed as the dominant phyla, and diversity of *Gammaproteobacteria* was particularly high in chromium-contaminated sites. Uniquely enough, total bacterial abundance in these sites was an order of magnitude higher than background sediments. The effects of local discharges from a wastewater treatment plant on the microbial communities of a small, low-gradient stream in South Australia were studied by Wakelin and co-workers (Wakelin et al. 2008). Using a combination of phylogenetic and functional-gene-based marker approaches, the authors demonstrated significant shifts in bacterial community composition influenced by the changes in physicochemical properties of the stream, especially high nutrient loading, associated with inputs from wastewater effluents. The highly affected stretches were found to be dominated by members of *Gammaproteobacteria*. More recently, García-Armisen and co-workers utilized 16S rRNA tag pyrosequencing approach to assess temporal fluctuations and resilience of indigenous bacterial communities of the Zenne River in Brussels, Belgium, a river highly disturbed by urban sewage inputs (García-Armisen et al. 2014). The bacterial communities upstream of the entry point of the wastewater effluents were observed to be markedly different than in downstream communities, which had overlapping similarities with the effluent microbial communities. Analysis of similarities (ANOSIM) revealed that the allochthonous microbes introduced to the river communities by wastewater discharges had significant impacts over the seasonal variations in community composition.

Other contaminants such as sulfide, heavy metals, and persistent hydrocarbons are of equal concern in the context of affecting microbial populations in river and stream ecosystems. In a recent study, Chen and co-workers estimated microbial diversity in the Shenzhen River in southern China, which is polluted with volatile sulfide compounds and undergoing bioremediation by external addition of nitrate (Chen et al. 2013). Using high-throughput sequencing of the hypervariable region V6 of the 16S rRNA marker, the authors found that bacterial richness and diversity decreased in areas with nitrate consumption due to the overabundance of certain functional groups, e.g., nitrate reducers. *Beta-*, *Gamma-*, and *Epsilonproteobacteria* were dominant in nitrate-treated sediments, while *Firmicutes*, *Chloroflexi*, and *Deltaproteobacteria* were dominant in the untreated sediments. Estimation of temporal bacterial diversity in the sediments of Clark Fork River, Montana, USA, a river exposed to complex mixtures of metals, by Bouskill and co-workers revealed a relatively stable, metal-tolerant community (Bouskill et al. 2010). Predominant community members belonged to *Beta-* and *Gammaproteobacteria*, and statistically significant relationship between community composition and metals was established, indicating natural selection towards a metal-resistant microbial population. Another study by Rastogi and co-workers gained insights into the microbial populations in sediments of a river polluted with significant amounts of toxic heavy metals (Rastogi et al. 2011). A combination of high-density 16S microarray (PhyloChip) and clone libraries specific to various functional groups of bacteria captured several chemolithoautotrophic groups of bacteria, suggesting a noticeable metabolic diversity among the indigenous microbial populations. Pratt and co-workers conducted a study that investigated the microbial communities in polyaromatic-hydrocarbon-contaminated riverbank sediments using phospholipid-derived fatty acid analysis (PLFA) (Pratt et al. 2012). The authors found that the subsurface sediments displayed highest proportions of PLFA biomarkers specific to anaerobic, heterotrophic bacteria, especially sulfate reducers, whereas methanotrophs were abundant in surface sediments. Another study undertaken recently by Hamonts and co-workers looked into the microbial community structure in eutrophic sediments of the Zenne River located in Vilvoorde, Belgium (Hamonts et al. 2014). These sediments were also contaminated with chlorinated aliphatic hydrocarbons (CAH) discharged into the riverbed from groundwater baseflow. A combined PCR-denaturing gradient gel electrophoresis (DGGE) analysis revealed that the dominant members were CAH-respiring *Dehalococcoides* and other functional groups such as methanogens and sulfate-reducing bacteria. The authors also observed that organic carbon content and CAH content were significant factors influencing the variance in community composition.

Freshwater lake ecosystems are presumably more vulnerable to different aquatic contaminants partly due to local pollution events from anthropogenic sources. Among other inland waters, lakes play a key role in global carbon budget and serve as sentinels indicating environmental changes through differential rates of carbon processing (Williamson et al. 2008). Needless to say, microbial activities greatly impact metabolic responses in lake ecosystems. Additionally, certain lakes provide special microhabitats supporting specific microbiota, e.g., variable organic matter content in the neuston layer (the air-water interface) of high alpine lakes (Hörtnagl et al. 2010) and stable redox transition layers due to permanently stratified water columns in meromictic lakes (Casamayor et al. 2008). In a nutshell, lake ecosystems can potentially harbor highly diverse microbial populations, and multi-lake studies have demonstrated that the difference in bacterial community composition between lakes can be quite large (Newton et al. 2011).

Effects of mercury contamination on the microbial communities in a eutrophic lake along spatiotemporal scale were investigated by Macalady et al. (2000). Although mercury methylation and organic carbon content in the sediment were primary factors influencing community composition, PLFA analysis revealed dominance of sulfate-reducing genera such as *Desulfobacter* and *Desulfovibrio*, suggesting significant roles of this functional group in mercury methylation. Gough and co-workers studied the microbial communities in anoxic sediments of Lake DePue, located in Illinois, USA, with a prolonged exposure history to heavy metals (Gough and Stahl 2011). Using terminal restriction fragment length polymorphism (T-RFLP) analysis, the authors observed prevalence of mesophilic Crenarchaeota populations which showed positive correlations with increasing metal concentrations. Lake DePue was the site for another microbial community study by Kang and co-workers, who used functional gene microarray (GeoChip) targeting multiple functional genes (Kang et al. 2013). The results showed that certain sulfate reduction genes (*dsr*) and a suite of metal-resistant genes were abundant and exhibited strong correlations with heavy metal concentrations, suggesting selection of metal-resistant and sulfate-reducing communities. Multiple studies have been conducted investigating microbial communities in Lake Geneva which, located in Switzerland and France, is one of the largest freshwater reservoirs in Europe and continually receives a huge volume of treated wastewater. Haller and co-workers used 16S rRNA clone library approach to characterize microbial communities from contaminated sediments of the Bay of Vidy near the northern shore of Lake Geneva (Haller et al. 2011). The bacterial community was found to be dominated by members of *Betaproteobacteria* closely related to *Dechloromonas*. Among other groups, sulfate- and iron-reducing clades belonging to *Deltaproteobacteria* were also abundant in the contaminated sediments. More recently, Sauvain and co-workers conducted microbial community profiling from similar locations in Lake Geneva and observed a high abundance of endospore-forming *Firmicutes*, closely related to *Clostridium* (Sauvain et al. 2014). This discovery has demonstrated the potential of spore-forming *Firmicutes* to be used as proxies for assessing stress conditions in lake environments (Wunderlin et al. 2014).

3.2.2 Groundwater Habitats

Groundwater makes up about one-third of world's total freshwater and is significant regarding global water use, accounting for approximately one-quarter of the total water withdrawal (Giordano 2009). In addition, groundwater harbors one of the largest populations of prokaryotes in the biosphere, with estimates of bacterial abundances in the range of $3.8–6 \times 10^{30}$ cells, constituting approximately 40 % of total prokaryotic biomass of the Earth (Whitman et al. 1998). Depending on the depth and topography, groundwater microorganisms can adapt to various growth conditions. On one end of this spectrum are deep anoxic habitats entirely devoid of organic carbon sources where geogenic hydrogen serves as the source of chemical energy for microbial growth. Microbial assemblages in such systems are usually dominated by methanogenic archaea, outnumbering bacteria substantially (Chapelle et al. 2002). Contrarily, shallow (<50 m below surface) subsurface systems that feature hypoxic to oxic zones and are comparatively richer in organic carbon host comparatively larger bacterial populations. Other than defined phylotypes, microorganisms affiliated with exotic phylogenetic lineages that are small enough to pass through filters of 0.2 μm pore size have also been reported to be associated with groundwater habitats (Miyoshi et al. 2005). However, till date there is no clear consensus regarding types of bacteria that can be considered typical for pristine groundwater systems (Pernthaler 2013).

Since groundwater serves as the major source of potable water in many parts of the world, particularly in developing countries, there are increasing concerns regarding groundwater pollution and associated water quality. Protection and restoration of groundwater, over the last decade, have garnered immense attention. Multiple factors such as precipitation, aquifer mineralogy, topography, and anthropogenic activities introduce contaminants in groundwater on various spatiotemporal scales. Groundwater pollutants can be of diverse nature, ranging from toxic heavy metals and metalloids to an array of persistent organic compounds. Microorganisms have been observed to be involved in biogeochemical cycling of almost all forms of groundwater contaminants and are recognized as key contributors towards bioremediation of contaminated subsurface systems. Consequently, immense efforts have been devoted to understanding microbial communities in contaminated groundwater systems (Griebler and Lueders 2009; Franzmann et al. 2002).

Heavy metals and metalloids, being more persistent than organic pollutants, are well-known contaminants of groundwater and have been subjected to exhaustive research in terms of different remediation strategies (Hashim et al. 2011). Heavy metals occur in the Earth's crust and get solubilized in groundwater due to changes in soil chemistry. On the other hand, these inorganic contaminants also get introduced to groundwater from landfill and mining leachate, sewage, and seepages from industrial waste lagoons. Certain guild of microorganisms, however, well adapt in heavy metal contaminated groundwater, withstanding the toxicity and also at times being able to harness energy from the redox reactions involving such metals (Colin et al. 2012). Hemme and co-workers used high-throughput-sequencing-based metagenomic analysis to assess the effects of high levels of uranium on groundwater microbial communities (Hemme et al. 2010). Their study indicated that prolonged exposure (more than 50 years) to heavy metals resulted in significant loss of species and metabolic diversity in the microbial community, which was dominated by *Beta-* and *Gammaproteobacteria*. Additionally, overabundance of stress response and resistance genes, likely propelled by the contaminants, was a striking observation of the functional profiling. The cumulative outcomes of this study led the authors to infer that lateral gene transfer could be a crucial adaptive mechanism, furnishing the surviving microbial communities with genes conferring resistance to stress incurred from different contaminants. More recently, Lin and co-workers examined vertical stratification of microbial diversity in subsurface sediments of the infamous Hanford Site, the most nuclear-waste-polluted site in the USA (Lin et al. 2012). The researchers analyzed 8,000 16S rRNA marker sequences from 21 samples collected from sediments of different geological strata including natural redox transition zones. Their study revealed 13 novel phylogenetic orders within *Deltaproteobacteria*, a class rich in microbes involved in redox transformations of heavy metals and radionuclides. Handley and co-workers employed PhyloChip analysis to determine diversity and abundance patterns in an acetate-amended, uranium-contaminated aquifer in Rifle, Colorado, USA.

Although overall community diversity was observed to be lower in acetate-stimulated communities relative to background sediments, abundance of operational taxonomic units (OTU) representing or closely related to *Delta-* and *Epsilonproteobacteria* was significantly higher (Handley et al. 2012). Sutton and co-workers characterized bacterial populations associated with arsenic mobilization in contaminated aquifers of Bangladesh (Sutton et al. 2009). The authors observed that microbial communities in the deep water Pleistocene aquifers were dominated by arsenic-tolerant bacteria, rather than arsenate- or Fe(III)-reducing bacteria, suggesting that arsenic mobilization through microbially mediated arsenate reduction may not occur at depth. Another study by Ghosh and co-workers investigated the temporal dynamics and community structure of arsenite-oxidizing bacteria from groundwater aquifers in Bengal Delta Plain (BDP) contaminated with high level of arsenic (Ghosh et al. 2014). Based on deep phylogeny analyses of libraries comprised of 460 clones representing arsenite oxidase subunit A (*aio*A) and 16S rRNA, the authors found that BDP aquifers harbor indigenous novel arsenite-oxidizing bacteria and their temporal dynamics were strongly influenced by seasonal fluctuations of arsenic concentration which was also linked with seasonal precipitation.

Petroleum hydrocarbons, polychlorinated organics, and antibiotics constitute another class of organic contaminants, frequently impacting shallow aquifers and indigenous microbial assemblages (Kleinsteuber et al. 2012). Dojka and co-workers studied the microbial diversity in aquifers contaminated with jet-fuel and chlorinated solvents using a 16S rRNA clone library and T-RFLP approach (Dojka et al. 1998). Among a total of 812 clones screened, 104 unique sequence types without any significant similarities with known taxonomic divisions were obtained. Additionally, sequence types characteristic of *Syntrophus* spp. and *Methanosaeta* spp. were observed. These findings led to the conclusion that terminal step of hydrocarbon degradation in the methanogenic zone of the aquifer was acetoclastic methanogenesis. In another study, Kleikemper and co-workers examined the diversity of sulfate-reducing bacteria (SRB) in an aquifer contaminated with petroleum hydrocarbons (Kleikemper et al. 2002). Using single well "push-pull" tests, the authors observed a certain degree of variability in the rate coefficients of sulfate reduction indicating activities of specific genera of SRB, whereas culture-independent analyses revealed a high SRB diversity. Another study used pyrosequencing for the characterization of the metagenome of a hydrocarbon-contaminated site in order to gain a comprehensive understanding of the microbial community composition and metabolic potentials (Abbai et al. 2012). Phylogenetic composition of the metagenome revealed that the community was dominated by heterotrophic members of *Alpha-*, *Beta-*, and *Gammaproteobacteria*. Metabolic profiling showed a higher percentage of genes associated with aromatic hydrocarbon degradation pathways, suggesting microbial biodegradation of organic contaminants. Among comparatively recent studies, Fahrenfeld and co-workers assessed the effects of biostimulants on microbial community composition in enrichment cultures obtained from a 2,4,6-trinitritoluene (TNT)-contaminated aquifer (Fahrenfeld et al. 2013). The researchers established anoxic microcosms amended with different organic substrates. Subsequent DGGE analysis showed the dominance of *Gamma-* and *Betaproteobacteria*, along with Gram-positive *Negativicutes* and *Clostridia*, suggesting their possible involvement in TNT biodegradation. In another contemporary study, Larentis and co-workers investigated the diversity and activity of toluene-degrading microorganisms from a tar-oil-contaminated aquifer (Larentis et al. 2013). The results pointed towards a novel, competitive niche partitioning behavior by aerobic and anaerobic hydrocarbon-degrading communities.

Nitrate-contaminated groundwater, aggravated by increasing nitrate input from intensive agriculture, livestock production, and other non-point sources, has been a major problem in many parts of the world, leading to inclusion of nitrate regulations and special protection for aquifers in the Safe Drinking Water Act. Various microorganisms, however, can effectively utilize nitrate either as a terminal electron acceptor during anaerobic respiration or as nitrogen source in the biomass. Consequently, bioremediation of nitrate-contaminated aquifers has received considerable scientific attention. Fields and co-workers investigated the prevalent microbial communities in acidified, nitrate-contaminated groundwater samples from The Field Research Center, Oak Ridge, Tennessee, USA (Fields et al. 2005). Using a 16S rRNA clone library approach, the authors analyzed 876 clones to determine the community structure in the contaminated samples relative to the background samples. The analysis revealed that microbial communities in the contaminated samples were less diverse and dominated by members of *Betaproteobacteria*, closely related to *Azoarcus*, a genus known for denitrifiers. Akob and co-workers used DNA-based stable isotope probing (SIP) to glean an understanding of the shifts in terminal electron-accepting processes during biostimulation of heterotrophic bacteria in subsurface sediments contaminated with uranium and nitrate (Akob et al. 2011). Using ^{13}C-ethanol-amended microcosms, the researchers observed that nitrate reduction was preferred by microbial communities (dominated by members of *Betaproteobacteria*) as the principal electron-accepting process compared to U(VI) and Fe(III) reduction. In another study, Torrentó and co-workers investigated the effects of addition of pyrite to nitrate contaminated groundwater and sediment samples (Torrentó et al. 2011). This study showed that pyrite addition led to enhanced rates of microbial denitrification coupled to Fe(II) oxidation and concomitant nitrate removal from both groundwater and sediment. Analysis of the microbial communities by a combined DGGE and quantitative-PCR (qPCR) approaches based on 16S rRNA marker and *nos*Z (nitrous oxide reductase) gene revealed the dominance of autotrophic, denitrifying bacterial populations closely related to *Xanthomonadaceae*. However, heterotrophic denitrifiers were also stimulated and contributed towards nitrate removal in spite of no external addition of organic carbon. The authors concluded that the heterotrophs probably used organic carbon from dead microbial biomass.

3.2.3 Wetland Habitats

Wetlands are widely recognized as unique ecosystems and a natural resource. Other than the presence of water, wetlands are usually characterized by unique soils different from upland soils and presence of vegetation adapted to saturated conditions (Scholz and Lee 2005). Wetlands often become landscape sinks, augmenting nutrients from enriched groundwater or surface runoff and thus possess an unusual capacity of removing contaminants from regional water bodies (Zedler and Kercher 2005). They play a particularly crucial role in biogeochemical nitrogen cycling by releasing gaseous nitrogen into the atmosphere by anaerobic, microbially regulated processes such as denitrification and anaerobic ammonium oxidation (anammox). Most wetlands are characterized by the prevalence of anoxic conditions as well as higher concentrations of nitrate and ammonium, creating ideal environments for anaerobic, nitrate-reducing microorganisms to thrive. A majority of these microorganisms can also utilize a range of organic pollutants as source of carbon, through respiratory or fermentative pathways. The capability of most wetlands to support such microbial communities has enhanced their importance in bioremediation. Brofft and co-workers investigated microbial diversity in a wetland adjacent to and impacted by an exposed reject coal pile in the US Department of Energy's Savannah River Site in South Carolina, USA, based on 16S rRNA clone libraries (Brofft et al. 2002). The results indicated that majority of the community members were represented by acidophilic organisms involved in sulfur and iron biogeochemical cycling. Bacterial clone libraries were dominated by sequences closely related to *Acidiphilium*, *Ferromicrobium*, and *Leptospirillum*, while the archaeal community consisted mainly of sequences related to the genus *Thermoplasma*. These libraries also exhibited particularly novel 16S rRNA marker sequences not retrieved from other acid mine drainage habitats. More recently, Gilbert and co-workers studied microbial communities and their abilities to tolerate high concentrations of 2,4-dichlorophenoxyacetic acids in three urban wetlands with different runoff regimes (Gilbert et al. 2012). A combined PCR and DGGE analysis revealed that despite having different input regimes and contaminant exposure histories, microbial community diversity and function in all three wetlands were generally similar.

Artificial wetlands, often referred to as constructed wetlands, are used for treating various forms of polluted waters, e.g., industrial and domestic wastewater (Scholz and Lee 2005). Since constructed wetlands are continuously fed by wastewater, microbial communities in these systems tend to be quite diverse, consisting of autochthonous and allochthonous groups of microorganisms, influent wastewater being the source for the latter group. A wide functional diversity also exists in constructed wetland microorganisms, including denitrifiers, anammox organisms, sulfate reducers, methanogens, and methanotrophs (Truu et al. 2009). These dynamic and heterogeneous systems have proven to be cost-effective alternatives to traditional wastewater treatment systems (Scholz and Lee 2005). Notwithstanding the importance of underlying microbial processes in constructed wetlands and related efficiency of contaminant removal, much remains to be known about the indigenous microbial communities, and a "black box" approach is usually applied during the design and operation of these systems (Faulwetter et al. 2009). Ibekwe and co-workers characterized the microbial communities in a constructed wetland used to treat dairy wastewater based on DGGE and sequencing of PCR-amplified fragments of the gene carrying the α-subunit of the ammonia monooxygenase gene (*amoA*) and observed dominance of phylotypes belonging to either *Firmicutes* or *Proteobacteria*, closely related to the types found in gastrointestinal tracts of mammals (Ibekwe et al. 2003). Additionally, phylogenetic analysis involving *amoA* gene showed a higher percentage of *Nitrospira*-like sequences. Microbial assemblages in a constructed wetland regularly fed by acidic drainage water from coal mines were assessed by Nicomrat and co-workers based on 16S rRNA PCR-DGGE approach (Nicomrat et al. 2006). Not surprisingly, the community was found to be less diverse and dominated by members of the genus *Acidithiobacillus*, widely recognized for flourishing in iron- and sulfur-rich, acid mine drainage environments. Among more recent studies, Zhu and co-workers investigated the diversity and abundance of anammox bacteria in lab-scale, vertical-flow wetland units fed by sludge (Zhu et al. 2011). Using a combined clone library and qPCR-based approach based on 16S rRNA, the authors observed a shift in community composition from a single *Candidatus Brocadia fulgida* to multiple dominant phylotypes such as *Candidatus Jettenia*, *Brocadia*, and *Anammoxoglobus*. In another study, dominance of *Proteobacteria* was observed in mesocosm-scale treatment wetlands for removal of heavy metals (arsenic and zinc) based on 16S rRNA clone library approach, and the highest bacterial diversity and richness was observed in wetlands planted with large wetland grass *Phragmites australis* (Arroyo et al. 2013). Most recently, Bai and co-workers used metagenomics for deep profiling of microbial diversity and functions in reed rhizosphere soils of a constructed wetland (Bai et al. 2014). Among different domains, bacteria were observed to be dominant in both soil and water, suggesting they were primarily responsible for pollutant removal from the system. *Proteobacteria* was the most abundant phylum and *Gammaproteobacteria* was the dominant class. Overall, microbial richness and diversity in rhizosphere soils was higher than in influent water. Functional annotation revealed a number of biodegradation pathways of xenobiotic compounds as well as biological manganese oxidation, implicating crucial roles played by the rhizosphere microbiota in removal of both nutrients and non-essential contaminants from wastewater.

3.3 Pollutants in Marine Ecosystem

Approximately 99.7 % of water found on planet Earth is contained in the oceans and seas. Contrary to freshwater counterparts, these habitats are typically saline in nature, although ranges of salinity can vary depending on the type of marine habitat. Marine environments are rich in microbial diversity which encompasses both prokaryotic and eukaryotic organismal groups. While many types of pollutant enter the marine ecosystem and predominantly affect habitats such as estuaries, coastal zones, coral reefs, and mangrove wetlands, in this chapter we have focused on two key pollutants that enter marine ecosystems globally and how prokaryotic communities, in particular bacteria, are able to cope up and degrade these pollutants.

3.3.1 Biodegradation of Oil Pollutants in Marine Environment

On an annual scale, approximately 1.3 million tonnes of petroleum enter the marine environment (NRC 2003). Additionally incidents such as the release of approximately 600,000 tonnes of crude oil after the Deepwater Horizon disaster in the Gulf of Mexico or the oil spill from Prestige oil tanker incident off the coast of northwest Spain can accelerate entry of petroleum products into the marine environment (ITOPF 2006; Crone and Tolstoy 2010). The fate of such spill can have immense consequences for marine biota (de la Huz et al. 2005).

Crude oil is a heterogenous mixture of hydrocarbons consisting predominantly of alkanes, cycloalkanes, and mono-aromatic and polycyclic aromatic forms (Harayama et al. 1999; Marshall and Rodgers 2004). Hydrocarbon molecules are relatively stable in marine environment, and their presence for prolonged period has influenced varieties of microbes through evolution to actively use these compounds as a major or sole source of carbon and energy. To date 175 genera of bacteria and several haloarchaeal genera have been identified that can grow on or transform hydrocarbons (Prince et al. 2010; Al-Mailem et al. 2010). Thus, hydrocarbon-contaminated marine environments are remediated predominantly by microbially mediated processes whereby crude oil is ultimately biodegraded into carbon dioxide and water.

Polycyclic aromatic hydrocarbons (PAHs) present in oil tend to have low water solubility and hydrophobicity, thus get associated with organic and inorganic suspended particles rapidly and ultimately get deposited in marine sediment. Since solubility of PAH decreases with increasing molecular weight, bioaccumulation of the same in marine organisms residing in sediment is generally greater for low molecular weight and more water-soluble PAHs (Singh 2012). In the literature, varying level of PAH has been reported in marine sediments, e.g., 48,000 ng/g have been found in Venice Lagoon (La Rocca et al. 1996), while in Todos Santos Bay, Mexico, the level was 800 ng/g (Macias-Zamora et al. 2002). However, many of the PAHs and other hydrocarbons can be degraded by prokaryotes, mainly bacteria which are present in marine sediment as well as in the water column.

Based on extensive research, it is known that following petroleum spill in the marine environment, there is a substantial increase in abundance of *Alcanivorax* spp., a Gram-negative bacterium belonging to the phylum *Proteobacteria*, which has the ability to degrade straight chain and branched alkanes, followed by *Cycloclasticus* spp., another member belonging to *Proteobacteria*, which can breakdown polyaromatic hydrocarbons (Harayama et al. 2004; Teira et al. 2007).

Using functional genomics and biochemical and physiological approaches, it has been shown that the oil-degrading bacterium *Alcanivorax borkumensis* possess multiple alkane-metabolism pathways including key enzymes such as alkane hydroxylases and three cytochrome P450-dependent alkane monooxygenases (Schneiker et al. 2006). Additionally, *A. borkumensis* possess a multitude of other adaptive capabilities to access hydrocarbons from marine environment including the ability to synthesize emulsifiers, form biofilms, scavenge nutrients, and resist to ultraviolet light (Schneiker et al. 2006; Sabirova et al. 2008). Besides *Cycloclasticus*, several other PAH-degrading bacterial genera such as *Vibrio*, *Marinobacter*, *Pseudoalteromonas*, *Marinomonas*, and *Halomonas* have been isolated from sediments of San Diego Bay, and many of them can grow on phenanthrene, a common polycyclic aromatic hydrocarbon prevalent in marine environment (Melcher et al. 2002). A study by Hilyard et al. (2008) based on DGGE approach coupled with classical enrichment techniques has shown that sediments in Elizabeth River (Virginia, USA) can harbor diverse group of bacteria such as *Bacteroidetes* and *Planctomycetes* that possess capabilities to degrade multiple polycyclic aromatic hydrocarbon (PAH) compounds such as naphthalene, phenanthrene, or pyrene.

Hedlund et al. (1999) isolated a novel bacterium, *Neptunomonas naphthovorans*, from creosote-contaminated Puget Sound sediment off the coast of USA which showed ability to utilize PAH compounds (2-methylnaphthalene and 1-methylnaphthalene), and they also identified naphthalene dioxygenase iron-sulfur protein (ISP) gene in the same bacterium which plays major role in PAH degradation. In 2001, Hedlund and Staley reported another bacterium, *Vibrio cyclotrophicus*, isolated from the same environment which showed metabolic capacity to utilize several two- and three-ring polycyclic aromatic hydrocarbons (PAHs) as substrates, including naphthalene, 2-methylnaphthalene, and phenanthrene. The authors based on gas chromatography experiments also showed that this bacterium was able to degrade several other PAHs but it was unable to use them as sole sources of carbon and energy (Hedlund and Staley 2001). When PAH-contaminated marine sediments of Baltic Sea were spiked additionally with phenanthrene and bromodeoxyuridine, followed by analysis of bromodeoxyuridine-labeled DNA, a remarkable diversity of putative PAH degraders belonging to the genera *Exiguobacterium*, *Shewanella*, *Methylomonas*, *Pseudomonas*, *Bacteroides*, as well as *Deltaproteobacteria* and *Gammaproteobacteria* were encountered (Edlund and Jansson 2008). Additionally, the authors found a significant fall in phenanthrene concentration in the spiked sediments which was linked to bacterial metabolism based on quantification of activity of genes (*xyl*E and *phn*Ac dioxygenase genes) involved in PAH degradation. A significant number of papers have been published in the last 10 years which have looked into aspects of isolation of novel bacterial strains from diverse environments including sediments contaminated with PAH compounds (e.g., review by Lu et al. 2011).

It is also well known that in cold marine environments, the psychrophilic alkane-degrading bacterium *Oleispira* is commonly associated with oil spills instead of *Alcanivorax* spp. (Coulon et al. 2007) and studies have shown the bacterium *Thalassolituus* spp. can outcompete *Alcanivorax* spp. in temperate environments (McKew et al. 2007).

Based on the genome sequencing of *Oleispira antarctica*, it has been shown that this bacterium has undergone massive gene transfer events and its large genome hosts an array of enzymes required for oil biodegradation including alkane monooxygenases as well as osmoprotectants, siderophores, and micronutrient scavenging pathways (Kube et al. 2013).

The ability of bacteria to degrade PAH and other hydrocarbons is dependent on several factors including composition of oil, degree of weathering, temperature, and ambient nutrient concentrations (Singh 2012). Besides metabolic capabilities, the success of degradation of oil pollutants by bacterial communities in marine environment is also controlled by the availability of nutrients such as nitrate and phosphate. Hence, degradation of petroleum including polyaromatic hydrocarbon compounds in natural marine environment is limited by nutrient availability, and when nutrients are additionally added to the ecosystem, the breakdown process of these persistent hydrocarbons is greatly enhanced. Several studies have also shown that microbes in marine environment have also evolved mechanisms to access PAHs which have low solubility and high levels of adsorption. It has been shown that some PAH-degrading microbial organisms have high-affinity uptake systems that efficiently reduce the PAH concentration close to the cell surface, thereby enhancing diffusive flux (Bastiaens et al. 2000). In the bacterium *Pseudomonas putida*, production of extracellular polymeric substances aids towards attachment of the cell to solid PAHs (Rodrigues et al. 2005). Production of biosurfactant molecules such as glycolipids, rhamnolipid, trehalose lipids, and lipopeptides by microbes including bacteria is known to enhance bioavailability of PAH compounds (e.g., Perfumo et al. 2010). Besides, interaction between different groups of organisms could also influence success rate of aerobic and anaerobic degradation of hydrocarbons including PAHs.

3.3.2 Biodegradation of Polychlorinated Biphenyls in Marine Environment

Polychlorinated biphenyls (PCBs), a class of compounds containing aromatic biphenyl rings, are extensively used for industrial and commercial applications (Pieper 2005). PCBs are considered as carcinogens because some of them have shown to induce hepatic tumor development in rats (Pieper 2005). Additional noncancerous effects in the endocrine system, reproductive system, and immune and nervous system have been also documented (Chen et al. 1980). There are evidences that PCB contamination in aquatic ecosystems is widespread including those in estuarine and marine sediments (Berkaw et al. 1996). Degradation of PCBs in aquatic environment is a slow process, and therefore they are highly ubiquitous contaminant in coastal ecosystems. For example, high concentration of PCB (6.570 ng/g; dry weight) was found in marine sediments from Chennai harbor with distinct PCB distribution pattern observed across other coastal harbor areas (Rajendran et al. 2005).

Biodegradation of PCBs in marine environment can be facilitated by microbes, in particular bacteria, through aerobic degradation (Focht 1995) and anaerobic dechlorination (e.g., Sokol et al. 1994). A study undertaken by De et al. (2006) has reported the isolation of a bacterium *Pseudomonas aeruginosa* CH07 (NRRL B-30604) from a contaminated coastal site in India which can degrade polychlorinated biphenyls of different configurations aerobically based on measurement using gas liquid chromatography. The authors have shown that the bacterium was sufficiently able to degrade highly chlorinated PCB congeners, CB-180 and CB-181 while on the other hand, it was unable to grow on biphenyl as the sole carbon source. Based on enrichment of marine sediments collected from the Croatian Adriatic coast, it has been shown using gas chromatography-mass spectrometry measurements that bacteria present in sediments can degrade PCBs under aerobic conditions (Kolar et al. 2007). Moreover, the identified bacterial taxa belonging to *Rhodococcus* and *Sphingomonas* were able to grow in the presence of biphenyl as the sole carbon source, and strains of *Rhodococcus* showed substantial PCB-degrading activity indicating the widespread presence of these bacteria in marine environment contaminated with pollutants (Kolar et al. 2007).

Bacteria which are capable of degrading PCBs have been also found in cold marine environments such as from the Antarctica region. Marine bacteria isolated from Antarctica seawater by culture-dependent methods and confirmed by 16S rRNA sequencing belonging to the genera *Pseudoalteromonas*, *Psychrobacter*, and *Arthrobacter* were found to degrade PCB congeners (35.6–82.8 %) at temperature ranging between 4 and 15 °C (Michaud et al. 2007). The authors found differences in the removal patterns of PCB congeners in relation to the phylogenetic affiliation: *Arthrobacter* isolate showed similar biodegradation efficiencies when growing at 4 and 15 °C, while *Pseudoalteromonas* showed higher degradation of PCBs at 15 °C. But in case of *Psychrobacter*, no biodegradation was detected at 4 °C. This study clearly highlights the occurrence of PCB-degrading bacteria in Antarctica seawater and provides an opportunity for utilization of these novel bacteria towards bioremediation of PCB.

Two psychrotrophic bacterial strains (B11 and B15) isolated from seawater of Ross Sea (Antarctica) representing marine variants of the bacterium *Rhodococcus fascians* have been found to effectively utilize PCBs as sole carbon source at 4 and 20 °C (De Domenico et al. 2004). The authors have pointed bacteria isolates from cold environments such as marine realm of Antarctic could be better suited for *in situ* bioremediation of persistent PCBs in both temperate and cold marine environments than their mesophilic counterparts.

It has been shown in published literature that bacteria which can degrade PCBs are known to possess an enzyme known as biphenyl dioxygenase (*bph* locus)

(Seeger et al. 2010). Research undertaken on several biphenyl 2,3-dioxygenases (*bph*As) of mostly terrestrial bacterial origin has revealed considerable differences in their PCB selectivity as well as their preference of the oxidized ring (McKay et al. 1997; Seeger et al. 1999, 2001).

In the modern era of genomics, whole genome of PCB-degrading bacterial strains, majorly of terrestrial origin, has been sequenced. For example, *bph* genes encoding enzymes of the biphenyl catabolic pathway have been identified in *Burkholderia xenovorans* LB400 and *Rhodococcus jostii* RHA1 genomes, and they are located on mobile genetic elements (Chain et al. 2006; McLeod et al. 2006). In the bacterium *B. xenovorans* LB400, *bph* genes were found to be located in a genomic island on the megaplasmid indicating its origin through horizontal gene transfer (Chain et al. 2006). But in case of *R. jostii* RHA1, the *bph* genes were found to be encoded by two plasmids (McLeod et al. 2006). Incidentally, two of the recently sequenced PCB-degrading bacterial genomes, namely, *Cupriavidus* sp. SK4 and *Dehalococcoides mccartyi* SG1, have been also isolated from sludge system (Vilo et al. 2014; Wang et al. 2014), thereby highlighting the dearth of information on PCB-degrading molecular pathways in bacteria of typically marine origin.

In this modern era of next-generation sequencing (NGS), numerous studies have attempted to understand how structure and function of microbial communities can be altered as a result of oil spill in marine environments. The Deepwater Horizon oil spill in the Gulf of Mexico is one such example which resulted in a deep-sea hydrocarbon plume that caused major shifts in indigenous microbial community composition with ecological consequences which are still unknown to date and yet to be adequately quantified. Several studies have investigated effects of Deep Water Horizon oil spill on microbial communities using an array of techniques including next-generation sequencing (e.g., Hazen et al. 2010; Valentine et al. 2010; Kessler et al. 2011). For example, in a recent study, Mason et al. (2012) have investigated the functional role of *Oceanospirillales* and other active members of the indigenous microbial community following the Gulf of Mexico incident based on deep sequencing of community DNA and RNA along with single-cell genomics. The authors found that genes which are responsible for motility, chemotaxis, and aliphatic hydrocarbon degradation were significantly enriched and expressed in hydrocarbon plume samples compared with uncontaminated seawater collected from plume depth. Additionally, they showed that genes coding for degradation of more recalcitrant compounds, such as benzene, toluene, ethylbenzene, total xylenes, and polycyclic aromatic hydrocarbons, were identified in the metagenomes; however, they were expressed at low levels, or not at all based on analysis of the metatranscriptomes. The authors using single-cell genomics approach of two *Oceanospirillales* single cells revealed that both cells possessed genes coding for *n*-alkane and cycloalkane degradation. The authors showed that members of *Oceanospirillales* can respond rapidly to aliphatic hydrocarbons as part of oil spill in marine environment. In another study, Liu and Liu (2013) investigated bacterial community structure in oil samples using next-generation sequencing of 16S rRNA including oil mousses collected on sea surface and salt marshes during the Deepwater Horizon oil spill and oil deposited in sediments adjacent to the wellhead 1 year after the spill. Based on molecular phylogeny of generated data, the authors found that *Erythrobacter*, *Rhodovulum*, *Stappia*, and *Thalassospira* belonging to *Alphaproteobacteria* were most prevalent in oil mousses, and the authors accounted their occurrence with high temperature and strong irradiance observed in surface water of the Gulf of Mexico. The authors reported that bacterial communities in oil-contaminated sediments were highly diversified and there was high abundance of the *Methylococcus*, *Methylobacter*, *Actinobacteria*, *Firmicutes*, and *Chloroflexi* bacteria resembling those found in certain cold-seep sediments with gas hydrates. In the overlying water of the oil-contaminated sediment, there was dominance of *Ralstonia*, a bacterium belonging to *Betaproteobacteria*, which can degrade small aromatics, and *Saccharophagus degradans* of *Gammaproteobacteria*, a cellulose degrader, suggesting that overlying water was affected by the oil-contaminated sediments, possibly due to the dissolution of small aromatics and biosurfactants produced during biodegradation.

As discussed in this chapter with respect to two major groups of pollutants in marine environment, bacterial communities have shown capabilities to be able to degrade and ultimately bioremediate environment from toxic pollutants. However, our understanding about the combined effects of these pollutants on microbial community present in marine ecosystems is not very clear to date and particularly from the context of toxicology. For example, the response of combination of pollutants along with heavy metal pollutants can have serious effect even on prokaryotic systems, and ultimately pollutants can persist in environment even for a longer time scale. Therefore, in future studies, laboratory-based manipulation experiments involving bacterial communities isolate from warm water and cold water marine environments should be investigated in the presence of a combination of persistent pollutants. Incorporation of modern methodologies such as transcriptomics and metabolomics can significantly increase our understanding of the effects of pollutants on microbial cell systems and ultimate long-term impacts on ecosystem processes.

References

Abbai N, Govender A, Shaik R, Pillay B (2012) Pyrosequence analysis of unamplified and whole genome amplified DNA from hydrocarbon-contaminated groundwater. Mol Biotechnol 50:39–48

Akkanen J, Vogt RD, Kukkonen JVK (2004) Essential characteristics of natural dissolved organic matter affecting the sorption of hydrophobic organic contaminants. Aquat Sci 66:171–177

Akob DM, Kerkhof L, Küsel K, Watson DB, Palumbo AV, Kostka JE (2011) Linking specific heterotrophic bacterial populations to bioreduction of uranium and nitrate in contaminated subsurface sediments by using stable isotope probing. Appl Environ Microbiol 77:8197–8200

Al-Mailem DM, Sorkhoh NA, Al-Awadhi H, Eliyas M, Radwan SS (2010) Biodegradation of crude oil and pure hydrocarbons by extreme halophilic archaea from hypersaline coasts of the Arabian Gulf. Extremophiles 14:321–328

Arroyo P, Ansola G, Miera LES (2013) Effects of substrate, vegetation and flow on arsenic and zinc removal efficiency and microbial diversity in constructed wetlands. Ecol Eng 51:95–103

Bai Y, Liang J, Liu R, Hu C, Qu J (2014) Metagenomic analysis reveals microbial diversity and function in the rhizosphere soil of a constructed wetland. Environ Technol 35:2521–2527

Bastiaens L, Springael D, Wattiau P, Harms H, de Wachter R, Verachtert H, Diels L (2000) Isolation of adherent polycyclic aromatic hydrocarbon (PAH)- degrading bacteria using PAH-sorbing carriers. Appl Environ Microbiol 66:1834–1843

Berkaw M, Sowers KR, May HD (1996) Anaerobic ortho dechlorination of polychlorinated biphenyls by estuarine sediments from Baltimore Harbor. Appl Environ Microbiol 62:2534–2539

Bouskill N, Barker-Finkel J, Galloway T, Handy R, Ford T (2010) Temporal bacterial diversity associated with metal-contaminated river sediments. Ecotoxicology 19:317–328

Branco R, Chung A-P, Veríssimo A, Morais PV (2005) Impact of chromium-contaminated wastewaters on the microbial community of a river. FEMS Microbiol Ecol 54:35–46

Brofft JE, McArthur JV, Shimkets LJ (2002) Recovery of novel bacterial diversity from a forested wetland impacted by reject coal. Environ Microbiol 4:764–769

Carpenter SR, Stanley EH, Vander Zanden MJ (2011) State of the world's freshwater ecosystems: physical, chemical, and biological changes. Annu Rev Environ Resour 36:75–99

Casamayor EO, García-Cantizano J, Pedrós-Alió C (2008) Carbon dioxide fixation in the dark by photosynthetic bacteria in sulfide-rich stratified lakes with oxic-anoxic interfaces. Limnol Oceanogr 53:1193–1203

Chain PS, Denef VJ, Konstantinidis KT, Vergez LM, Agulló L, Latorre-Reyes V, Hauser L, Córdova M, Gómez L, González M, Land M, Larimer F, LiPuma JJ, Mahenthiralingam E, Malfatti SA, Marx CJ, Parnell J, Ramette A, Richardson P, Seeger M, Smith D, Spilker T, Sul WJ, Tsoi TV, Ulrich LE, Zhulin I, Tiedje J (2006) *Burkholderia xenovorans* LB400 harbors a multi-replicon, 9.73-Mbp genome shaped for versatility. Proc Natl Acad Sci U S A 103:15280–15287

Chapelle FH, O'Neill K, Bradley PM, Methe BA, Ciufo SA, Knobel LL, Lovley DR (2002) A hydrogen-based subsurface microbial community dominated by methanogens. Nature 415:312–315

Chen PH, Wong CK, Rappe C, Nygren M (1980) Polychlorinated biphenyls, dibenzofurans and quaterbiphenyls in toxic rice bran oil and in the blood and tissue of patients with PCB poisoning (Yu-Cheng) in Taiwan. Environ Health Perspect 59:59–65

Chen L, Wang L-Y, Liu S-J, Hu J-Y, He Y, Zhou H-W, Zhang X-H (2013) Profiling of microbial community during in situ remediation of volatile sulfide compounds in river sediment with nitrate by high throughput sequencing. Int Biodeterior Biodegrad 85:429–437

Colin VL, Villegas LB, Abate CM (2012) Indigenous microorganisms as potential bioremediators for environments contaminated with heavy metals. Int Biodeterior Biodegrad 69:28–37

Coulon F, McKew BA, Osborn AM, McGenity TJ, Timmis KN (2007) Effects of temperature and biostimulation on oil-degrading microbial communities in temperate estuarine waters. Environ Microbiol 9:177–186

Crone TJ, Tolstoy M (2010) Magnitude of the 2010 Gulf of Mexico oil leak. Science 330:634

De J, Ramaiah N, Sarkar A (2006) Aerobic degradation of highly chlorinated polychlorobiphenyls by a marine bacterium Pseudomonas CH07. World J Microbiol Biotechnol 22:1321–1327

De Domenico M, Lo Giudice A, Michaud L, Saitta M, Bruni V (2004) Diesel oil and PCB-degrading psychrotrophic bacteria isolated from Antarctic seawaters (Terra Nova Bay, Ross Sea). Polar Res 23:141–146

de la Huz R, Lastra M, Junoy J, Castellanos C, Vieitez JM (2005) Biological impacts of oil pollution and cleaning in the intertidal zone of exposed sandy beaches: preliminary study of the "Prestige" oil spill. Estuar Coast Shelf Sci 65:19–29

Dojka MA, Hugenholtz P, Haack SK, Pace NR (1998) Microbial diversity in a hydrocarbon- and chlorinated-solvent-contaminated aquifer undergoing intrinsic bioremediation. Appl Environ Microbiol 64:3869–3877

Edlund A, Jansson JK (2008) Use of bromodeoxyuridine immunocapture to identify psychrotolerant phenanthrene-degrading bacteria in phenanthrene-enriched polluted Baltic Sea sediments. FEMS Microbiol Ecol 65:513–525

Fahrenfeld N, Zoeckler J, Widdowson M, Pruden A (2013) Effect of biostimulants on 2,4,6-trinitrotoluene (TNT) degradation and bacterial community composition in contaminated aquifer sediment enrichments. Biodegradation 24:179–190

Faulwetter JL, Gagnon V, Sundberg C, Chazarenc F, Burr MD, Brisson J, Camper AK, Stein OR (2009) Microbial processes influencing performance of treatment wetlands: a review. Ecol Eng 35:987–1004

Fields MW, Yan T, Rhee S-K, Carroll SL, Jardine PM, Watson DB, Criddle CS, Zhou J (2005) Impacts on microbial communities and cultivable isolates from groundwater contaminated with high levels of nitric acid–uranium waste. FEMS Microbiol Ecol 53:417–428

Focht DD (1995) Strategies for the improvement of aerobic metabolism of polychlorinated biphenyls. Curr Opin Biotechnol 6:341–346

Franzmann PD, Robertson WJ, Zappia LR, Davis GB (2002) The role of microbial populations in the containment of aromatic hydrocarbons in the subsurface. Biodegradation 13:65–78

García-Armisen T, İnceoğlu Ö, Ouattara NK, Anzil A, Verbanck MA, Brion N, Servais P (2014) Seasonal variations and resilience of bacterial communities in a sewage polluted urban river. PLoS One 9, e92579

Ghosh D, Routh J, Bhadury P (2014) Diversity of arsenite oxidizing bacterial communities in arsenic-rich deltaic aquifers in West Bengal, India. Front Microbiol. doi:10.3389/fmicb.2014.00602

Gilbert N, Fulthorpe R, Kirkwood AE (2012) Microbial diversity, tolerance, and biodegradation potential of urban wetlands with different input regimes. Can J Microbiol 58:887–897

Giordano M (2009) Global groundwater? Issues and solutions. Annu Rev Environ Resour 34:153–178

Gough HL, Stahl DA (2011) Microbial community structures in anoxic freshwater lake sediment along a metal contamination gradient. ISME J 5:543–558

Griebler C, Lueders T (2009) Microbial biodiversity in groundwater ecosystems. Freshw Biol 54:649–677

Hahn M (2006) The microbial diversity of inland waters. Curr Opin Biotechnol 17:256–261

Haller L, Tonolla M, Zopfi J, Peduzzi R, Wildi W, Poté J (2011) Composition of bacterial and archaeal communities in freshwater sediments with different contamination levels (Lake Geneva, Switzerland). Water Res 45:1213–1228

Hamonts K, Ryngaert A, Smidt H, Springael D, Dejonghe W (2014) Determinants of the microbial community structure of eutrophic, hyporheic river sediments polluted with chlorinated aliphatic hydrocarbons. FEMS Microbiol Ecol 87:715–732

Handley KM, Wrighton KC, Piceno YM, Andersen GL, DeSantis TZ, Williams KH, Wilkins MJ, N'Guessan AL, Peacock A, Bargar J, Long PE, Banfield JF (2012) High-density PhyloChip profiling of stimulated aquifer microbial communities reveals a complex response to acetate amendment. FEMS Microbiol Ecol 81:188–204

Harayama S, Kishira H, Kasai Y, Shutsubo K (1999) Petroleum biodegradation in marine environments. J Mol Microbiol Biotechnol 1:63–70

Harayama S, Kasai Y, Hara A (2004) Microbial communities in oil-contaminated seawater. Curr Opin Biotech 15:205–214

Hashim MA, Mukhopadhyay S, Sahu JN, Sengupta B (2011) Remediation technologies for heavy metal contaminated groundwater. J Environ Manag 92:2355–2388

Hazen TC, Dubinsky EA, DeSantis TZ, Andersen GL, Piceno YM, Singh N et al (2010) Deep-Sea oil plume enriches indigenous oil-degrading bacteria. Science 330:204–208

Hedlund BP, Staley JT (2001) *Vibrio cyclotrophicus* sp. nov., a polycyclic aromatic hydrocarbon (PAH)-degrading marine bacterium. Int J Syst Evol Microbiol 51:61–66

Hedlund BP, Geiselbrecht AD, Bair TJ, Staley JT (1999) Polycyclic aromatic hydrocarbon degradation by a new marine bacterium, *Neptunomonas naphthovorans* gen. nov., sp. nov. Appl Environ Microbiol 65:251–259

Hemme CL, Deng Y, Gentry TJ, Fields MW, Wu L, Barua S, Barry K, Tringe SG, Watson DB, He Z, Hazen TC, Tiedje JM, Rubin EM, Zhou J (2010) Metagenomic insights into evolution of a heavy metal-contaminated groundwater microbial community. ISME J 4:660–672

Hilyard EJ, Jones-Meehan JM, Spargo BJ, Hill RT (2008) Enrichment, isolation and phylogenetic identification of polycyclic aromatic hydrocarbon degrading bacteria from Elizabeth River sediments. Appl Environ Microbiol 74:1176–1182

Hiorns WD, Methé BA, Nierzwicki-Bauer SA, Zehr JP (1997) Bacterial diversity in Adirondack mountain lakes as revealed by 16S rRNA gene sequences. Appl Environ Microbiol 63:2957–2960

Hörtnagl P, Pérez MT, Zeder M, Sommaruga R (2010) The bacterial community composition of the surface microlayer in a high mountain lake. FEMS Microbiol Ecol 73:458–467

Ibekwe AM, Grieve CM, Lyon SR (2003) Characterization of microbial communities and composition in constructed dairy wetland wastewater effluent. Appl Environ Microbiol 69:5060–5069

ITOPF (2006) Oil tanker spill statistics: 2005. International Tanker Owners Pollution Federation, London

Kang S, Van Nostrand JD, Gough HL, He Z, Hazen TC, Stahl DA, Zhou J (2013) Functional gene array-based analysis of microbial communities in heavy metals-contaminated lake sediments. FEMS Microbiol Ecol 86:200–214

Kessler JD, Valentine DM, Redmond MC, Du MR, Chan EW, Mendes SD et al (2011) A persistent oxygen anomaly reveals the fate of spilled methane in the deep Gulf of Mexico. Science 331:312–315

Kleikemper J, Schroth MH, Sigler WV, Schmucki M, Bernasconi SM, Zeyer J (2002) Activity and diversity of sulfate-reducing bacteria in a petroleum hydrocarbon-contaminated aquifer. Appl Environ Microbiol 68:1516–1523

Kleinsteuber S, Schleinitz K, Vogt C (2012) Key players and team play: anaerobic microbial communities in hydrocarbon-contaminated aquifers. Appl Microbiol Biotechnol 94:851–873

Kolar AB, Hrsak D, Fingler S, Cetkovic H, Petric I, Kolic NU (2007) PCB-degrading potential of aerobic bacteria enriched from marine sediments. Biodegradation 60:16–24

Kube M, Chernikova TN, Al-Ramahi Y et al (2013) Genome sequence and functional genomic analysis of the oil-degrading bacterium Oleispira antarctica. Nat Commun. doi:10.1038/ncomms3156

La Rocca C, Conti L, Crebelli R, Crochi B, Iacovella N, Rodriguez F, Turrio-Baldassarri L, Di Domenico A (1996) PAH content and mutagenicity of marine sediments from the Venice lagoon. Ecotoxicol Environ Saf 33:236–245

Larentis M, Hoermann K, Lueders T (2013) Fine-scale degrader community profiling over an aerobic/anaerobic redox gradient in a toluene-contaminated aquifer. Environ Microbiol Rep 5:225–234

Lin CHM, Pedersen JA, Suffet IH (2003) Influence of aeration on hydrophobic organic contaminant distribution and diffusive flux in estuarine sediments. Environ Sci Technol 37:3547–3554

Lin X, Kennedy D, Fredrickson J, Bjornstad B, Konopka A (2012) Vertical stratification of subsurface microbial community composition across geological formations at the Hanford Site. Environ Microbiol 14:414–425

Liu Z, Liu J (2013) Evaluating bacterial community structures in oil collected from the sea surface and sediment in the northern Gulf of Mexico after the Deepwater Horizon oil spill. Microbiol Open 2:492–504

Lu X-Y, Zhang T, Fang HH-P (2011) Bacteria-mediated PAH degradation in soil and sediment. Appl Microbiol Biotechnol 89:1357–1371

Macalady JL, Mack EE, Nelson DC, Scow KM (2000) Sediment microbial community structure and mercury methylation in mercury-polluted Clear Lake, California. Appl Environ Microbiol 66:1479–1488

Macias-Zamora JV, Mendoza-Vega E, Villaescusa-Celaya JA (2002) PAHs composition of surface marine sediments: a comparison to potential local sources in Todos Santos Bay, BC, Mexico. Chemosphere 46:459–468

Marshall AG, Rodgers RP (2004) Petroleomics: the next grand challenge for chemical analysis. Acc Chem Res 37:53–59

Mason OU, Hazen TC, Borglin S, Chain PSG, Dubinsky EA, Fortney JL, Han J, Holman NH-Y, Hultman J, Lamendella R, Mackelprang R, Malfatti S, Tom LM, Tringe SG, Woyke T, Zhou J, Rubin EM, Jansson JK (2012) Metagenome, metatranscriptome and single-cell sequencing reveal microbial response to Deepwater Horizon oil spill. ISME J 6:1715–1727

McKay DB, Seeger M, Zielinski M, Hofer B, Timmis KN (1997) Heterologous expression of biphenyl dioxygenase-encoding genes from a gram-positive broad-spectrum polychlorinated biphenyl degrader and characterization of chlorobiphenyl oxidation by the gene products. J Bacteriol 179:1924–1930

McKew BA, Coulon F, Osborn AM, Timmis KN, McGenity TJ (2007) Determining the identity and roles of oil-metabolizing marine bacteria from the Thames estuary, UK. Environ Microbiol 9:165–176

McLeod MP, Warren RL, Hsiao WW, Araki N, Myhre M, Fernandes C, Miyazawa D, Wong W, Lillquist AL, Wang D, Dosanjh M, Hara H, Petrescu A, Morin RD, Yang G, Stott JM, Schein JE, Shin H, Smailus D, Siddiqui AS, Marra MA, Jones SJ, Holt R, Brinkman FS, Miyauchi K, Fukuda M, Davies JE, Mohn WW, Eltis LD (2006) The complete genome of *Rhodococcus* sp. RHA1 provides insights into a catabolic powerhouse. Proc Natl Acad Sci U S A 103:15582–15587

Melcher RJ, Apitz SE, Hemmingsen BB (2002) Impact of irradiation and polycyclic aromatic hydrocarbon spiking on microbial populations in marine sediment for future aging and biodegradability studies. Appl Environ Microbiol 68:2858–2868

Michaud L, Di Marco G, Bruni V, Lo Giudice A (2007) Biodegradative potential and characterization of psychrotolerant polychlorinated biphenyl-degrading marine bacteria isolated from a coastal station in the Terra Nova Bay (Ross Sea, Antarctica). Mar Pollut Bull 54:1754–1761

Miyoshi T, Iwatsuki T, Naganuma T (2005) Phylogenetic characterization of 16S rRNA gene clones from deep-groundwater microorganisms that pass through 0.2-micrometer-pore-size filters. Appl Environ Microbiol 71:1084–1088

National Research Council (NRC) (2003) Oil in the Sea III: inputs, fates, and effects. National Academy Press, Washington, DC

Newton RJ, Jones SE, Eiler A, McMahon KD, Bertilsson S (2011) A guide to the natural history of freshwater lake bacteria. Microbiol Mol Biol Rev 75:14–49

Nicomrat D, Dick W, Tuovinen O (2006) Assessment of the microbial community in a constructed wetland that receives acid coal mine drainage. Microb Ecol 51:83–89

Perfumo A, Smyth TJP, Marchant R, Banat IM (2010) Production and roles of biosurfactants and bioemulsifiers in accessing hydrophobic substrates. In: Timmis KN, McGenity TJ, van der Meer JR, de Lorenzo V (eds) Handbook of hydrocarbon and lipid microbiology. Springer, Berlin, pp 1501–1512

Pernthaler J (2013) Freshwater microbial communities. In: Rosenberg E, DeLong E, Lory S, Stackebrandt E, Thompson F (eds) The prokaryotes. Springer, Berlin, pp 97–112

Pieper DH (2005) Aerobic degradation of polychlorinated biphenyls. Appl Microbiol Biotechnol 67:170–191

Pratt B, Riesen R, Johnston C (2012) PLFA analyses of microbial communities associated with PAH-contaminated riverbank sediment. Microb Ecol 64:680–691

Prince RC, Gramain A, McGenity TJ (2010) Prokaryotic hydrocarbon degraders. In: Timmis KN, McGenity TJ, van der Meer JR, de Lorenzo V (eds) Handbook of hydrocarbon and lipid microbiology. Springer, Berlin, pp 1671–1692

Proia L, Cassió F, Pascoal C, Tlili A, Romaní A (2012) The use of attached microbial communities to assess ecological risks of pollutants in river ecosystems: the role of heterotrophs. In: Guasch H, Ginebreda A, Geiszinger A (eds) Emerging and priority pollutants in rivers. The handbook of environmental chemistry. Springer, Berlin, pp 55–83

Rajendran RB, Imagawa T, Tao H, Ramesh R (2005) Distribution of PCBs, HCHs and DDTs, and their ecotoxicological implications in Bay of Bengal, India. Environ Int 31:503–512

Rastogi G, Barua S, Sani R, Peyton B (2011) Investigation of microbial populations in the extremely metal-contaminated Coeur d'Alene River sediments. Microb Ecol 62:1–13

Rodrigues AC, Wuertz S, Brito AG, Melo LF (2005) Fluorene and phenanthrene uptake by *Pseudomonas putida* ATCC 17514: kinetics and physiological aspects. Biotechnol Bioeng 90:281–289

Sabirova JS, Chernikova TN, Timmis KN, Golyshin PN (2008) Niche-specificity factors of a marine oil-degrading bacterium *Alcanivorax borkumensis* SK2. FEMS Microbiol Lett 285:89–96

Safe Drinking Water Act P. L. 93-523

Sauvain L, Bueche M, Junier T, Masson M, Wunderlin T, Kohler-Milleret R, Gascon Diez E, Loizeau J-L, Tercier-Waeber M-L, Junier P (2014) Bacterial communities in trace metal contaminated lake sediments are dominated by endospore-forming bacteria. Aquat Sci 76:33–46

Schneiker SVA, dos Santos PM, Bartels D, Bekel T, Brecht M, Buhrmester J, Chernikova TN, Denaro R, Ferrer M, Gertler C, Goesmann A, Golyshina OV, Kaminski F, Khachane AN, Lang S, Linke B, McHardy AC, Meyer F, Nechitaylo T, Pühler A, Regenhardt D, Rupp O, Sabirova JS, Selbitschka W, Yakimov MM, Timmis KN, Vorhölter FJ, Weidner S, Kaiser O, Golyshin PN (2006) Genome sequence of the ubiquitous hydrocarbon-degrading marine bacterium *Alcanivorax borkumensis*. Nat Biotechnol 24:997–1004

Scholz M, Lee BH (2005) Constructed wetlands: a review. Int J Environ Stud 62:421–447

Schwarzenbach RP, Escher BI, Fenner K, Hofstetter TB, Johnson CA, von Gunten U, Wehrli B (2006) The challenge of micropollutants in aquatic systems. Science 313:1072–1077

Seeger M, Zielinski M, Timmis KN, Hofer B (1999) Regiospecificity of dioxygenation of dito pentachlorobiphenyls and their degradation to chlorobenzoates by the *bph*-encoded catabolic pathway of *Burkholderia* sp. strain LB400. Appl Environ Microbiol 65:3614–3621

Seeger M, Cámara B, Hofer B (2001) Dehalogenation, denitration, dehydroxylation and angular attack on substituted biphenyls and related compounds by a biphenyl dioxygenase. J Bacteriol 183:3548–3555

Seeger M, Hernandez M, Mendez V, Ponce B, Cordova M, Gonzalez M (2010) Bacterial degradation and bioremediation of chlorinated herbicides and biphenyls. J Soil Sci Plant Nutr 10:320–332

Singh SN (2012) Microbial degradation of xenobiotics. Springer, Berlin, 486 p (eBook)

Small C, Cohen JE (2004) Continental physiography, climate, and the global distribution of human population. Curr Anthropol 45:269–277

Sokol RC, Bethoney CM, Rhee G-Y (1994) Effect of hydrogen on the pathway and products of PCB dechlorination. Chemosphere 29:1735–1742

Sutton NB, van der Kraan GM, van Loosdrecht MCM, Muyzer G, Bruining J, Schotting RJ (2009) Characterization of geochemical constituents and bacterial populations associated with As mobilization in deep and shallow tube wells in Bangladesh. Water Res 43:1720–1730

Teira E, Lekunberri I, Gasol JM, Nieto-Cid M, Alvarez-Salgado XA, Figueiras FG (2007) Dynamics of the hydrocarbon-degrading *Cycloclasticus* bacteria during mesocosm-simulated oil spills. Environ Microbiol 9:2551–2562

Torrentó C, Urmeneta J, Otero N, Soler A, Viñas M, Cama J (2011) Enhanced denitrification in groundwater and sediments from a nitrate-contaminated aquifer after addition of pyrite. Chem Geol 287:90–101

Truu M, Juhanson J, Truu J (2009) Microbial biomass, activity and community composition in constructed wetlands. Sci Total Environ 407:3958–3971

Valentine DL, Kessler JD, Redmond MC, Mendes SD, Heintz MB, Farwell C et al (2010) Propane respiration jump starts microbial response to a deep oil spill. Science 330:208–211
Vilo C, Benedik MJ, Ilori M, Dong Q (2014) Draft genome sequence of Cupriavidus sp. strain SK-4, a di-ortho-substituted biphenyl utilizing bacterium isolated from polychlorinated biphenyl contaminated sludge. Genome Announc 2:e00474-14
Wakelin SA, Colloff MJ, Kookana RS (2008) Effect of wastewater treatment plant effluent on microbial function and community structure in the sediment of a freshwater stream with variable seasonal flow. Appl Environ Microbiol 74:2659–2668
Wang S, Chng KR, Wu C, Wilm A, Nagarajan N, He J (2014) Draft genome sequence of polychlorinated biphenyl-dechlorinating *Dehalococcoides mccartyi* strain SG1, which carries a circular putative plasmid. Genome Announc 2:e00901–e00914
Wetzel RG (2000) Freshwater ecology: changes, requirements, and future demands. Limnology 1:3–9
Whitman WB, Coleman DC, Wiebe WJ (1998) Prokaryotes: the unseen majority. Proc Natl Acad Sci U S A 95:6578–6583
Williamson CE, Dodds W, Kratz TK, Palmer MA (2008) Lakes and streams as sentinels of environmental change in terrestrial and atmospheric processes. Front Ecol Environ 6:247–254
Wunderlin T, Corella J, Junier T, Bueche M, Loizeau J-L, Girardclos S, Junier P (2014) Endospore-forming bacteria as new proxies to assess impact of eutrophication in Lake Geneva (Switzerland–France). Aquat Sci 76:103–116
Zedler JB, Kercher S (2005) Wetland resources: status, trends, ecosystem services, and restorability. Annu Rev Environ Resour 30:39–74
Zhu G, Wang S, Feng X, Fan G, Jetten MSM, Yin C (2011) Anammox bacterial abundance, biodiversity and activity in a constructed wetland. Environ Sci Technol 45:9951–9958
Zwart G, Hiorns WD, Methé BA, van Agterveld MP, Huismans R, Nold SC, Zehr JP, Laanbroek HJ (1998) Nearly identical 16S rRNA sequences recovered from lakes in North America and Europe indicate the existence of clades of globally distributed freshwater bacteria. Syst Appl Microbiol 21:546–556
Zwart G, Crump BC, van Agterveld MP, Hagen F, Han S-K (2002) Typical freshwater bacteria: an analysis of available 16S rRNA gene sequences from plankton of lakes and rivers. Aquat Microb Ecol 28:141–155

Chapter 4
Role of Biosensors in Environmental Monitoring

Lata S.B. Upadhyay and Nishant Verma

4.1 Introduction

Environmental pollution is considered to be our world's most dangerous and constant threat. Our world is exposed to hazardous pollutants and chemicals from different sources every day. Slowly, our ecosystem is being brought down by the impending danger of pollutants. The predicted increase in temperature up to 6 °C till year 2100 is expected to affect the organisms and ecological processes in both terrestrial and aquatic ecosystems. These problems require immediate scientific attention to find an appropriate and cost-effective solution. With increasing economical and industrial development, a lot of pressure has been put on the environment in terms of pollutions. Since there exists equilibrium between water, air, and soil phases, contamination of any of these phases will ultimately spread to all the phases. Climate models predict an increase in the frequency and magnitude of extreme weather events, such as drought periods, intense rainfalls, changing the levels, and bioavailability of contaminants in freshwaters as a result of runoff events from the surrounding soils. Metal contamination is an environmental problem in both developing and developed countries throughout the world. Pollution by metals is a matter of great concern because of its toxicity to living organisms and their persistence in the environment. Monitoring of pollutants in water, air, and soil is an essential component in understanding and managing the risks associated with the human health and local ecosystems. These environmental issues are now gaining greater attention in technical, scientific, and popular press. A number of legislation has been put forward to monitor the release and the levels of different hazardous chemicals in the environment. Conventional methods such as chromatographic and spectroscopic techniques are usually performed with better accuracy and sensitivity, but require expensive instruments and expert personnel and are

L.S.B. Upadhyay (✉) • N. Verma
National Institute of Technology, Raipur 492010, Chhattisgarh, India
e-mail: contactlata@yahoo.com; nishantbiotech1985@gmail.com

L.B. Sukla et al. (eds.), *Environmental Microbial Biotechnology*, Soil Biology 45,
DOI 10.1007/978-3-319-19018-1_4

tedious and time consuming (Prieto-Simón et al. 2006; Rodriguez-Mozaz et al. 2009). So, a need arises for the development and deployment of simple, portable, cost-effective, and precise analytical devices.

Biosensor, according to the International Union of Pure and Applied Chemistry (IUPAC), can be described as a self-contained integrated device which is capable of providing specific quantitative or semiquantitative analytical information with the help of a biological recognition element spatially contacted with a transduction element (Thévenot et al. 2001). They are the reagentless analytical devices which are characterized by their high specificity and sensitivity toward the substance and thus become an active area of research for food and clinical and environmental monitoring (Bachan Upadhyay and Verma 2013). A typical biosensor has three main features: a biorecognizing element, a transducer, and a signal processing system. Here, the biorecognizing element interacts with the analyte of interest in the sample and induces some physicochemical changes. Those changes are then converted into signals by the transducer element, which are further converted into a readable form by the signal processing system.

4.2 Biosensor Classification

General classification of biosensors is based on the type of biological recognition element being used during their designing process or the mode of signal transduction they perform. The biological recognition element may further be classified as a biocatalytic recognition element (composed of a single biocatalytic element or multiple biocatalytic elements) and a bio-complexing recognition element. In biocatalytic element, a reaction is catalyzed by the immobilized molecule, with continuous consumption and release of substrate/product (analyte). The analyte consumption is then monitored either by measuring the consumption of co-substrate or formation of the reaction product. For example, the first biosensor developed by Clarks and Lyons was an immobilized enzyme electrode, i.e., a glucose oxidase immobilized electrode, in which the enzyme glucose oxidase catalyzes the conversion of glucose to glucuronic acid and hydrogen peroxide. Two strategies were employed to determine the glucose concentration in the sample. In the first strategy, the consumption of oxygen during the reaction was monitored to determine the glucose concentration in samples (Clark and Lyons 1962). Alternatively, oxidation of one of the reaction product, hydrogen peroxide, at electrode was monitored to quantify glucose (Updike and Hicks 1967). Later, a number of enzyme biosensors with different modifications have been developed and reported (Liu and Lin 2006; Wang et al. 2009; Deng et al. 2009). Some other biocatalytic elements, such as whole cell (bacteria, fungi, yeast, eukaryotic cells), cell organelles, and tissue (animal and plant), are also used in biosensor, but enzymes were the first biorecognition elements to be used in biosensor and still remain the most widely used bio-element.

Second is the bio-complexing recognition element, where some type of interactions occurs between the analyte and immobilized macromolecules, and an equilibrium response is formed which is then monitored by the detector. The most common example is antigen–antibody interaction, where an antigen binds to a specific antibody, and this complex formation is usually determined by employing enzyme-coupled antibodies or antigens. This particular category of biosensor is called antibody-based biosensor or immunosensor. However, these immunosensors should possess some high standards such as high specificity in complex medium, well-characterized binding properties and high stability (Saerens et al. 2008). Such bio-complexing element can also be operated as receptor/antagonist/agonist, where ion channels, membrane receptors, or binding proteins have been used as molecular recognition elements. For example, acetylcholine receptor, a first group of transmembrane receptors, has been used in sensing nerve gases, while the nicotinic acetylcholine receptor has been used for sensing naturally occurring toxins (Collings and Caruso 1997).

On the basis of signal transduction, biosensors can be broadly classified as electrochemical biosensors, optical biosensors, calorimetric biosensors, and acoustic biosensors. Electrochemical biosensors may be further classified as amperometric, potentiometric, and conductometric biosensors depending on the type of electrochemical parameters measured such as current, potential difference, and conductivity. Optical biosensors are based on the monitoring of optical phenomenon such as absorption, reflection, fluorescence, chemiluminescence, and phosphorescence. Calorimetric biosensors are based on the property of biological reactions involving exothermic and endothermic reaction, in which the heat exchange is measured using some sensitive thermistor. However, the application of these calorimetric biosensors is restricted by the cost of operation and very long experimental procedures (Ramanathan and Danielsson 2001). Acoustic biosensors are based on the detection of change of mass density and viscoelastic, electric, or dielectric properties of a membrane, which is made of a chemically interactive material in contact with a piezoelectric material (Reyes De Corcuera and Cavalieri 2003).

In recent years, the application of nanomaterials in biosensors has advanced, and a lot of nanomaterials are being fabricated to increase the potential applications of biosensors in different fields (Zhang et al. 2009).

4.3 Biosensor Response

Enzyme biosensors have been extensively used to determine the environmental pollutants due to their unique specificity and sensitivity. There are two modes by which the biosensors monitor the pollutants. First is the *direct monitoring*, in which the immobilized enzyme consumed the analyte along with a co-substrate (if any) and yield product(s). The biosensor response is then achieved in terms of co-substrate consumption or product yield. However, the applications of these biosensors have been limited by the small number of pollutants that can act as the substrate for the immobilized enzyme and their high detection limits. The other mode is called *indirect monitoring*, in which the analyte or pollutant interacts with

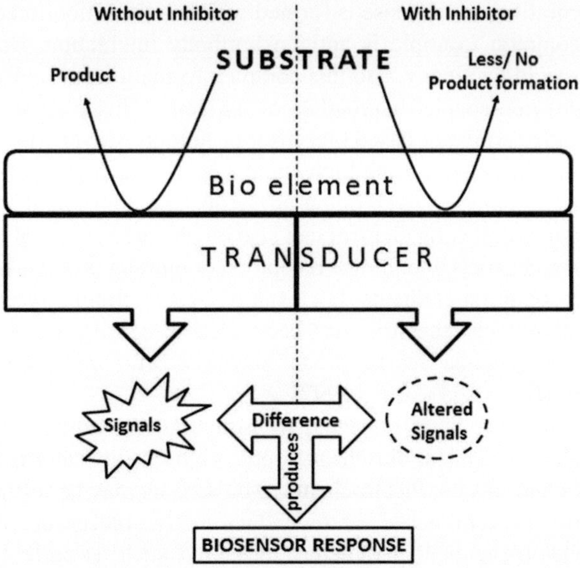

Fig. 4.1 Diagrammatic representation of indirect monitoring of enzyme inhibition-based biosensor

the immobilized enzyme and interferes with its catalytic activity. Here, the pollutant acts as an inhibitor of the enzyme activity. The biosensor response is generally determined by assessing the deviation in enzyme activity, after it is exposed to the inhibitor (Fig. 4.1). The concentration of pollutants is determined by measuring the percentage of inhibition, which is calculated as follows:

$$I(\%) = \left\{1 - \frac{E_t}{E_i}\right\} \times 100$$

Where, I (%) is the percentage of inhibition, E_i is the activity of immobilized enzyme before exposure to the inhibitor, and E_t is the activity of immobilized enzyme after the exposure to inhibitor.

The advantage of this indirect monitoring over direct monitoring is the high sensitivity of enzyme inhibition-based biosensors, as most of the enzymes are susceptible to a very low concentration of inhibitors. However, the major disadvantage of this indirect monitoring is the presence of more than one inhibitor in the sample which may inhibit the enzyme activity and, therefore, produces unexpected results. Moreover, in addition to the inhibitor, these biosensors also require some substrate during the monitoring process which complicates the overall design of the biosensor.

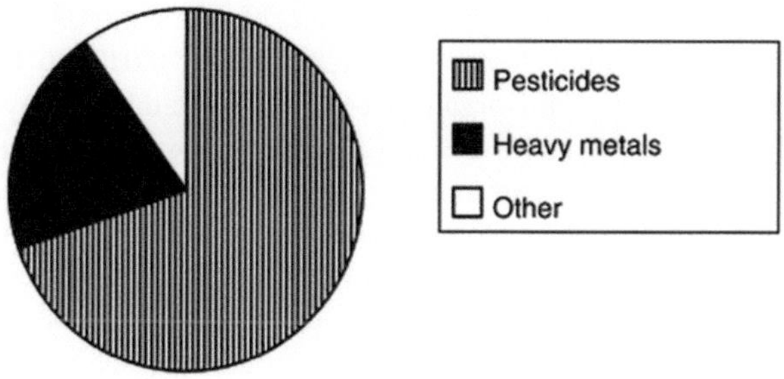

Fig. 4.2 Distribution of inhibitors as determined by the enzymatic biosensors [reprinted from Amine et al. (2006) with permission from Elsevier]

4.4 Environmental Applications

In the context of environmental monitoring, a large number of biosensors have been developed for the detection of a variety of pollutants such as pesticides, phenols, heavy metals, polluting gases, surfactants, and polychlorinated biphenyls (PCBs). Most of these biosensors are based on enzyme inhibition which employed some common enzymes such as cholinesterase (acetylcholinesterase and butyrylcholinesterase), horseradish peroxidase (HRP), urease, and tyrosinase. A large number of these biosensors have been extensively used for the analysis of pesticides (71 %), followed by the heavy metal determination (21 %) as shown in Fig. 4.2 (Amine et al. 2006).

4.4.1 Biosensors for Pesticides Determination

Due to their high pesticidal activity, pesticides are widely used around the world for agricultural and industrial purposes. They are generally used to control the weeds, insects, and fungi and, therefore, increase the crop production. However, with extensive use, the food, soil, and water become contaminated with pesticide residues and its metabolites and, thus, represent one of the major environmental pollutants (Mostafa 2010). Pesticides can be classified into three different groups: insecticides, herbicides, and fungicides, which differ in their polarity, volatility, and persistence (Hernández et al. 2005). Among different enzymes, acetylcholine esterase (AChE) has been widely used in developing biosensor for the detection of pesticides, mainly organophosphorus and carbamate pesticides. The organophosphorus compounds partly inhibit the activity of acetylcholine esterase by phosphorylating its serine group and therefore provides the basis for its detection by various transducers. A number of pesticides such as carbaryl, carbofuran, paraoxon, chlorpyrifos ethyl oxon, malaoxon, aldicarb, trichlorfon, and diisopropyl fluorophosphates have been reported to be successfully detected by the AChE inhibition biosensors (Dzyadevych et al. 2005). A further modification in AChE biosensor is the coupling of a second enzyme (bienzymatic system), choline oxidase, which provides a higher sensitivity, wide linear range, and lower detection limit for the analysis of organophosphorus compounds. Snejdarkova et al. (2003) reported a much lower detection limit of 0.01 ppb for carbamate pesticide and 1.3×10^{-3} ppb for organophosphate pesticide, employing AChE–choline oxidase bienzymatic system. Besides this, few tyrosinase inhibition-based biosensors have been also demonstrated for the determination of pesticides such as triazine (Campanella et al. 2006), paraoxon, and methyl parathion (de Albuquerque and Ferreira 2007). Another approach for pesticides detection is the direct monitoring by the use of an enzyme, organophosphorus hydrolase (OPH). OPH is a biological catalyst that hydrolyzes a range of organophosphotriester according to the following equation.

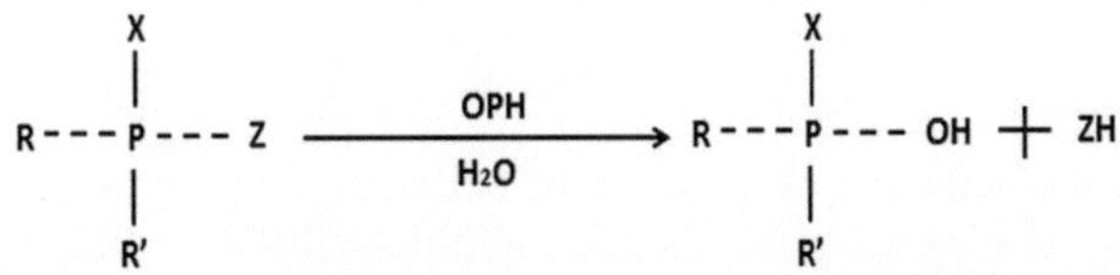

Where, X is oxygen or sulfur, R is an alkoxy group ranging in size from methoxy to butoxy, R′ is an alkoxy or phenyl group, and Z is a phenoxy group, a thiol moiety, a cyanide, or fluoride group. Because of the hydrolyzing ability, the OPH enzyme can be incorporated with different transducers to monitor a number of organophosphorus pesticides such as parathion, paraoxon, diazinon, and chemical warfare agents such as soman, sarin, and VX (Mulchandani et al. 1999). A flow injection amperometric OPH biosensor has been developed for the rapid and direct detection of paraoxon with detection limit of 0.1 μM (Wang et al. 2003). A potentiometric biosensor has also been reported for the detection of OP nerve agents with 2 μM detection limit (Mulchandani et al. 1999). For enhanced sensitivity, selectivity, and reliability, a dual amperometric/potentiometric biosensor was also developed that provides distinct determination of different groups of OP pesticides (Schöning et al. 2003).

In recent years, a number of immunosensors have been reported to be powerful analytical devices for pesticide detection. It may be categorized as a labeled biosensor that utilizes an enzyme to quantify the amount of pesticides, or a label-free biosensor which detects the binding of pesticides on a transducer surface. With the help of anti-atrazine monoclonal antibodies, atrazine pesticide can easily be detected in the agricultural field and water samples, by a label-free electrochemical immunosensor (Tran et al. 2012; Liu et al. 2014b). A novel multianalyte immunosensor was developed for the detection of pesticides such as endosulfan and paraoxon with a high sensitivity and detection limit of 0.05 ppb and 2 ppb, respectively (Liu et al. 2014a). Table 4.1 summarizes some of the biosensor reported for pesticide determination.

Table 4.1 Biosensors for pesticide detection

Pesticide	Enzyme	Transducer	Detection limit	Reference
Paraoxon	AChE	Amperometric	1.91×10^{-8} M	Andreescu (2002)
Carbaryl	AChE	Amperometric	1×10^{-8} M	Bucur et al. (2006)
Carbofuran	AChE	Amperometric	8×10^{-10} M	Bucur et al. (2006)
Chlorpyrifos ethyl oxon	AChE	Amperometric	1.24×10^{-9} M	Andreescu (2002)
Malaoxon	AChE	Amperometric	0.6–1.2 μM	Jeanty et al. (2002)
Aldicarb	AChE	Amperometric	24 ppb	Arduini et al. (2006)
Trichlorfon	AChE	Amperometric	0.5 nM	Li et al. (2007)
Diisopropyl fluorophosphates	AChE	Conductometric	5×10^{-11} M	Dzyadevych et al. (2005)
Triazine	Tyr	Amperometric	0.5×10^{-6} mM	Campanella et al. (2006)
Paraoxon	Tyr	Amperometric	0.5×10^{-6} mM	Campanella et al. (2007)
Methyl parathion	Tyr	Amperometric	6–100 ppb	de Albuquerque and Ferreira (2007)
Parathion	OPH	Amperometric	10 nM	Tang et al. (2014)
Paraoxon	OPH	Optical	10^{-9} M	Cao et al. (2004)
Diazinon	OPH	Optical	800 ppt	White and Harmon (2005)
Atrazine	–	IS	0.016 ng/mL	Liu et al. (2014b)
Atrazine	–	IS	0.2 ng/L	Tran et al. (2012)
Endosulfan	–	IS	2 ppb	Liu et al. (2014a)

AChE acetylcholinesterase, *Tyr* tyrosinase, *OPH* organophosphorus hydrolase, *IS* immunosensor

4.4.2 Biosensors for Heavy Metal Determination

Heavy metals such as copper, cadmium, mercury, and zinc impose a threat to the environment, owing to their high toxicity and ability to be accumulated in the living organisms (Rodriguez-Mozaz et al. 2009). Heavy metal biosensors using immobilized urease and glucose oxidase are widely used for the detection of metals such as mercury, cadmium, chromium, lead, and copper salts. Mercury has a higher affinity for the cysteine residue of urease and, therefore, could be detected with a limit as lower as 10 nM (Tsai and Doong 2005). A renewable potentiometric biosensor based on urease inhibition has been developed that can detect mercury ions with a detection limit of 0.05 μM (Yang et al. 2006). An amperometric urease inhibition biosensor was developed by immobilizing urease in poly(vinylferrocenium) film, which has a much lower detection limit of 7.4×10^{-6} M for mercury (Kuralay et al. 2007). Other heavy metals such as cadmium, copper, lead, and nickel are also found to inhibit the activity of several enzymes and, thus, could be monitored by enzymatic biosensors (Tsai et al. 2003; Guascito et al. 2008; Ghica et al. 2013; Ilangovan et al. 2014). However, these enzyme inhibition-based biosensors are not specific and usually applied for the analysis of total inhibition effect caused by heavy metals in the sample. To overcome this, some recombinant bacterial sensors have also been developed, in which a specific metal-inducible promoter is attached to a reporter gene which codes for a luminescent protein. A cell-free biosensor (*E. coli* S30 extract) was developed by using merR as a regulatory protein, and LucFF of firefly luciferase as a reporter gene, for the detection of mercury at a concentration of 10^{-9} M (Pellinen et al. 2004). Some metal-binding proteins such as cytochrome c3 and metallothionein *SmtA* have been purified from bacteria and were used for the determination of heavy metals employing amperometric and capacitance methods (Bontidean et al. 2000; Michel et al. 2006). Another approach being used for the heavy metal detection is *immunoassay* with the advantages of having high sensitivity, selectivity, and species specificity. Blake and coworkers reported a KinExA-automated immunoassay for the detection of uranium in a range of 1–5 mm. They also reported the same immunoassay format for the detection of different metals such as cadmium, cobalt, lead, and uranium (Blake et al. 2001). Table 4.2 lists some of the biosensors reported for heavy metal determination.

Table 4.2 Biosensors for heavy metal detection

Heavy metals	Biorecognition element	Transducer	Detection limit	Reference
Cu (II) Cd (II)	Urease	Optical	10 μM	Tsai et al. (2003)
Hg (II)	Urease	Potentiometric	0.05 μM	Yang et al. (2006)
Cu (II) Cd (II) Pb (II)	Urease	Conductometric	0.1–10 mM	Ilangovan et al. (2014)
Hg (II)	Urease	Amperometric	7.4×10^{-6} M	Kuralay et al. (2007)
Hg (II) Ag (I) Cu (II) Cd (II)	Glucose oxidase	Amperometric	2.5 μM 0.05 μM 5 μM 5 μM	Guascito et al. (2008)
Hg (II)	Glucose oxidase	Amperometric	0.49 μg/L	Liu et al. (2009)
Cu (II) Cd (II) Co (II) Ni (II)	Glucose oxidase	Amperometric	0.2 μM 2.4 μM 2.1 μM 3.3 μM	Ghica et al. (2013)
Hg (II)	*Escherichia coli* S30 extract	Luminescence	5×10^{-9} M	Pellinen et al. (2004)
Hg (II) Cu (II) Zn (II) Cd (II)	Metallothionein *SmA* (*E. coli*)	Potentiometric	10^{-15} M	Bontidean et al. (2000)
Cr (VI)	Cytochrome C_3 (*D. norvegicum*)	Amperometric	0.2 mg/L	Michel et al. (2006)

4.4.3 Other Environmental Pollutants

In addition to the determination of pesticides and heavy metals, the biosensors have been also used for the analysis of some other environmental pollutants, which includes:

(i) **Phenolic compounds**
Phenols and its compounds are widely used in the mining, plastic, paint, and pharmaceutical industries and, therefore, are priority-considered pollutant in water owing to its toxicity at a very low concentration. Phenols also have the potential to cause persistence and bioaccumulation effect in animals and plants (Davì and Gnudi 1999). For phenolic determination, a number of biosensors have been developed; most of them were based on the enzyme tyrosinase. Tyrosinase (E.C.1.14.18.1) is a copper-containing enzyme that catalyzes the oxidation of phenol to catechols and finally to quinines along with the depletion of oxygen. Therefore, phenols can easily be monitored by measuring the oxygen depletion using oxygen electrodes. A phenol biosensor was developed by immobilizing tyrosinase on modified magnetic nanoparticles which can detect phenol at a concentration of 10^{-7} M. A tyrosinase/laccase bienzymatic biosensor was developed using titania gel matrix that has a higher sensitivity for different phenolic compounds such as methylcatechol, dimethoxyphenol, and chlorophenol (Kochana et al. 2008). Apart from tyrosinase, polyphenol oxidase and horseradish peroxidase were also used for developing phenol biosensor. A polyphenol oxidase-based biosensor was employed to be used for the determination of phenol, *p*-cresol, *m*-cresol, and catechol with a detection limit of 0.96, 1.38, 1.5, and 2.03 AM^{-1}/cm^{-2} (Xue and Shen 2002). Horseradish peroxidase-modified electrodes were used to design amperometric biosensor for the detection of phenol and its derivatives, with a detection limit of 0.5 mM, 0.22 μM, and 0.93 mM (Korkut et al. 2008).

(ii) **Surfactants**
The surfactants are the basic "active" components in most of the detergent products and represent a wide group of organic pollutants. In contrast to their efficiency as detergents, these surfactants possess resistance to biodegradation due to the presence of ramified hydrocarbons (Badía and Díaz García 1998). The anionic surfactants are most widely used as compared to cationic surfactants, and therefore, a number of biosensors have been reported only for the anionic surfactants such as alkane sulfonates, sulfosuccinates, methyl ester sulfonates, sodium dodecyl sulfate (SDS), and volgonat. A microbial-based biosensor was developed by using strains of genera *Pseudomonas* and *Achromobacter* which have the potential to degrade anionic surfactants. The detection was based on the measurement of decreased dissolved oxygen concentration as a result of SDS-activated cell respiration. The developed biosensor can detect SDS with a detection limit lower to 1 μM (Taranova et al. 2002). Similarly, an amperometric biosensor was constructed employing

Pseudomonas rathonis T for the detection of SDS and volgonat in which the microorganism contains the plasmid for surfactant degradation with detection limit in the range of 0.25–0.75 mg/L (Reshetilov et al. 1997). Some biodegradable surfactants such as linear alkyl benzene sulfonates (LASs) which took several days for degradation can easily be degraded and monitored on real-time basis (Nomura et al. 1998).

(iii) **Pathogenic microorganism**

Pathogenic form of microorganism such as of bacteria, virus, and protozoa poses a serious threat to the public health and, therefore, must be eliminated from potable and polluted water. The detection of these microorganisms is not only important from an environmental viewpoint but also for clinical diagnosis and food quality control. The conventional methods of pathogen identification include the cultivation of bacteria, which is tedious and time consuming (Eriksson et al. 2009); therefore, biosensors have played increasingly important roles in their detection. A variety of immunosensors utilizing the phenomenon of fluorescence, surface plasmon resonance, quartz crystal microbalance, and impedance (Barreiros dos Santos et al. 2013) have been developed for the pathogen detection in different biological samples (Heyduk and Heyduk 2010; Baccar et al. 2010; Guo et al. 2012; Barreiros dos Santos et al. 2013). However, DNA detection is more specific than immunosensor with enhanced sensitivity. Aptamer is a kind of synthetic oligonucleotide which binds to certain target molecules with high specificity and, therefore, has been used in the pathogen detection biosensor (Wang et al. 2012). Mycobacteria, the etiological agent of diseases tuberculosis and leprosy, can be detected in the environmental samples by a microfluidic culture-based biosensor (Jing et al. 2007). The biosensor relies on the paraffinophilic nature, which is unique to mycobacteria. Recently, a potentiometric biosensor was developed for the detection of living organism that can detect a single CFU/mL of *Staphylococcus aureus*, an assay close to the real time (Hernández et al. 2014).

(iv) **Biological oxygen demand**

Biological oxygen demand (BOD) is one of the most important and widely used indexes for the characterization of organic pollution of water. It is the amount of oxygen required by aerobic microorganism for degrading organic matters present in wastewater. The conventional method of BOD estimation is a 5-day-long procedure and also requires experience and skills (Wang et al. 2010). So biosensors provide an alternative approach for BOD estimation with the advantages of being a rapid and reliable method. BOD biosensors generally consist of a biomembrane (mostly microbes) and an oxygen electrode, and it measures the respiration activity of cells. The first microbial biosensor for BOD estimation was reported by Karube et al. (1977); after that, a number of biosensors have been developed with enhanced features (Dhall et al. 2008). For long-time operations, yeasts, specifically of genus *Debaryomyces*, are the most preferable biomaterials for the development of

BOD biosensors as they are resistant to the negative environmental factors (Arlyapov et al. 2013). To avoid the disadvantages associated with microbial membrane-type biosensor such as low stability, a novel biosensor was developed by using immobilized microbial cell (IMC) beads freely suspended in the aqueous solution (Wang et al. 2010). A cell-based biosensor was also developed for monitoring BOD level in wastewater samples from factory processing concentrated rubber latex, with a response time of 10–15 min (Kumlanghan et al. 2008), while an amperometric biosensor was developed for quick and reliable estimation of BOD from beverage industrial wastewater with a detection limit of 1 mg/mL (Dhall et al. 2008). Unlike living cells, a thermally killed complex of microbial culture (at 300 °C) can also be used in the development of BOD biosensor, with advantages of having better sensitivity, stability, and reproducibility (Tan and Lim 2005).

4.5 Conclusion

In the recent years, there has been a boost in the development of biosensors for their potential applications in environmental analysis. Despite their huge potential, there commercial impact is not up to the mark. Some issues still raise a barricade in their broader area of application which may be lower sensitivity, high response time, and low lifetime stability. However, these limitations could be improved by better immobilization techniques and better understanding of interference mechanism of environmental pollutants and advanced receptors and therefore will be helpful in designing more precise and reliable devices.

References

Amine A, Mohammadi H, Bourais I, Palleschi G (2006) Enzyme inhibition-based biosensors for food safety and environmental monitoring. Biosens Bioelectron 21:1405–1423

Andreescu S (2002) Screen-printed electrode based on AChE for the detection of pesticides in presence of organic solvents. Talanta 57:169–176

Arduini F, Ricci F, Tuta CS et al (2006) Detection of carbamic and organophosphorous pesticides in water samples using a cholinesterase biosensor based on Prussian Blue-modified screen-printed electrode. Anal Chim Acta 580:155–162

Arlyapov VA, Yudina NY, Asulyan LD et al (2013) BOD biosensor based on the yeast Debaryomyces hansenii immobilized in poly(vinyl alcohol) modified by N-vinylpyrrolidone. Enzyme Microb Technol 53:257–262

Bachan Upadhyay LS, Verma N (2013) Enzyme inhibition based biosensors: a review. Anal Lett 46:225–241

Baccar H, Mejri MB, Hafaiedh I et al (2010) Surface plasmon resonance immunosensor for bacteria detection. Talanta 82:810–814

Badía R, Díaz García M (1998) Room temperature phosphorescence flow-through biosensing of anionic surfactants. Anal Chim Acta 371:73–80

Barreiros dos Santos M, Agusil JP, Prieto-Simón B et al (2013) Highly sensitive detection of pathogen Escherichia coli O157:H7 by electrochemical impedance spectroscopy. Biosens Bioelectron 45:174–180

Blake DA, Jones RM, Blake RC et al (2001) Antibody-based sensors for heavy metal ions. Biosens Bioelectron 16:799–809

Bontidean I, Lloyd JR, Hobman JL et al (2000) Bacterial metal-resistance proteins and their use in biosensors for the detection of bioavailable heavy metals. J Inorg Biochem 79:225–229

Bucur B, Fournier D, Danet A, Marty J-L (2006) Biosensors based on highly sensitive acetylcholinesterases for enhanced carbamate insecticides detection. Anal Chim Acta 562:115–121

Campanella L, Dragone R, Lelo D et al (2006) Tyrosinase inhibition organic phase biosensor for triazinic and benzotriazinic pesticide analysis (part two). Anal Bioanal Chem 384:915–921

Campanella L, Lelo D, Martini E, Tomassetti M (2007) Organophosphorus and carbamate pesticide analysis using an inhibition tyrosinase organic phase enzyme sensor; comparison by butyrylcholinesterase + choline oxidase opee and application to natural waters. Anal Chim Acta 587:22–32

Cao X, Mello SV, Leblanc RM et al (2004) Detection of paraoxon by immobilized organophosphorus hydrolase in a Langmuir–Blodgett film. Colloids Surf A Physicochem Eng Aspects 250:349–356

Clark LC, Lyons C (1962) Electrode systems for continuous monitoring in cardiovascular surgery. Ann N Y Acad Sci 102:29–45

Collings AF, Caruso F (1997) Biosensors: recent advances. Rep Prog Phys 60:1397–1445

Davì ML, Gnudi F (1999) Phenolic compounds in surface water. Water Res 33:3213–3219

De Albuquerque YDT, Ferreira LF (2007) Amperometric biosensing of carbamate and organophosphate pesticides utilizing screen-printed tyrosinase-modified electrodes. Anal Chim Acta 596:210–221

Deng S, Jian G, Lei J et al (2009) A glucose biosensor based on direct electrochemistry of glucose oxidase immobilized on nitrogen-doped carbon nanotubes. Biosens Bioelectron 25:373–377

Dhall P, Kumar A, Joshi A et al (2008) Quick and reliable estimation of BOD load of beverage industrial wastewater by developing BOD biosensor. Sensors Actuators B Chem 133:478–483

Dzyadevych SV, Soldatkin AP, Arkhypova VN et al (2005) Early-warning electrochemical biosensor system for environmental monitoring based on enzyme inhibition. Sensors Actuators B Chem 105:81–87

Eriksson R, Jobs M, Ekstrand C et al (2009) Multiplex and quantifiable detection of nucleic acid from pathogenic fungi using padlock probes, generic real time PCR and specific suspension array readout. J Microbiol Methods 78:195–202

Ghica ME, Carvalho RC, Amine A, Brett CMA (2013) Glucose oxidase enzyme inhibition sensors for heavy metals at carbon film electrodes modified with cobalt or copper hexacyanoferrate. Sensors Actuators B Chem 178:270–278

Guo X, Lin CS, Chen SH et al (2012) A piezoelectric immunosensor for specific capture and enrichment of viable pathogens by quartz crystal microbalance sensor, followed by detection with antibody-functionalized gold nanoparticles. Biosens Bioelectron 38:177–183

Guascito MR, Malitesta C, Mazzotta E, Turco A (2008) Inhibitive determination of metal ions by an amperometric glucose oxidase biosensor. Sensors Actuators B Chem 131:394–402

Hernández F, Sancho JV, Pozo OJ (2005) Critical review of the application of liquid chromatography/mass spectrometry to the determination of pesticide residues in biological samples. Anal Bioanal Chem 382:934–946

Hernández R, Vallés C, Benito AM et al (2014) Graphene-based potentiometric biosensor for the immediate detection of living bacteria. Biosens Bioelectron 54:553–557

Heyduk E, Heyduk T (2010) Fluorescent homogeneous immunosensors for detecting pathogenic bacteria. Anal Biochem 396:298–303

Ilangovan R, Daniel D, Krastanov A et al (2014) Enzyme based biosensor for heavy metal ions determination. Biotechnol Biotechnol Equip 20:184–189

Jeanty G, Wojciechowska A, Marty J-L, Trojanowicz M (2002) Flow-injection amperometric determination of pesticides on the basis of their inhibition of immobilized acetylcholinesterases of different origin. Anal Bioanal Chem 373:691–695

Jing G, Polaczyk A, Oerther DB, Papautsky I (2007) Development of a microfluidic biosensor for detection of environmental mycobacteria. Sensors Actuators B Chem 123:614–621

Karube I, Matsunaga T, Mitsuda S, Suzuki S (1977) Microbial electrode BOD sensors. Biotechnol Bioeng 19:1535–1547

Kochana J, Nowak P, Jarosz-Wilkołazka A, Bieroń M (2008) Tyrosinase/laccase bienzyme biosensor for amperometric determination of phenolic compounds. Microchem J 89:171–174

Korkut S, Keskinler B, Erhan E (2008) An amperometric biosensor based on multiwalled carbon nanotube-poly(pyrrole)-horseradish peroxidase nanobiocomposite film for determination of phenol derivatives. Talanta 76:1147–1152

Kumlanghan A, Kanatharana P, Asawatreratanakul P et al (2008) Microbial BOD sensor for monitoring treatment of wastewater from a rubber latex industry. Enzyme Microb Technol 42:483–491

Kuralay F, Özyörük H, Yıldız A (2007) Inhibitive determination of Hg2+ ion by an amperometric urea biosensor using poly(vinylferrocenium) film. Enzyme Microb Technol 40:1156–1159

Li X-H, Xie Z-H, Min H et al (2007) Amperometric biosensor based on immobilization acetylcholinesterase on manganese porphyrin nanoparticles for detection of trichlorfon with flow-injection analysis system. Electroanalysis 19:2551–2557

Liu G, Lin Y (2006) Amperometric glucose biosensor based on self-assembling glucose oxidase on carbon nanotubes. Electrochem Commun 8:251–256

Liu J, Xu X, Tang L, Zeng G (2009) Determination of trace mercury in compost extract by inhibition based glucose oxidase biosensor. Trans Nonferrous Met Soc China 19:235–240

Liu G, Guo W, Song D (2014a) A multianalyte electrochemical immunosensor based on patterned carbon nanotubes modified substrates for detection of pesticides. Biosens Bioelectron 52:360–366

Liu X, Li W-J, Li L et al (2014b) A label-free electrochemical immunosensor based on gold nanoparticles for direct detection of atrazine. Sensors Actuators B Chem 191:408–414

Michel C, Ouerd A, Battaglia-Brunet F et al (2006) Cr(VI) quantification using an amperometric enzyme-based sensor: interference and physical and chemical factors controlling the biosensor response in ground waters. Biosens Bioelectron 22:285–290

Mostafa GA (2010) Electrochemical biosensors for the detection of pesticides. Open Electrochem J 2:22–42

Mulchandani P, Mulchandani A, Kaneva I, Chen W (1999) Biosensor for direct determination of organophosphate nerve agents. 1. Potentiometric enzyme electrode. Biosens Bioelectron 14:77–85

Nomura Y, Ikebukuro K, Yokoyama K et al (1998) Application of a linear alkylbenzene sulfonate biosensor to river water monitoring. Biosens Bioelectron 13:1047–1053

Pellinen T, Huovinen T, Karp M (2004) A cell-free biosensor for the detection of transcriptional inducers using firefly luciferase as a reporter. Anal Biochem 330:52–57

Prieto-Simón B, Campàs M, Andreescu S, Marty J-L (2006) Trends in flow-based biosensing systems for pesticide assessment. Sensors 6:1161–1186

Ramanathan K, Danielsson B (2001) Principles and applications of thermal biosensors. Biosens Bioelectron 16:417–423

Reshetilov AN, Semenchuk IN, Iliasov PV, Taranova LA (1997) The amperometric biosensor for detection of sodium dodecyl sulfate. Anal Chim Acta 347:19–26

Reyes De Corcuera J, Cavalieri R (2003) Biosensors. In: Heldmen DR, Moraru CI (eds) Encyclopedia of agriculture and food biology engineering, 2nd edn. CRC, Boca Raton, pp 119–123

Rodriguez-Mozaz S, Marco MP, de Alda MJL, Barceló D (2009) Biosensors for environmental applications: future development trends: pure and applied chemistry. Pure Appl Chem 76:723–752

Saerens D, Huang L, Bonroy K, Muyldermans S (2008) Antibody fragments as probe in biosensor development. Sensors 8:4669–4686

Schöning MJ, Krause R, Block K et al (2003) A dual amperometric/potentiometric FIA-based biosensor for the distinctive detection of organophosphorus pesticides. Sensors Actuators B Chem 95:291–296

Snejdarkova M, Svobodova L, Nikolelis DP et al (2003) Acetylcholine biosensor based on dendrimer layers for pesticides detection. Electroanalysis 15:1185–1191

Tan TC, Lim EWC (2005) Thermally killed cells of complex microbial culture for biosensor measurement of BOD of wastewater. Sensors Actuators B Chem 107:546–551

Tang X, Zhang T, Liang B et al (2014) Sensitive electrochemical microbial biosensor for p-nitrophenylorganophosphates based on electrode modified with cell surface-displayed organophosphorus hydrolase and ordered mesopore carbons. Biosens Bioelectron 60:137–142

Taranova L, Semenchuk I, Manolov T et al (2002) Bacteria-degraders as the base of an amperometric biosensor for detection of anionic surfactants. Biosens Bioelectron 17:635–640

Thévenot DR, Toth K, Durst RA, Wilson GS (2001) Electrochemical biosensors: recommended definitions and classification. Biosens Bioelectron 16:121–131

Tran HV, Yougnia R, Reisberg S et al (2012) A label-free electrochemical immunosensor for direct, signal-on and sensitive pesticide detection. Biosens Bioelectron 31:62–68

Tsai H, Doong R (2005) Simultaneous determination of pH, urea, acetylcholine and heavy metals using array-based enzymatic optical biosensor. Biosens Bioelectron 20:1796–1804

Tsai H-C, Doong R-A, Chiang H-C, Chen K-T (2003) Sol–gel derived urease-based optical biosensor for the rapid determination of heavy metals. Anal Chim Acta 481:75–84

Updike SJ, Hicks GP (1967) The enzyme electrode. Nature 214:986–988

Wang J, Krause R, Block K et al (2003) Flow injection amperometric detection of OP nerve agents based on an organophosphorus–hydrolase biosensor detector. Biosens Bioelectron 18:255–260

Wang Z, Liu S, Wu P, Cai C (2009) Detection of glucose based on direct electron transfer reaction of glucose oxidase immobilized on highly ordered polyaniline nanotubes. Anal Chem 81:1638–1645

Wang J, Zhang Y, Wang Y et al (2010) An innovative reactor-type biosensor for BOD rapid measurement. Biosens Bioelectron 25:1705–1709

Wang Y-X, Ye Z-Z, Si C-Y, Ying Y-B (2012) Application of aptamer based biosensors for detection of pathogenic microorganisms. Chin J Anal Chem 40:634–642

White BJ, Harmon HJ (2005) Optical solid-state detection of organophosphates using organophosphorus hydrolase. Biosens Bioelectron 20:1977–1983

Xue H, Shen Z (2002) A highly stable biosensor for phenols prepared by immobilizing polyphenol oxidase into polyaniline–polyacrylonitrile composite matrix. Talanta 57:289–295

Yang Y, Wang Z, Yang M et al (2006) Inhibitive determination of mercury ion using a renewable urea biosensor based on self-assembled gold nanoparticles. Sensors Actuators B Chem 114:1–8

Zhang X, Guo Q, Cui D (2009) Recent advances in nanotechnology applied to biosensors. Sensors (Basel) 9:1033–1053

Chapter 5
Microbial Biosurfactant for Hydrocarbons and Heavy Metals Bioremediation

Arun Kumar Pradhan and Nilotpala Pradhan

5.1 Introduction

Biosurfactants are produced by diverse bacterial genera (Desai and Banat 1997). As they are produced by a wide variety of microorganisms, there is immense diversity in structure, function, and genetic basis. These factors provide the advantage for the producing microorganism over other microorganism in an ecosystem. The genes responsible for the diversity are also evolved differently. The hydrophobic lipid fraction forming the tail of the molecule is more similar across the microbial community. The polar fraction of the molecule generally depicted as head group are more diverse and mostly composed of diverse glycan, proteins, and other polymeric substances of microbial origin. Diverse groups can be categorized in categories like glycolipids, lipoproteins, phospholipids, fatty acid salts, polymeric biosurfactants, etc. (Fig. 5.1).

Biosurfactants are surface active compounds produced by various microorganisms like bacteria and fungi as secondary metabolites. Biosurfactants (saponin) can also be produced by certain plants part like quillaja bark (Hong et al. 2002), fruit pericarps of *Sapindus mukorossi* (Zhou et al. 2013). Due to presence of both hydrophilic and hydrophobic moieties, biosurfactants can interact with any type of surfaces and modify these surfaces. Various applications or utilities of biosurfactants are only possible due to its amphiphilic nature. Nowadays, biosurfactants are commonly used for the bioremediation purpose such as removal of heavy metals (Mulligan et al. 1999, 2001; Sriram et al. 2011a; Plociniczak et al. 2011), hydrocarbons/crude oils (Kumar et al. 2006a, b, 2008; Abdolhamid

A.K. Pradhan (✉) • N. Pradhan
Academy of Scientific and Innovative Research, CSIR – Institute of Minerals and Materials Technology, Bhubaneswar 751013, Odisha, India

Bioresources Engineering Department, CSIR – Institute of Minerals and Materials Technology, Bhubaneswar 751013, Odisha, India
e-mail: arunpradhan83@gmail.com; npradhan@immt.res.in

L.B. Sukla et al. (eds.), *Environmental Microbial Biotechnology*, Soil Biology 45,
DOI 10.1007/978-3-319-19018-1_5

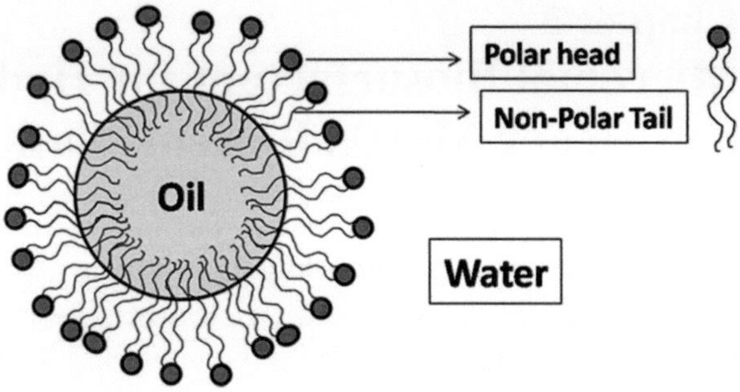

Fig. 5.1 Attachment of biosurfactants on oil droplet

et al. 2009; Chandraja et al. 2014) from contaminated soil. Besides that biosurfactants can be used in diverse fields like cosmetic industries (Desai and Banat 1997; Pradhan et al. 2013, 2014a), pharmaceutical industries (Desai and Banat 1997), antifouling agent (Sriram et al. 2011b; Pradhan et al. 2013, 2014a), and anticancer agent (Zaragoza et al. 2009; Janek et al. 2013; Pradhan et al. 2014b) in biomedical field, etc.

5.2 Biosurfactant Producing Microorganisms

Generally, bacteria produce biosurfactants as their secondary metabolites. Most common biosurfactant producing bacteria reported are *Pseudomonas* sp. (Hong et al. 2005; Chang et al. 2008; Jayabarath et al. 2009; Mani et al. 2011; Milena et al. 2012) and *Bacillus* sp. (Christova et al. 2004; Priya and Usharani 2009; Kosaric 2001; Stankovic et al. 2012; Pradhan et al. 2013, 2014a, b). Other reported bacteria are *Rhodococcus* (Kosaric 2001), *Corynebacterium* (Thavasi et al. 2010), *Sphingomonas* (Chang et al. 2008), *Serratia* (Anyanwu et al. 2011), and *Lactococcus* (Rodrigues et al. 2006).

Few species of fungus can also produce biosurfactants. *Torulopsis petrophilum* culturing with animal tissue produce glycolipid type of biosurfactant (Folch et al. 1957) and *Candida bombicola* produces sophorolipid type of biosurfactants (Casas et at. 1999). *Candida lipolytica* produces protein–lipid–polysaccharide type of complex biosurfactant (Sarubbo et al. 2007). Another fungus, *Pichia anomala*, produces sophorolipid utilizing 4 % soybean oil as carbon source (Thaniyavarn et al. 2008).

5.3 Chemical Structure of Biosurfactants

Chemically microbial biosurfactants are glycan or peptide derivatives of lipid molecules. So biosurfactants are broadly categorized as glycolipid, lipopeptide, or protein–lipid–polysaccharide types (Table 5.1).

Table 5.1 List of different types of biosurfactants and related microorganisms

Type	Examples	Microorganisms	References
Glycolipid	Rhamnolipid	*Pseudomonas aeruginosa*	Hong et al. (2005), Chang et al. (2008), Milena et al. (2012), Mani et al. (2011)
		Sphingomonas sp. 12A	Chang et al. (2008)
	Sophorolipid	*Candida bombicola*	Casas and García-Ochoa (1999)
		Pichia anamola	Thaniyavarn et al. (2008)
	Cellobiose lipids	*Cryptococcus humicola* *Pseudozyma fusiformata*	Kulakovskaya et al. (2009)
	Trehalose lipids	*Rhodococcus*	Franzetti et al. (2010a)
	Emulsan (lipopolysaccharide)	*Acinetobacter calcoaceticus*	Panilaitis et al. (2002), Julia et al. (1989)
	Liamocins and exophilins (mannitol-lipids)	*Aureobasidium pullulans*	Price et al. (2013)
Lipopeptide	Surfactin	*Bacillus subtilis*	Priya and Usharani (2009), Tang et al. (2010)
		Bacillus amyloliquefaciens	Liu et al. (2012)
	Subtilin	*Bacillus subtilis*	Kuboi et al. (1994), Noudeh et al. (2010)
		Bacillus tequilensis	Pradhan et al. (2014a)
Protein–lipid–polysaccharide complex		*Candida lipolytica*	Sarubbo et al. (2007)

5.3.1 *Rhamnolipid*

Rhamnolipid is the most common type of biosurfactants produced by *Pseudomonas aeruginosa* (Hong et al. 2005; Chang et al. 2008; Milena et al. 2012; Mani et al. 2011). *Sphingomonas* sp. 12A also produces rhamnolipid which is a glycolipid type of biosurfactant (Chang et al. 2008). Its lipid part is associated with rhamnose (glycan). If one rhamnose is attached to lipid moiety, it is called as mono-rhamnolipid (Fig. 5.2a). If two rhamnoses are attached with one lipid moiety, it is called as di-rhamnolipid (Fig. 5.2b).

Rhamnolipid can be used as antibacterial, antifungal, and emulsifying agents. It has the capacity to remove many toxic metals form contaminated soil efficiently (Table 5.2).

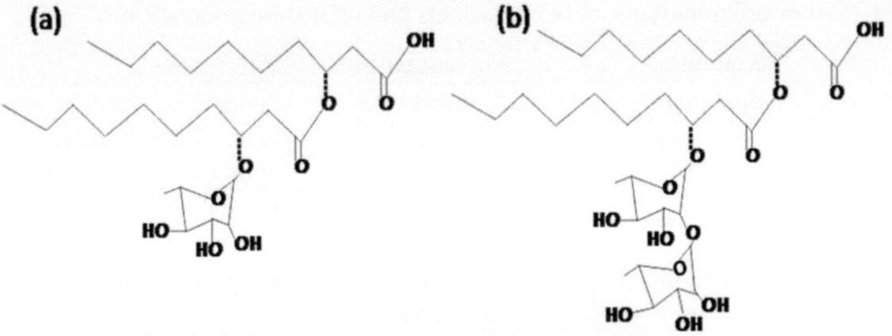

Fig. 5.2 Structure of (**a**) mono-rhamnolipid and (**b**) di-rhamnolipid

Table 5.2 List of heavy metals removed by biosurfactants

Heavy metals	Biosurfactants (concentration)	% of removal	References
Cd	Rhamnolipid (80 ppm)	53	Elouzi et al. (2012)
	Rhamnolipid (5 mM)	92	Tan et al. (1994)
	Rhamnolipid (0.1 %)	92	Juwarkar et al. (2007)
	Rhamnolipid (40 ppm)	95	Bondarenko et al. (2010)
Pb	Rhamnolipid (80 ppm)	62	Elouzi et al. (2012)
	Rhamnolipid (0.1 %)	88	Juwarkar et al. (2007)
U	Emulsan (0.9 μm U/mg)	90	Zosim et al. (1983)
Ba	Rhamnolipid (80 ppm)	28	Elouzi et al. (2012)
Ni	Rhamnolipid (80 ppm)	56	
Zn	Rhamnolipid (80 ppm)	20	
	Rhamnolipid (12 %)	19.5	Mulligan et al. (1999)
	Sophorolipid (4 %) + HCl (0.7 %) (five batch washes)	100	
	Rhamnolipid (0.5 %) (single wash)	18	Mulligan et al. (2001)
	Sophorolipid (4 %) (single wash)	60	
Sr	Rhamnolipid (80 ppm)	7	Elouzi et al. (2012)
Cu	Rhamnolipid (12 %)	25	Mulligan et al. (1999)
	Rhamnolipid (2 %) + NaOH (1 %)		
	Surfactin (0.1 %) + NaOH (1 %) (five batch washes)	70	
	Rhamnolipid (0.5 %) (single wash)	65	Mulligan et al. (2001)
	Sophorolipid (4 %) (single wash)	25	

5.3.2 Sophorolipid

Candida bombicola (Casas and García-Ochoa 1999) and *Pichia anamola* (Thaniyavarn et al. 2008) produce sophorolipid having lipid and sophorose moieties. Sophorolipid is reported to be useful in removing heavy metals like Cu and Zn from contaminated soil (Mulligan et al. 1999, 2001). Sophorolipid's cyclic structure and all operational conditions are well studied (Casas and García-Ochoa 1999).

5.3.3 Cellobiose Lipids

Cellobiose lipid can behave as biosurfactant in which cellobiose is hydrophilic part whereas lipid is the hydrophobic part. *Cryptococcus humicola* and *Pseudozyma fusiformata* produce cellobiose lipid (Kulakovskaya et al. 2009). This is a fungal biosurfactant having fungicidal activity.

5.3.4 Trehalose Lipids

Trehalose is one of the biosurfactants mainly produced by various species of *Rhodococcus*. Trehalose lipid is having a number of chemical structures like trehalose mono-, di-, and tri-mycolates, trehalose tetraesters, and succinoyl trehalose lipids (Franzetti et al. 2010a). Trehalose lipids are used to degrade hydrocarbons from contaminated soil.

5.3.5 Emulsan

Acinetobacter calcoaceticus produces lipo-heteropolysaccharides having emulsifying capacity called as emulsans (Julia et al. 1989; Panilaitis et al. 2002). Emulsans can be used for the biodegradation of crude oil (Julia et al. 1989). It is reported that uranium could be removed from contaminated soil by emulsan treatment (Zosim et al. 1983).

5.3.6 Surfactin

Commonly *Bacillus* sp. produces lipopeptide type of biosurfactants. Surfactin is one of lipopeptide in which peptide moiety is in cyclic form (Priya and Usharani 2009; Tang et al. 2010; Liu et al. 2012). It is used to recover oil from crude oil and removal of heavy metals from soil (Table 5.2). It is also having antifungal activity (Liu et al. 2012), antibacterial property, and cytotoxic effect (Fig. 5.3).

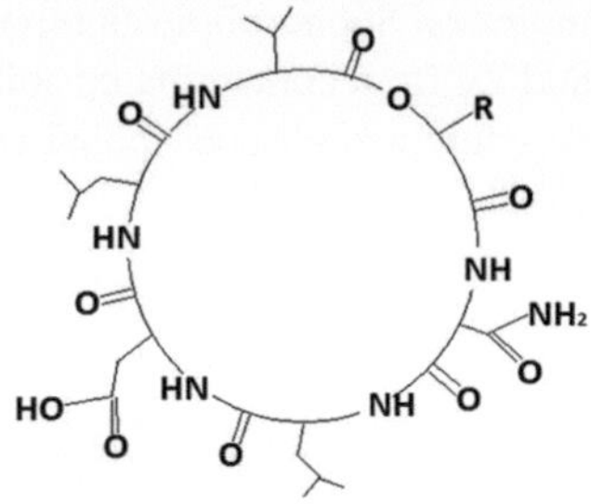

Fig. 5.3 Structure of surfactin

5.3.7 *Subtilin*

Subtilin, a lipopeptide type of biosurfactants, is produced by mostly *Bacillus subtilis* (Kuboi et al. 1994; Noudeh et al. 2010). It is both antibacterial and antifungal agent. Subtilin is mostly effective to gram positive bacteria like *Lactobacillus*, *Streptococcus*, *Mycobacterium*, *Staphylococcus*, *Micrococcus*, etc. and fungus. It can be used as natural bio-preservatives for food.

5.4 Biosurfactant Mediated Bioremediation

5.4.1 *Biodegradation of Oils/Hydrocarbons by Biosurfactants*

Biosurfactant contains both hydrophobic and hydrophilic moieties and so increases the solubility of any compounds like organics or inorganics by reducing the surface tension of solvents (Bustamante et al. 2012). Using this unique property of biosurfactant, one of the global problems like spillage of crude oil of oil industries can be solved (Abdolhamid et al. 2009). Hexadecane can be degraded by *Pseudomonas* sp. with the secretion of rhamnolipid biosurfactants (Cameotra and Singh 2009). A suitable procedure is established to estimate hydrocarbon degradation in contaminated water and soil, i.e., gas chromatograph coupled with flame ionization detector (FID) (Kumar et al. 2006b; Obayori et al. 2009).

5.4.2 *Degradation of Crude Oils*

There is an increase in degradation of crude oil in contaminated soil at oilfield with increasing the concentration of rhamnolipid (0.05 %, 0.5 %) (Abdolhamid et al. 2009). Rhamnolipid and SDS remove 80 % of crude oil where as lecithin removes 42 % from contaminated soil (Bustamante et al. 2012; Urum and Pekdemir 2004). *Gordonia* sp. strain BS29 produces biosurfactant which is able to remove crude oil by soil washing (Franzetti et al. 2010b). Biosurfactant produced by *Acinetobacter calcoaceticus* RAG-1, emulsan, removes 50–90 % of crude oil from soil (Julia et al. 1989).

5.4.3 Degradation of Hydrocarbons

Pseudomonas sp. uptakes hexadecane to degrade 70–75 % within 7 days (Cameotra and Singh 2009). Surfactin at a concentration of 40 mg/L degrades 90 % of diesel (Bustamante et al. 2012). *P. aeruginosa* biosurfactant (glycolipid) removed 56 % aliphatic hydrocarbons from hydrocarbon contaminated sandy loam soil (Desai and Banat 1997). Coculture of hydrocarbon degrading bacteria (*Pseudomonas putida*) and biosurfactant producing bacteria (*Pseudomonas aeruginosa*) results in degradation of diesel oil, a representative of hydrocarbon like pyrene, hexadecane, ethanol, phenol, naphthalene, phenanthrene, indole, toluene, etc. (Kumar et al. 2006b, 2008). With the help of bacterial adhesion to hydrocarbons (BATH) assay, they reported that biosurfactants produced by one bacterium have two important roles. One is emulsification of hydrocarbons and other is to change adhesion of hydrocarbons for the easy attachment of these to cell surfaces which results in degradation of hydrocarbons.

Lawniczak et al. (2013) reported that with biosurfactant from *Lactobacillus pentosus*, the biodegradation efficiencies of autochthonous microflora are 59 % and 63 % when initial concentrations of octane in soil are 700 and 70,000 mg/kg, respectively. Without biosurfactant, the efficiencies are only 1 and 25 % for same initial concentration of octane.

Similarly, *Candida tropicalis* degrades hexadecane 93 % in presence of rhamnolipid whereas it is 78 % in absence of rhamnolipid (Zeng et al. 2011).

5.4.4 Oil Spillage Degradation

Oil spillage is one of the most important global problems related to soil as well as water bodies. Oil spillage makes the water unsuitable for all types of aquatic animals and plants. Similarly, growth of plants can be reduced in oil contaminated soil. So, to maintain the balance of ecosystem, it needs degradation. Most of oil degrading microbes attaches to oil contaminated water forming a mat like structure in Persian Gulf and Shetland Isles (Desai and Banat 1997). These bacteria produce bioemulsifiers or biosurfactants which reduce surface tension or interfacial tension of water and oil. Then these droplets of oil which are the result of emulsion are cleared by bacteria easily. The presence of both hydrophilic and hydrophobic moieties makes the emulsion possible.

5.4.5 Microbial Enhanced Oil Recovery

Few amount of oil remains un-retrieved inside the oil well at the time of oil recovery. Nowadays, microorganisms are used to produce compounds that mobilize the attached oil and assists oil flow. So a greater amount of oil can be recovered from the oil well. According to Lazar et al. (2007), the important mechanisms of MEOR are oil solubilization, emulsion, interfacial tension alternation, porosity, and permeability alternation. MEOR can be used to clean large oil tank sludge with the help of biosurfactants.

5.4.6 Degradation of Polycyclic Aromatic Hydrocarbons

Polycyclic aromatic hydrocarbons (PAH) (phenanthrene, naphthalene) is one of the major contaminant for soil. Researchers are also trying to degrade PAH with the application of biosurfactants. *Pseudomonas putida* ATCC 17484 degrades 68 % of phenanthrene without biosurfactant, but degradation efficiency was increased to 91 % with rhamnolipid (initial concentration of phenanthrene is 500 mg/kg) (Gottfried et al. 2010). Treatment of PAH with rhamnolipid results in 61 % degradation of PAH within 90 days (12.85 g/kg is the initial concentration of PAH in soil) (Zhang et al. 2010). Anthracene was treated with biosurfactant produced by *Sphingomonas* sp. and *Pseudomonas* sp. monocultures, and 52 % of degradation of anthracene was observed within 18 days whereas degradation was 32 % without biosurfactants (Cui et al. 2008). The uptake of phenanthrene by ryegrass can be increased from 77 to 435 mg/kg in presence of rhamnolipid (Zhu and Zhang 2008).

All these above studies confirm that biosurfactants of different nature degrade PAH and make the soil fertile.

5.4.7 Probable Mechanism for Degradation of Oils/Hydrocarbons/PAH

Biosurfactants increase the hydrophobicity of bacterial cell surfaces so that adhesion and degradation of hydrophobic compounds (oils or hydrocarbons) will become easier (Wouter and Dick 2002; Plaza et al. 2005). Biosurfactant treated hydrocarbons become more bioavailable to hydrocarbon degrading microorganisms due to emulsification (Carrillo et al. 1996; Wei et al. 2005).

5.5 Removal of Heavy Metals from Soil/Water by Biosurfactants

Heavy metal (Cd, Hg, Pb, U, Ni, Ba, Sr, Zn, etc.) contamination is one of the most critical problems for soil and water, because these heavy metals are toxic to microflora (Juwarkar et al. 2007) present in soil as well as water. Microbial consortia and their secondary products like biosurfactants are most useful to solve this type of critical environmental problem (Miller 1995; Desai and Banat 1997). Biosurfactants are used for the bioremediation of soil site polluted with toxic heavy metals like Pb, Cd, U, etc. (Miller 1995). Rhamnolipid at a concentration of 80 ppm removes Pd, Cd, Ni, Ba, Zn, and Sr by 62, 53, 56, 28, 20, and 7 %, respectively (Elouzi et al. 2012). Approximately, 90 % of uranium removed from 0.9 μm of U per 1 mg Emulsan (Zosim et al. 1983). Mulligan et al. (1999, 2001) reported that 4 % sophorolipid removed 15.8 % Cu and 12 % rhamnolipid removed 19.5 % Zn. They reviewed about the removal of Cu and Zn with the help of rhamnolipid, sophorolipid, and surfactin in different conditions (Table 5.2). Lipopeptide produced by *Candida lipolytica* removes approximately 96 % of Zn and Cu (Rufino et al. 2012). Lipopetide is also effective to reduce heavy metal like Fe, Pb, and Cd.

Effect of pH on biosurfactant at the time of heavy metal removal from soil is well studied and reported that pH 9 is the optimum pH for removal of Cd, Co, Ni, Zn, Cu, Pd from contaminated soil (Singh and Cameotra 2013).

5.5.1 Probable Mechanisms for Bioremediation of Heavy Metals from Water/Soil

Due to presence of polar head of biosurfactant, its micelles can form complex with metals present in soil and detaches from soil as shown in Fig. 5.4 (Plociniczak et al. 2011).

Metals are generally positively charged species. Anionic biosurfactants like rhamnolipid, sophorolipid, surfactin, etc. chelate cations like metals and makes bio-unavailable for microflora or plants in soil.

5.6 Conclusion

Finally, it is accomplished that biosurfactants (rhamnolipid, sophorolipid, emulsan, surfactin) are able to degrade hydrocarbons like PAH, hexadecane, octane, diesel oils and remove heavy metals like Cd, Pd, Zn, U, Ba, Sr, Ni, Cu, etc. So we can say that this green surfactant is one of the best tools for bioremediation of heavy metals and hydrocarbons from soil and water. Due to its nontoxic and biodegradable natures, it is preferred over chemical surfactants. Further study is needed for

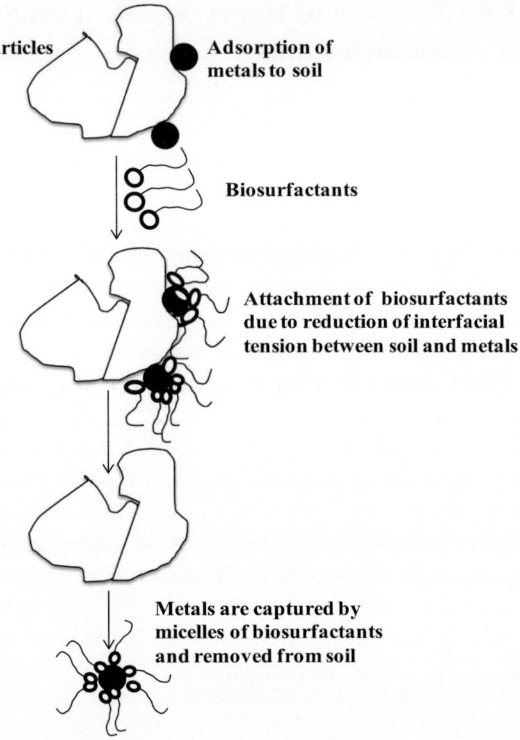

Fig. 5.4 Probable mechanism of removal of heavy metals from contaminated soil by biosurfactants (adapted from Plociniczak et al. 2011)

establishment of exact mechanism of bioremediation by biosurfactants. It also requires the study about large scale in situ bioremediation by biosurfactants. Polar heads of biosurfactant micelles modify the bacterial surfaces as well as soil/metal interface in such a way that hydrocarbons are emulsified and bioavailable for bacteria whereas metals are captured by micelles and desorbed from soil particles. Thus, major environmental problems like hydrocarbons and toxic heavy metals contamination to soil/water may be resolved using biosurfactants.

Acknowledgement One of the authors (AKP) is grateful to the Council of Scientific and Industrial Research (CSIR), India, for the award of Senior Research Fellowship (SRF).

References

Abdolhamid HR, Sadiq al-baghdadi MAR, Hinshiri AKE (2009) Evaluation of bio-surfactants enhancement on bioremediation process efficiency for crude oil contaminated soil at oilfield: strategic study. Ovidius Univ Ann Chem 20:25–30

Anyanwu CU, Obi SKC, Okolo BN (2011) Lipopeptide biosurfactant production by *Serratia marcescens* NSK-1 strain isolated from petroleum-contaminated soil. J Appl Sci Res 7:79–87

Bondarenko O, Rahman PKSM, Rahman TJ, Kahru A, Ivask A (2010) Effects of rhamnolipids from *Pseudomonas aeruginosa* DS10-129 on luminescent bacteria: toxicity and modulation of cadmium bioavailability. Microb Ecol 59:588–600

Bustamante M, Durán N, Diez MC (2012) Biosurfactants are useful tools for the bioremediation of contaminated soil: a review. J Soil Sci Plant Nutr 12:667–687

Cameotra SS, Singh P (2009) Synthesis of rhamnolipid biosurfactant and mode of hexadecane uptake by *Pseudomonas* species. Microb Cell Fact 8:16

Carrillo PG, Mardaraz C, Alvarez SI, Giuliett AM (1996) Isolation and selection of biosurfactant producing bacteria. World J Microbiol Biotechnol 12:82–84

Casas JA, García-Ochoa F (1999) Sophorolipid production by *Candida bombicola*: medium composition and culture methods. J Biosci Bioeng 88:488–494

Chandraja CV, Sridevi K, Gopu KST, Praseetha PK (2014) Bioremediation of crude oil and hydrocarbons by actinomycetes with the enhanced production of bioactive compounds. J Chem Pharm Res 6:1066–1071

Chang ZC, Zeng C, Wan X, Chen D, Zhang YJ, Shen P (2008) Effect of rhamnolipids on degradation of anthracene by two newly isolated strains, *Sphingomonas* sp. 12A and *Pseudomonas* sp. 12B. J Microbiol Biotechnol 18:63–66

Christova N, Tuleva B, Nikolova DB (2004) Enhanced hydrocarbon biodegradation by a newly isolated *Bacillus subtilis* strain. Z Naturforsch 59:205–208

Cui C-Z, Zeng C, Wan X, Chen D, Zhang J-Y, Shen P (2008) Effect of rhamnolipids in degradation of anthracene by two newly isolated strains, *Sphingomonas* sp. 12A and *Pseudomonas* sp. 12B. World J Microbiol Biotechnol 18:63–66

Desai JD, Banat IM (1997) Microbial production of surfactants and their commercial potential. Microbiol Mol Biol Rev 61:47–64

Elouzi AA, Akasha AA, Elgerbi AM, El Gammudi BA (2012) Using of microbial bio-surfactants (rhamnolipid) in heavy metals removal from contaminated water. J Sebha Univ Pure Appl Sci 11:85–95

Folch J, Lees M, Stanly SGH (1957) A simple method for the isolation and quantification of total lipids from animal tissues. J Biol Chem 226:497–509

Franzetti A, Gandolfi I, Bestetti G, Banat IM (eds) (2010a) (Bio) surfactant and bioremediation, successes and failures. Research Signpost, Trivandrum, pp 145–156

Franzetti A, Gandolfi I, Bestetti G, Smyth TJP, Banat IM (2010b) Production and applications of trehalose lipid biosurfactants. Eur J Lipid Sci Technol 112:617–627

Gottfried A, Singhal N, Elliot R, Swift S (2010) The role of salicylate and biosurfactant in inducing phenanthrene degradation in batch soil slurries. Appl Microbiol Biotechnol 86:1563–1571

Hong KJ, Tokunaga S, Kajiuchi T (2002) Evaluation of remediation process with plant-derived biosurfactant for recovery of heavy metals from contaminated soils. Chemosphere 49:379–387

Hong WY, Liang CC, Shu CJ (2005) Rhamnolipid production by indigenous *Pseudomonas aeruginosa* J4 originating from petrochemical wastewater. Biochem Eng J 27:146–154

Janek T, Krasowska A, Radwanska A, Lukaszewicz M (2013) Lipopeptide biosurfactant Pseudofactin II induced apoptosis of melanoma A 375 cells by specific interaction with the plasma membrane. PLoS One 8:1–9

Jayabarath J, Shyam S, Arulmurugan R, Giridhar R (2009) Bioremediation of heavy metals using biosurfactants. Int J Biotechnol Appl 1:50–54

Julia MF, Gutnick DL, Westlake DWS (1989) Effect of emulsan on biodegradation of crude oil by pure and mixed bacterial cultures. Appl Environ Microbiol 55:36–42

Juwarkar AA, Nair A, Dubey KV, Singh SK, Devotta S (2007) Biosurfactant technology for remediation of cadmium and lead contaminated soils. Chemosphere 68:1996–2002

Kosaric N (2001) Biosurfactants and their application for soil bioremediation. Food Technol Biotechnol 39:295–304

Kuboi R, Maruki T, Tanaka H, Komasawa H (1994) Fermentation of *Bacillus subtilis* ATCC 6633 and production of subtilin in polyethylene glycol/phosphate aqueous two-phase systems. J Ferment Bioeng 78:431–436

Kulakovskaya T, Shashkov A, Kulakovskaya E, Golubev W, Zinin A, Tsvetkov Y, Grachev A, Nifantiev N (2009) Extracellular cellobiose lipid from yeast and their analogues: structures and fungicidal activities. J Oleo Sci 58:133–140

Kumar M, Leon V, Materano ADS, Ilzins OA (2006a) Enhancement of oil degradation by co-culture of hydrocarbon degrading and biosurfactant producing bacteria. Pol J Microbiol 55:139–146

Kumar M, Leon V, Materano ADS, Ilzins OA, Castro IG, Fuenmayor SL (2006b) Polycyclic aromatic hydrocarbon degradation by biosurfactant-producing *Pseudomonas* sp. IR1. Z Naturforsch 61:203–212

Kumar M, Leon V, Materano ADS, Ilzins OA, Luis L (2008) Biosurfactant production and hydrocarbon-degradation by halotolerant and thermotolerant *Pseudomonas* sp. World J Microbiol Biotechnol 24:1047–1057

Lawniczak L, Marecik R, Chrzanowski L (2013) Contributions of biosurfactants to natural or induced bioremediation. Appl Microbiol Biotechnol 97:2327–2339

Lazar I, Petrisor IG, Yen TF (2007) Microbial enhanced oil recovery (MEOR). Pet Sci Technol 25:1353–1366

Liu X, Ren B, Gao H, Liu M, Dai H et al (2012) Optimization for the production of surfactin with a new synergistic antifungal activity. PLoS One 7:1–9

Mani KA, Hamid SS, Ramalingam S, Kaliamoorthi R (2011) Effect of rhamnolipid potential on biodegradation of endosulfan by *Pseudomonas aeruginosa* in batch studies. J Biosci Tech 2:268–278

Milena GR, Gordana GC, Miroslav MV, Ivanka K (2012) Production and characterization of rhamnolipids from *Pseudomonas aeruginosa*. J Serb Chem Soc 77:27–42

Miller RM (1995) Biosurfactant-facilitated remediation of metal-contaminated soils. Environ Health Perspect 103:59–62

Mulligan CN, Yong RN, Gibbs BF (1999) On the use of biosurfactants for the removal of heavy metals from oil contaminated soil. Process Saf Prog 18:50–54

Mulligan CN, Yong RN, Gibbs BF (2001) Heavy metal removal from sediments by biosurfactants. J Hazard Mater 85:111–125

Noudeh GD, Noodeh AD, Moshafi MH, Afzadi MA, Pardakhti A, Salandari M (2010) Investigating the effects of various additives on surface activity and emulsification index of biosurfactant resulting from broth media of *Bacillus subtilis* PTCC 1023. Afr J Microbiol Res 4:1981–1990

Obayori OS, Ilori MO, Adebusoye SA, Oyetibo GO, Omotayo AE, Amund OO (2009) Degradation of hydrocarbons and biosurfactant production by *Pseudomonas* sp. strain LP1. World J Microbiol Biotechnol 25:1615–1623

Panilaitis B, Johri A, Blank W, Kaplan D, Fuhrman J (2002) Adjuvant activity of emulsan, a secreted lipopolysaccharide from *Acinetobacter calcoaceticus*. Clin Diagn Lab Immunol 9:1240–1247

Plaza GA, Ulfig K, Brigmon RL (2005) Surface active properties of bacterial strains isolated from petroleum hydrocarbon-bioremediated soil. Pol J Microbiol 54:161–167

Plociniczak MP, Płaza GA, Seget ZP, Cameotra SS (2011) Environmental applications of biosurfactants: recent advances. Int J Mol Sci 12:633–654

Pradhan AK, Pradhan N, Mall G, Panda HT, Sukla LB, Panda PK, Mishra BK (2013) Application of lipopeptide biosurfactant isolated from a halophile: *Bacillus tequilensis* CH for inhibition of biofilm. Appl Biochem Biotechnol 171:1362–1375

Pradhan AK, Pradhan N, Mohapatra P, Kundu CN, Panda PK, Mishra BK (2014a) Cytotoxic effect of microbial biosurfactants against human embryonic kidney cancerous cell: HEK-293 and their possible role in apoptosis. Appl Biochem Biotechnol. doi:10.1007/s12010-014-1168-8

Pradhan AK, Pradhan N, Sukla LB, Panda PK, Mishra BK (2014b) Inhibition of pathogenic bacterial biofilm by biosurfactant produced by *Lysinibacillus fusiformis* S9. Bioprocess Biosyst Eng 37:139–149

Price NP, Manitchotpisit P, Vermillion KE, Bowman MJ, Leathers TD (2013) Structural characterization of novel extracellular liamocins (mannitol oils) produced by *Aureobasidium pullulans* strain NRRL 50380. Carbohydr Res 5:24–32

Priya T, Usharani G (2009) Comparative study for biosurfactant production by using *Bacillus subtilis* and *Pseudomonas aeruginosa*. Bot Res Int 2:284–287

Rodrigues LR, Teixeira JA, Mei HC, Oliveira R (2006) Physicochemical and functional characterization of a biosurfactant produced by *Lactococcus lactis* 53. Colloids Surf B 49:79–86

Rufino RD, Luna JM, Campos-Takaki GM, Ferreira SRM, Sarubbo LA (2012) Application of the biosurfactant produced by *Candida lipolytica* in the remediation of heavy metals. Chem Eng Trans 27:61–66

Sarubbo LA, Farias CB, Campos-Takaki GM (2007) Co-utilization of canola oil and glucose on the production of a surfactant by *Candida lipolytica*. Curr Microbiol 54:68–73

Singh AK, Cameotra SS (2013) Efficiency of lipopeptide biosurfactants in removal of petroleum hydrocarbons and heavy metals from contaminated soil. Environ Sci Pollut Res. doi:10.1007/s11356-013-1752-4

Sriram MI, Gayathiri S, Gnanaselvi U, Jenifer PS, Raj SM, Gurunathan S (2011a) Novel lipopeptide biosurfactant produced by hydrocarbon degrading and heavy metal tolerant bacterium *Escherichia fergusonii* KLU01 as a potential tool for bioremediation. Bioresour Technol 102:9291–9295

Sriram MI, Kalishwaralal K, Deepak V, Gracerosepat R, Srisakthi K, Gurunathan S (2011b) Biofilm inhibition and antimicrobial action of lipopeptide biosurfactant produced by heavy metal tolerant strain *Bacillus cereus* NK1. Colloids Surf B Biointerfaces 85:174–181

Stankovic S, Mihajlovic S, Draganic V, Dimkic I, Vukotic G, Beric T, Fira D (2012) Screening for the presence of biosynthetic genes for antimicrobial lipopeptides in natural isolates of *Bacillus* sp. Arch Biol Sci 64:1425–1432

Tan H, Champion JT, Artiola JF, Brusseau ML, Miller RM (1994) Complexation of cadmium by a rhamnolipid biosurfactant. Environ Sci Technol 28:2402–2406

Tang JS, Zhao F, Gao H, Dai Y, Yao ZH, Hong K et al (2010) Characterization and online detection of surfactin isomers based on HPLC-MSn analyses and their inhibitory effects on the overproduction of nitric oxide and the release of TNF-α and IL-6 in LPS-induced macrophages. Mar Drugs 8:2605–2618

Thaniyavarn J, Chianguthai T, Sangvanich P, Roongsawang N, Washio K (2008) Production of sophorolipid biosurfactant by *Pichia anomala*. Biosci Biotechnol Biochem 72:2061–2068

Thavasi R, Jayalakshmi S, Banat IM (2010) Biosurfactants from marine bacterial isolates. In: Vilas AM (ed) Current research, technology and education topics in applied microbiology and microbial biotechnology. Formatex, Badajoz, pp 1367–1373

Urum K, Pekdemir T (2004) Evaluation of biosurfactants for crude oil contaminated soil washing. Chemosphere 57:1139–1150

Wei QF, Mather RR, Fotheringham AF (2005) Oil removal from used sorbents using a biosurfactant. Bioresour Technol 96:331–334

Wouter HN, Dick BJ (2002) Rhamnolipid stimulate uptake of hydrophobic compounds by *Pseudomonas aeruginosa*. Appl Environ Microbiol 68:4502–4508

Zaragoza A, Aranda FJ, Espuny MJ, Teruel JA, Marques A, Manresa A, Ortiz A (2009) Effects of a bacterial trehalose lipid on phosphatidylglycerol membranes. Langmuir 25:7892–7898

Zeng G, Liu Z, Zhong H, Li J, Yuan X, Fu H, Ding Y, Wang J, Zhou M (2011) Effect of monorhamnolipid on the degradation of n-hexadecane by *Candida tropicalis* and the association with cell surface properties. Appl Microbiol Biotechnol 90:1155–1161

Zhang J, Yin R, Lin X, Liu W, Chen R, Li X (2010) Interactive effect of biosurfactant and microorganism to enhance phytoremediation for removal of aged polycyclic aromatic hydrocarbons from contaminated soils. J Health Sci 56:257–266

Zhou W, Wang X, Chen C, Zhu L (2013) Enhanced soil washing of phenanthrene by a plant-derived natural biosurfactant, Sapindus saponin. Colloids Surf A Physicochem Eng Asp 425:122–128

Zhu L, Zhang M (2008) Effect of rhamnolipids on the uptake of PAHs by ryegrass. Environ Pollut 156:46–52

Zosim Z, Gutnick D, Rosenberg E (1983) Uranium binding by emulsan and emulsanosols. Biotechnol Bioeng 25:1725–1735

Chapter 6
Anaerobic Treatment of Organic Saline Waste/Wastewater: Overcome Salinity Inhibition by Addition of Compatible Solutes

Ioannis Vyrides

6.1 Introduction

Saline wastewater is discharged by many industries and represents about 5 % of annual total wastewater (Le Borgne et al.2008). Seafood processing and aquaculture industries produce wastewater with high salinity and high concentrations of organics (Intrasungkha et al.1999). Oil refineries, chemical factories, tanneries, the textile industry and the food industry all discharge wastewater with large amounts of salt (Belkin et al.1993; Feijoo et al.1995; Dalmacija et al.1996; Lefebvre and Moletta 2006). Municipal sewage plants located near coastal regions may contain significant amounts of salt due to saline infiltration (Gomec et al.2001; Ozalp et al.2003).

The disposal of organic saline wastewater to the sea can cause severe environmental problems such as eutrophication, a decrease in oxygen concentration and toxicity towards aquatic life. Moreover, the high organic content in the saline wastewater could infiltrate to the plants and natural waters, and as a result, the removal of the organic content is a top priority. In a recent review (Lefebvre and Moletta 2006) on the treatment of organic saline wastewater, the authors stated the following steps as an optimum solution for saline wastewater treatment: homogenisation, preparation of the wastewater, organic matter removal and salt removal.

However, organics in highly saline wastewaters are poorly biodegraded by conventional wastewater treatment plants with non-adapted biomass due to the toxic effect of sodium in the wastewater. High concentration of salts (10 g NaCl/L) can cause cell plasmolysis and the death of some microorganisms due to the dramatic increase in osmotic pressure (Kargi and Dincer 1999; Kempf and Bremer 1998). To overcome this problem, many industries dilute the saline wastewater with

I. Vyrides (✉)
Department of Environmental Science and Technology, Cyprus University of Technology, 95 Eirinis Str., 50329, 3603 Lemesos, Cyprus
e-mail: ioannis.vyrides@cut.ac.cy

L.B. Sukla et al. (eds.), *Environmental Microbial Biotechnology*, Soil Biology 45,
DOI 10.1007/978-3-319-19018-1_6

fresh water or other wastewater to reduce salinity. This process is not always effective due to variations in the composition of the saline wastewater, so the amount of dilution has to be constantly adjusted to compensate for these fluctuations. Another problem is that the dilution may be 10–20 times to make biological treatment feasible. This practice is unsustainable due to continuous pressure on industry to reduce fresh water consumption (Vallero et al.2002). As a result, biological treatment of undiluted saline wastewater is extremely desirable for the environment, and it could be an economically viable solution as well (Tuin et al.2006).

The most common strategy for the biological treatment of a toxic wastewater is the gradual adaptation of sludge/biomass to this wastewater (Vyrides and Stuckey 2009a). However, the drawback of this method is that the adaptation period can take place during a considerable amount of time (more than a year). In addition, during fluctuation of the toxic wastewater, the biological system may recover relatively slow. Moreover, washout of the biomass from the reactor results in a failure of the system as the new biomass needs long time to readapt to the toxic wastewater.

Apart from this, the use of halophilic methanogens as an inoculum has also been reported as an approach to deal with high salinity (Riffat and Krongthamchat 2007). However, this approach could be successfully applied on saline wastewater but could be difficult to be implemented to a bioreactor that received wastewater with periodically sudden short shock of saline wastewater. Moreover, laboratory scale reactors are easily inoculated which already adapted to toxic wastewater biomass, but in the case of a full-scale reactor, considerable amount of biomass is needed; thus, the biomass requires long adaptation period.

In addition, most of these studies only investigated a gradual increase in salinity despite the fact that it is often highly variable in many industrial wastewaters (Lefebvre and Moletta 2006). Moreover, in many industries, salinity only increases during certain short periods of the year. Furthermore, in some industries, salinity can suddenly increase due to process instability or changes, and this could be a serious drawback for biological treatment.

Based on the above considerations, this chapter reviews several studies that followed a new approach for the rapid adaptation of anaerobic biomass to a sudden increase in salinity. This is based on previous insights gained about how cells cope with stress conditions through the use of compatible solutes.

6.2 Mechanisms for Anaerobic Biomass Cope Under Stress Conditions

6.2.1 "Salt-In Strategy" and Synthesis of Compatible Solutes

There are two fundamental strategies for cells to survive under osmotic stress: the "salt-in" strategy and the compatible solute strategy. However, the second strategy that this review focuses on has been employed by the cell not only under osmotic stress but also under other extreme conditions such as (low and high temperatures).

(a) Cells increase the intracellular ion concentration (mainly potassium) in order to balance the external osmotic pressure, and all intracellular enzymes have to adapt to the new conditions. Anaerobic halophilic bacteria, whose entire physiology has been adapted to high saline environments, use this so-called salt-in strategy (Muller et al. 2005). An early study by Kugelman and McCarty (1965) found that low concentration of potassium can reduce the inhibition of sodium to methanogens. Kugelman and McCarty (1965) explained that due to reactivation of the injured enzymes by potassium, few studies have employed the insight of this strategy and found that the presence of potassium can decrease the sodium inhibition on anaerobic biomass. Feijoo et al. (1995) found that the use of seawater wastewater instead of synthetic wastewater resulted in higher performance for anaerobic biomass. An anaerobic toxicity assay of sludge from mussel-processing wastewater showed that when the effluent of the anaerobic filter as assay medium was used, the tolerance to sodium was highly increased compared to distilled water (Soto et al. 1993). It is most likely the increase of the performance of biomass in a low concentration of potassium under high salinity to be due to "salt-in strategy" and not to the reactivation of the injured enzymes as was initially proposed by Kugelman and McCarty (1965).

(b) Many microorganisms accumulate organic solutes called "compatible solutes". The high external osmotic pressure is balanced within the cytoplasm by organic compatible solutes without the need for special adaptation of the intracellular enzymes, and compatible solutes also serve as protein stabilisers in the presence of high ionic strength inside the cell (Muller et al. 2005). These solutes can be synthesised by the cell, or provided by the medium, but for most species, uptake from the medium is energetically more favourable than synthesis. Methanogenic archaea isolated from moderately saline environments showed an accumulation of β-glutamine, α-glutamate, *N*-acetyl β-lysine, and glycine betaine (Muller et al. 2005; Sowers and Gunsalus 1995). When glycine betaine was provided externally to *Methanosarcina thermophila*, it was taken up into the cell, and at the same time, the synthesis of compatible solutes such as *N*-acetyl β-lysine or α-glutamate was downregulated (Pfluger and Muller 2004). Lai and Gunsalus (1992) found that exogenous addition of glycine betaine to *Methanohalophilus* strains Z7401, and FDF2 significantly decreased intracellular accumulation of de novo compatible solutes, indicating that these halophilic methanogens preferentially accumulate glycine betaine rather than synthesise because it is bioenergetically more favourable. Roessler and Muller (2001) isolated a *Methanosarcina mazei* from a municipal sewage plant to study the transport mechanism of glycine betaine inside the cell. The authors reported that intracellular generation of compatible solute can be very slow (up to 4 days). However, the addition of glycine betaine in a medium (800 mM) resulted in 160-fold enrichment of glycine betaine inside the cell after only 120 min.

6.2.2 Extracellular Polysaccharides

Apart from compatible solutes, anaerobic biomass produces extracellular polysaccharides (EPSs) to help them survive under sodium toxicity (Vyrides and Stuckey 2009b). The EPSs consist of various organic substances such as polysaccharides, proteins, nucleic acids and lipids. The EPSs consisted of heterogeneous components that have various functions (Tsuneda et al.2003). EPSs have significant role in the aggregation of bacterial cells in flocs and biofilms (Zhang et al.1998). Moreover, they can be generated as a protective barrier around the bacteria especially under harsh conditions (Vyrides and Stuckey 2009b; Xin and Wang 2007). Vyrides and Stuckey (2009b) found that accumulation of EPSs under salinity was another strategy that anaerobic biomass employs simultaneously with the generation of compatible solutes. Higher amounts of salinity resulted in higher amounts of EPSs. Moreover, in this strategy, part of the energy was consumed for the production of EPSs so less substrate is available for methane production. The high molecular weight (MW) compounds of EPSs could be non-biodegradable/or slowly biodegradable by anaerobic biomass, and this can result in increased chemical oxygen demand (COD).

6.2.3 Species Evolution

Apart from the generation of compatible solutes and the accumulations of EPSs for anaerobic biomass under salinity, the microbial evolution could be an important factor for the adaptation of biomass to salinity. Lefebvre et al. (2007), in batch reactors, used not adapted to salinity biomass, found that increase in salinity little affects the microbial diversity of anaerobic biomass and that salinity had mostly an impact on biodegradation rate. The same group found in another studies (Lefebvre et al. 2004, 2006) that the diversity of a salt-tolerant ecosystem treating hypersaline industrial wastewater could be similar to that of a non-salt-tolerant one. Osaka et al. (2008) found no major changes of bacterial populations in a methanol-fed reactor at 40 g NaCl/L; however, the populations of the genera *Azoarcus* and *Methylophaga* increased when salinity concentration was at 10–30 g NaCl/L. Vyrides et al. (2010) found no archaeal evolution under salinity after 360 h in a batch bioreactor. Despite the low methane production under salinity, it seems that the archaea population is not destroyed but uses the substrate for the physiological adaptation (generation of compatible solutes and EPSs). The dominant species identified under these conditions were *Methanosarcina mazei* and *Methanosaeta* sp. probably due to the ability to adapt rapidly in saline conditions.

6.3 Strategies for Biological Treatment of Toxic Wastewater

6.3.1 The Use of the Salt-In Strategy

An early study by Kugelman and McCarty (1965) found that low concentration of potassium can reduce the inhibition of sodium to methanogens. Kugelman and McCarty (1965) explained that due to reactivation of the injured enzymes by potassium, few studies have employed the insight of this strategy and found that the presence of potassium can decrease the sodium inhibition on anaerobic biomass. Feijoo et al. (1995) found that the use of seawater wastewater instead of synthetic wastewater resulted in higher performance for anaerobic biomass. An anaerobic toxicity assay of sludge from mussel-processing wastewater showed that when the effluent of the anaerobic filter as assay medium was used, the tolerance to sodium was highly increased compared to distilled water (Soto et al. 1993). It is most likely the increase of the performance of biomass in a low concentration of potassium under high salinity to be due to "salt-in strategy" and not to the reactivation of the injured enzymes as was initially proposed by Kugelman and McCarty (1965).

6.3.2 Addition of Compatible Solutes as a Strategy to Overcome Salinity Inhibition

Another possible strategy for the biological treatment of toxic wastewater could be the addition of compatible solutes in the medium. In many industries, the extreme condition of the wastewater only increases during certain short periods of the year. Moreover, in some industries, toxicity can suddenly increase due to process instability, and this could be a serious drawback for biological treatment. By the addition of compatible solutes during this certain period, the biomass would be able to cope well without previous acclimation or the need to stop the process. Apart from the generation of compatible solutes during osmotic stress, it is possible to be produced during other environmental stress, and this will be discussed further in Sect. 6.4 of this chapter.

6.3.2.1 Compatible Solute Strategy for the Anaerobic Treatment of Waste

Oh et al. (2008) investigated the addition of different compatible solutes on anaerobic biomass that treated saline food waste. Glycine betaine to anaerobic biomass can enhance methane production by sixfold when it was treating saline food waste (11.6 g NaCl/L). The second most positive effect was found by choline where carnitine and trehalose had a positive effect on anaerobic biomass but not as pronounce as glycine betaine and choline. The optimum concentration of glycine betaine for maximum methane production was 1.5 g/L where 0.5 and 5 g/L of betaine show slightly higher methane compared with the control (no salinity). The time that glycine betaine was inserted in the anaerobic medium was examined. The addition of glycine betaine after 7 days from the incubation resulted in the highest methane production. The addition of glycine betaine in the beginning of the incubation also produces considerable amount of methane, however slightly less methane compared with the case that glycine betaine was added after 7 days. The addition of glycine betaine after 14 days from the incubation did not result to any positive result to anaerobic biomass. Oh et al. (2008) pointed out that glycine betaine added in the medium was partially transferred into the cell and remained constant over a period of 7 days and then suddenly declined.

Oh et al. (2008) found that the uptake of glycine betaine by anaerobic biomass occurs relatively slowly over 1–7 days. In addition, the amount of intracellular compatible solute concentration depends on the salinity to which the cell is subjected. Halotolerant methanogens and halophilic methanogens increase their compatible solutes up to a threshold level by an increase in external osmolarity. Moreover, transport studies with glycine betaine to *Methanosarcina thermophila* revealed a 70 % increased transport rate when the NaCl concentration was increased from 0.1 to 0.4 M NaCl (Pfluger and Muller 2004). As was reported above, compatible solutes not only balance the osmotic strength of the cytoplasm but also have a significant function in the maintenance of protein structure and strength, which indicates that they might be used to counterbalance other stresses as well (Muller et al. 2005).

Zhang et al. (2014a) investigated the anaerobic digestion of sludge from brackish/marine aquaculture recirculation system. The results of BMP tests show that addition of glycine betaine (GB) 0.50 g/L, trehalose (T) 0.50 g/L and the combination 0.25 gGB/L plus 0.25 gT/L enhanced BMP of the sludge by 9.0 %, 11.6 % and 10.3 %, respectively, compared with that of the control group without addition of compatible solutes.

6.3.2.2 Compatible Solutes Strategy for the Anaerobic Treatment of Waste

Despite these insight findings of cell mechanisms, there are very few studies in the literature on the use of compatible solutes as osmoprotectants for anaerobic biomass treating saline wastewater. These studies are summarised at Table 6.1.

More specifically, Yerkes et al. (1997) successfully found that the addition of small concentrations of betaine (1–10 mM) to *Methanosarcina* and *Methanosaeta* cultures in sucrose-fed batch assays, CSTRs, fluidized bed reactors and UASB reactors; 1 mM of betaine was found to be effective in reducing sodium toxicity, and higher methane was produced compared with the reactors where betaine was not added (Table 6.1). However, the COD effluent remained relatively unchanged.

Table 6.1 Effect of compatible solutes on anaerobic treatment of saline wastewater

Sludge type	Type of wastewater/ substrate	System	Salt concentration g/L	Type of compatible solutes	Performance	Time	References
Anaerobic sludge	Korean food waste	Batch	10	Addition of 1.5 g GB/L after 7 days	200 ml CH_4 with GB compared to 50 ml CH_4 of control	38 days	Oh et al. (2008)
Anaerobic sludge	Korean food waste	Batch	35	Addition of 1.5 g GB/L after 7 days	58 ml CH_4 with GB compared to 16 ml CH_4 of control	38 days	Oh et al. (2008)
Anaerobic sludge	Sludge from recirculating culture system	BMP test	17	Combination of 0.5 g/L GB and 0.5 g/L trehalose	11.6 %	12 days	Zhang et al. (2014a)
Anaerobic sludge	Acetate	Batch	35 g NaCl/L	1.0 mM trehalose	22 % SMA increase		Zhang et al. (2014b)
Anaerobic sludge	Acetate	Batch	35 g NaCl/L	1.0 mM GB	12 % SMA increase		Zhang et al. (2014b)
Anaerobic sludge	Sucrose	UASB	17.25 g Na^+/L	1 mM GB	120 ml/day at 17.25 g Na^+/L 530 ml/day at 17.25 g Na^+/L and 1 mM GB 950 ml/day without Na^+	31 days	Yerkes et al. (1997)
Anaerobic sludge	Glucose	Batch	35 g NaCl/L	1GB	42 ml at 35 g NaCl/L and 1 GB 19 ml at 35 g NaCl/L	520 h	Vyrides and Stuckey (2009b)
Anaerobic sludge	Glucose	Batch	35 g NaCl/L	b-glutamate	36 ml at 35 g NaCl/L and b-glutamate 19 ml at 35 g NaCl/L	126 days	Vyrides and Stuckey (2009b)
Anaerobic sludge	Synthetic sewage wastewater	Submerged anaerobic membrane bioreactor (SAMBR)	35 g NaCl/L	5 mM GB and 2 days batch mode	46 % higher DOC effluent removal	126 days	Vyrides et al. (2010)

(continued)

Table 6.1 (continued)

Sludge type	Type of wastewater/ substrate	System	Salt concentration g/L	Type of compatible solutes	Performance	Time	References
Anaerobic sludge	Synthetic sewage wastewater	Submerged anaerobic membrane bio-reactor (SAMBR)	35 g NaCl/L	Injection of 1 mM GB for 5 consecutive days	38 % higher DOC effluent removal	126 days	Vyrides et al. (2010)
Anaerobic sludge	Mustard tuber wastewater	Anaerobic sequencing batch biofilm reactor (ASBBR)	20 g Cl/L	0.5 mmol	4,461 with GB addition 6,100 without GB addition	22 days	He et al. (2012)
Anaerobic granular sludge	Methanol (2g COD/L) + sulphate COD/S ration 0.5	UASB	25 g NaCl/L	10 mmol/L glutamate, betaine, ectoine, choline, a mixture of compatible solutes and K^+ and Mg^{2+}	710 mg COD/L.m^3 A no salinity, 116 mg COD/L.m^3 at 25 g NaCl/L and 278 mg COD/L.m^3 with addition of choline No improvement in methane generation		Vallero et al. (2003)

GB glycine betaine

The same study found that when betaine was depleted from the system, the anaerobic biomass left unacclimated to sodium.

Vyrides and Stuckey (2009b) found that the addition of compatible solutes to the medium increases the performance of anaerobic biomass under saline conditions. Glycine betaine was the most effective in counteracting sodium toxicity compared to β-glutamate and then α-glutamate. Methanogens were the most positively affected microbial group when glycine betaine was present in the saline medium, while propionate utilisers were less positively affected. The same study found that anaerobic biomass exposed for 28 days to 35 g NaCl/L showed low acclimatisation to salinity. On the other hand, anaerobic biomass that had not been exposed to salinity before but had glycine betaine added to the saline medium increased its acclimation potential significantly. Non-replacement of the medium with the addition of a substrate-enhanced methane production, while the replacement of old media with the addition of new media and substrate resulted in significant inhibition possibly due to disruption of the compatible solute balance between the cell and the medium (Vyrides and Stuckey 2009b) (Table 6.1).

Moreover, Vyrides et al. (2010) have shown that by using continuous submerged anaerobic membrane bioreactor (SAMBR), the addition of 5 mM glycine betaine and operation in batch mode for 2 days can substantially enhance organic degradation, at 12 h HRT and up to 35 g NaCl/L. Also, injecting the SAMBR with 1 mM of glycine betaine for 5 days into operation could also substantially enhance the dissolved organic carbon (DOC) removal under these conditions. In the same study, using nuclear magnetic resonance (NMR) has identified trehalose as the compatible solute generated internally by anaerobic biomass after 96 h of exposure to 20 and 40 g NaCl/L. However, the addition of 1 mM trehalose, 1 mM β-glutamate, 1 mM α-glutamate, *N*-acetyl β-lysine and potassium when added independently to the medium with anaerobic biomass was found to slightly reduce sodium inhibition. In contrast, the addition of glycine betaine dramatically alleviated sodium inhibition. Methanogens are the most positively affected microbial group, while propionate utilisers are less positively affected. Concentrations as low as 0.1 mM of glycine betaine can also be advantageous for anaerobic biomass under 35 g NaCl/L. The beneficial effect of glycine betaine can be retained over time and enhances the adaptation of anaerobic biomass for the following four batch feedings at 35 g NaCl/L. Anaerobic biomass exposed for 28 days to 35 g NaCl/L showed low acclimatisation to salinity. On the other hand, anaerobic biomass that had not been exposed to salinity before but had glycine betaine added to the saline medium increased its acclimation potential significantly.

Zhang et al. (2014b) found that 1.0 mM trehalose increased specific methanogenic activity (SMA) from 0.094 to 0.115 g COD-CH_4/g VSS and by 22 % compared to the control. Glycine betaine (GB) additions of 1.0 mM increased SMA by 12 % (Table 6.1). He et al. (2012) treated a composite mustard tuber wastewater with high concentrations of salt (about 20 g Cl) and organics (about 8,000 mg L 1COD) by an anaerobic sequencing batch biofilm reactor (ASBBR). As a strategy to overcome the high salinity, betaine was added to the influent wastewater. The efficiency reached the highest when the optimal dosage of betaine was

0.5 mmol/L. The average effluent COD, after stable acclimation, was 4,461 mg/L. Relative to ASBBR without adding betaine, the activity of the sludge increased significantly. Meanwhile, the dehydrogenase activity of anaerobic microorganisms and the COD removal efficiency were increased by 18.6 % and 18.1 %, respectively.

Suwannoppadol et al. (2012) added grass clippings at the onset of anaerobic digestion of acetate containing a sodium concentration of 7.8 g Na^{+}/L. A total methane production about 8 L/L was obtained, whereas no methane was produced in the absence of grass leaves. In an attempt to narrow down which components of grass leaves caused decrease of sodium toxicity, different hypotheses were tested. Results revealed that betaine could be a significant compound in grass leaves causing reduction to sodium inhibition (Table 6.1).

6.3.2.3 Thermophilic Treatment of Saline Wastewater

Vallero et al. (2003) using a mixture of compatible solutes found no significant reduction of sodium toxicity in thermophilic sulphate reducing granular biomass in a UASB reactor and batch assays. The organic compatible solutes that were used possibly were degraded by microorganisms from anaerobic biomass prior to the uptake by other microorganisms that may use them as osmoprotectant. Apart from this, the morphological characteristic of biomass (granular) may inhibit the diffusion of the compatible solutes to the cell. Moreover, the compatible solutes that were added were found more in mesophilic microorganisms compared with the thermophilic microorganisms that were used in that study (Vallero et al. 2003) (Table 6.1).

6.4 Future Potential

Up to date, the use of compatible solutes in bioreactor to overcome salinity inhibition is still in its infancy level, and most studies have took place at a laboratory level. A serious reason for not using this strategy at large-scale bioreactor could be the fact that adding high concentration of compatible solutes is not economically viable. The use of low concentration of compatible solutes (Vyrides et al. 2010) or adding substrates such as grass clippings that contain compatible solutes (Suwannoppadol et al. 2012) could be a possible solution for the high economical cost of compatible solute addition to bioreactors.

Before the use of compatible solutes in a bioreactor, it is recommended to examine the effect of various compatible solutes in alleviating sodium inhibition, under high salinity and batch mode. It is likely a mixture of compatible solutes with potassium to be more effective than the use of solely one type of compatible solute. Moreover, it is important to find out the compatible solutes' optimum concentration on the anaerobic sludge under various salinity concentrations. From the current results so far, it is more likely the compatible solutes' positive effect to be more pronounced under higher salinity concentrations. In addition, during continuous bioreactor operation, the strategy (continuous or batch) and the dosage of adding compatible solutes need to be investigated. All in all, the applied compatible strategy needs to be in line with low-cost and easy implementation strategy.

As was reported in the introduction, there are various strategies to deal with saline wastewater. The most common is either the use of halophilic and halotolerant microorganisms or the use of acclimated to salinity anaerobic sludge. However, these strategies could not be applied under sudden increase of salinity during short period usually found at industrial wastewater. This increase can be deleterious for anaerobic biomass, and its recovery could take considerable amount of time (Vyrides and Stuckey 2009a). During sudden increase of salinity in wastewater, the application of compatible solutes could be a good strategy to overcome sodium inhibition on anaerobic biomass. Moreover, during rapid start-up of anaerobic bioreactor to treat saline wastewater, the use of compatible solutes in the initial stage could be used in order to reduce the acclimatisation time of anaerobic biomass to salinity.

So far, the compatible solute strategy is in its early stages. Still, there is big room for improvement and exploration, not only for anaerobic treatment of organic saline wastewater but also for the nitrification and denitrification of saline wastewater. Moreover, the use of compatible solutes on activated sludge to treat saline organic wastewater was not examined so far. The use of compatible solutes on bioreactor to overcome sudden variation of temperatures has not been investigated. Apart from this, compatible solute addition can be examined during low or high pH, drying, low oxygen for activated sludge, drying and starvation phase. In the near future, it is expected that the research in this field will be increased. However, the use of this strategy at a full-scale bioreactor has still a long way to be covered.

References

Belkin S, Brenner A, Abeliovich A (1993) Biological treatment of a high salinity chemical industrial waste-water. Water Sci Technol 27:105–112

Dalmacija B, Karlovic E, Tamas Z, Miskovic D (1996) Purification of high-salinity wastewater by activated sludge process. Water Res 30:295–298

Feijoo G, Soto M, Mendez R, Lema JM (1995) Sodium inhibition in the anaerobic-digestion process - antagonism and adaptation phenomena. Enzyme Microb Technol 17:180–188

Gomec CY, Gonuldinc S, Eldem N, Ozturk I (2004) Behaviour of an up-flow anaerobic sludge bed (UASB) reactor at extreme salinity. In: 4th International Black Sea conference on environmental protection technologies for coastal area. Varna, Bulgaria, pp 115–120. (I W a Publishing)

He Q, Kong XJ, Chai HX, Fan MY, Du J (2012) Efficiency influence of exogenous betaine on anaerobic sequencing batch biofilm reactor treating high salinity mustard tuber wastewater. Environ Technol 33(13-15):1695–1699

Intrasungkha N, Keller J, Blackall L (1999) Biological nutrient removal efficiency in treatment of saline wastewater. Water Sci Technol 39:183–190

Kargi F, Dincer AR (1999) Salt inhibition effects in biological treatment of saline wastewater in RBC. J Environ Eng-Asce 125:966–971

Kempf B, Bremer E (1998) Uptake and synthesis of compatible solutes as microbial stress responses to high-osmolality environments. Arch Microbiol 170:319–330

Kugelman IJ, McCarty PL (1965) Cation toxicity and stimulation in anaerobic waste treatment. J Water Pollut Control Fed 37:97–116

Lai MC, Gunsalus RP (1992) Glycine betaine and potassium-ion are the major compatible solutes in the extremely halophilic methanogen Methanohalophilus strain Z7302. J Bacteriol 174:7474–7477

Le Borgne S, Paniagua D, Vazquez-Duhalt R (2008) Biodegradation of organic pollutants by halophilic bacteria and archaea. J Mol Microbiol Biotechnol 15:74–92

Lefebvre O, Moletta R (2006) Treatment of organic pollution in industrial saline wastewater: a literature review. Water Res 40:3671–3682

Lefebvre O, Habouzit F, Bru V, Delgenes JP, Godon JJ, Moletta R (2004) Treatment of hypersaline industrial wastewater by a microbial consortium in a sequencing batch reactor. Environ Technol 25:543–553.

Lefebvre O, Quentin S, Torrijos M, Godon J, Delgenes P, Moletta R (2007) Impact of increasing NaCl concentrations on the performance and community composition of two anaerobic reactors. Appl Microbiol Biotechnol 75:61–69

Lefebvre O, Vasudevan N, Thanasekaran K, Moletta R, Godon JJ (2006) Microbial diversity in hypersaline wastewater: the example of tanneries. Extremophiles 10:505–513

Muller V, Spanheimer R, Santos H (2005) Stress response by solute accumulation in archaea. Curr Opin Microbiol 8:729–736

Oh G, Zhang L, Jahng D (2008) Osmoprotectants enhance methane production from the anaerobic digestion of food wastes containing a high content of salt. J Chem Technol Biotechnol 83:1204–1210

Osaka T, Shirotani K, Yoshie S, Tsuneda S (2008) Effects of carbon source on denitrification efficiency and microbial community structure in a saline wastewater treatment process. Water Res 42:3709–3718

Ozalp G, Gomec CY, Ozturk I, Gonuldinc S, Altinbas M (2003) Effect of high salinity on anaerobic treatment of low strength effluents. Water Sci Technol 48:207–212

Pfluger K, Muller V (2004) Transport of compatible solutes in extremophiles. J Bioenerg Biomembr 36:17–24

Riffat R, Krongthamchat K (2007) Anaerobic treatment of high-saline wastewater using halophilic methanogens in laboratory-scale anaerobic filters. Water Environ Res 79:191–198

Roessler M, Muller V (2001) Osmoadaptation in bacteria and archaea: common principles and differences. Environ Microbiol 3:743–754

Soto M, Mendez R, Lema JM (1993) Sodium inhibition and sulfate reduction in the anaerobic treatment of mussel processing wastewaters. J Chem Technol Biotechnol 58:1–7

Sowers KR, Gunsalus RP (1995) Halotolerance in Methanosarcina Spp - role of N-epsilon-acetyl-beta-lysine, alpha-glutamate, glycine betaine, and K+ as compatible solutes for osmotic adaptation. Appl Environ Microbiol 61:4382–4388

Suwannoppadol S, Ho G, Cord-Ruwisch R (2012) Overcoming sodium toxicity by utilizing grass leaves as co-substrate during the start-up of batch thermophilic anaerobic digestion. Bioresour Technol 125:188–192

Tsuneda S, Aikawa H, Hayashi H, Yuasa A, Hirata A (2003) Extracellular polymeric substances responsible for bacterial adhesion onto solid surface. FEMS Microbiol Lett 223:287–292

Tuin BJW, Geerts R, Westerink JB, Van Ginkel CG (2006) Pretreatment and biotreatment of saline industrial wastewaters. Water Sci Technol 53:17–25

Vallero M, Lettinga G, Lens P (2003) Assessment of compatible solutes to overcome salinity stress in thermophilic (55 C) methanol-fed sulfate reducing granular sludges. Water Sci Technol 48:195–202

Vallero MVG, Poll LWH, Lens PNL, Lettinga G (2002) Effect of high salinity on the fate of methanol during the start-up of thermophilic (55 degrees C) sulfate reducing reactors. Water Sci Technol 45:121–126

Vyrides I, Stuckey DC (2009a) Effect of fluctuations in salinity on anaerobic biomass and production of soluble microbial products (SMPs). Biodegradation 100:165–175

Vyrides I, Stuckey DC (2009b) Adaptation of anaerobic biomass to saline conditions: role of compatible solutes and extracellular polysaccharides. Enzyme Microb Technol 44:46–51

Vyrides I, Santos H, Mingote A, Ray MJ, Stuckey DC (2010) Are compatible solutes compatible with biological treatment of saline wastewater? Batch and continuous studies using submerged anaerobic membrane bioreactors (SAMBRs). Environ Sci Technol 44:7437–7442

Xin X, Wang Y (2007) Ultrastructural and intracellular chemical changes of a novel halophilic strain V430 of Staphylococcus saprophyticus under CaCl2 stress. Appl Biochem Biotechnol 142:298–306

Yerkes DW, Boonyakitsombut S, Speece RE (1997) Antagonism of sodium toxicity by the compatible solute betaine in anaerobic methanogenic systems. Water Sci Technol 36:15–24

Zhang XQ, Bishop PL, Kupferle MJ (1998) Measurement of polysaccharides and proteins in biofilm extracellular polymers. Water Sci Technol 37:345–348

Zhang X, Hu J, Spanjers H, van Lier JB (2014a) Performance of inorganic coagulants in treatment of backwash waters from a brackish aquaculture recirculation system and digestibility of salty sludge. Aquac Eng 61:9–16

Zhang X, Ferreira RB, Hu J, Spanjers H, van Lier JB (2014b) Improving methane production and phosphorus release in anaerobic digestion of particulate saline sludge from a brackish aquaculture recirculation system. Bioresour Technol 162:384–388

Chapter 7
Uranium Bioremediation: Approaches and Challenges

Celin Acharya

7.1 Introduction

Uranium is the heaviest naturally occurring element found in the earth's crust and is the 49th most abundant element. It is an alpha emitter and radioactive element which presents both radiotoxicity and chemotoxicity. It occurs in the environment either as natural deposit or as a result of anthropogenic activities including mining, enrichment for manufacturing of nuclear weapons, electricity generations in nuclear fuel power plant, nuclear weapon testing, usage of phosphate fertilizers, etc. All the major isotopes of uranium, i.e. ^{238}U, ^{235}U and ^{234}U are radioactive. Despite of its low radiotoxicity, uranium has gained notoriety as a radiological hazard. The chemical toxicity of dissolved uranium is of greatest environmental significance and poses a major concern for public health and safety. Ingestion of high concentrations of soluble uranium compounds manifests in chemotoxic effects on renal tissue leading to kidney failure (Gavrilescu et al. 2009).

The mobility of uranium in the environment depends on its speciation and redox state as well. Uranium is present as U (VI) under oxidizing conditions. It occurs either in the form of UO_2^{2+} predominantly at $pH \leq 2.5$ or hydroxyl complexes at $pH < 6.5$ or as highly mobile carbonated species at higher pH (at a relative atmospheric CO_2 partial pressure of 0.03 %) (Choppin et al. 2011). UO_2^{2+} ion shares several chemical and biological properties with alkaline earth ions (Priest 2001). Carbonate complexes account for 90–100 % of uranium dissolved in ocean water. In the absence of carbonate, uranyl dication (UO_2^{2+}) and its complexes are adsorbed by iron oxide/organics or it forms complexes with humic acid and fulvic acid present in soil (Hsi and Langmuir 1985). Under reducing conditions, uranium predominates as insoluble and immobile U (IV) as mineral uraninite, UO_2[s].

C. Acharya (✉)
Environmental Biotechnology Group, Molecular Biology Division, Bhabha Atomic Research Centre, Trombay, Mumbai 400085, Maharashtra, India
e-mail: celin@barc.gov.in

L.B. Sukla et al. (eds.), *Environmental Microbial Biotechnology*, Soil Biology 45,
DOI 10.1007/978-3-319-19018-1_7

Uranium contamination is a worldwide problem. Preventing the uranium contamination in the environment is quite challenging and requires a thorough understanding of the microbiological, ecological and biogeochemical features of the contaminated sites. Bioremediation of uranium is largely dependent on reducing its bioavailability in the environment. Biogeochemical interactions can control/accelerate uranium bioremediation by altering pH or redox conditions, changing mineral surfaces, generating ligands for complexation or altering microbial processes. This chapter focuses on various microbial interactions which contribute towards uranium bioremediation.

7.2 Microbial Processes and Their Implications for Uranium Bioremediation

7.2.1 Bioadsorption

Microbial cells (live or dead) sequester uranium through passive sorption resulting from interaction of the latter with charged constituents or ligands on cell surface. Phosphate and carboxylate groups on bacterial cell surfaces/lipopolysaccharides have been proposed to be predominant ligands for uranyl binding (Fowle et al. 2000; Kelly et al. 2002). The gram-negative bacterial strains (*Microbacterium* sp. and *Sphingomonas* sp.) isolated from radioactive subsurface depository have demonstrated the complexation of uranium with cellular organic phosphate (fructose 6 phosphate) present at the cell surface at pH 2 (Koban et al. 2004). Various cellular mechanisms employed by cyanobacterial strains for uranium detoxification including surface complexation by ligands harboured within the extracellular polysaccharides (EPS) have been reported (Acharya et al. 2009; Acharya and Apte 2013a).

The carboxyl and phosphate groups of the surface structures overlaying bacterial cell walls, i.e. S-layers, have been shown to be involved in complexing uranium (Merroun et al. 2005). Bioadsorption is one of the best suited processes for treating contaminated effluents with low to medium metal/radionuclide contamination as the binding to cell surface is extremely rapid, and it is convenient to remove the bound metal and regenerate biomass. Dead biomass is even a better bioadsorbent as the limitations exerted by metal toxicity are negated (Schiewer and Volesky 2000). A potential algal biosorbent developed using freshwater alga, *Chlorella vulgaris*, immobilized in silica could reduce the concentrations of toxic ions (Cu, Ni, Pb, Hg, Cd, Zn, As, Ag and U) from dilute solutions of 1–100 mg/L to 1 mg/L (Beveridge and Fyfe 1985).

7.2.2 Bioaccumulation

Microbial cells are capable of sequestering uranium by energy-consuming processes involving its transport to the interior of the cell. Uranium has no biological function, and its transportation into the microbial cell has been proposed to occur due to increased membrane permeability (Suzuki and Banfield 1999). The toxicity of cationic radionuclide species like UO_2^{2+} existing in low pH environments is established due to its high positive charge and small size which facilitates its internalization or accumulation. Uranyl ion has a high binding affinity ($K_{U\text{-}Rcell} \geq 10^{7.7}$) which is equivalent for other transition metals like Zn, Fe and Mn (Fortin et al. 2007). Once internalized, it may be bound, precipitated or localized within intracellular structures. In the absence of any evidence for the occurrence of uranium transporters, the intracellular accumulation of uranium in microorganisms is considered as metabolism-independent phenomenon. Majority of the published reports of intracellular accumulation have shown the formation of uranyl phosphates as a result of precipitation of uranium and phosphate within the cells (VanEngelen et al. 2010; Suzuki and Banfield 2004). Uranium immobilization by acid-soluble, surface-associated polyphosphates demonstrated a novel uranium sequestration phenomenon in a marine, filamentous cyanobacterium, *Anabaena torulosa* (Acharya et al. 2012; Acharya and Apte 2013b).

There is hardly any evidence on uranium bioaccumulation being a viable option for uranium bioremediation in contaminated lands and water.

7.2.3 Bioreduction

Microorganisms capable of oxidizing organic matters along with reduction of Fe (III) or Mn (IV) were first demonstrated by Lovley and Phillips (1988) and Myers and Nealson (1988). Dissimilatory reduction resulted from reduction coupled with energy metabolism under anaerobic conditions where the oxidized form of the metal served as terminal electron acceptor for respiration. *Geobacter metallireducens* was shown to reduce U(VI) to poorly soluble U(IV) phase while oxidizing acetate to CO_2, thereby deriving energy for its growth (Lovley et al. 1991). U (VI) was also demonstrated to be reduced to U (IV) uraninite by sulphate-reducing bacterium *Desulfovibrio desulfuricans* and *Desulfovibrio vulgaris* via c-type cytochrome activity (Lovley and Phillips 1992; Lovley et al. 1993). Reduction of U (VI) to uraninite (crystalline and ordered) or “monomeric” U (IV) (noncrystalline and disordered phase which is coordinated with carboxyl or phosphate ligands, Bernier-Latmani et al. 2010) is important for long-term bioremediation strategies. One of the biggest challenges for uranium bioreduction is the tendency of reduced minerals for reoxidation in the environment. Although both forms, i.e. uraninite and “monomeric” U (IV), are susceptible to reoxidation, the former is less susceptible than the latter due to its crystalline structure. Long-term experiments conducted over 3 months to 1 year resulted in nano-crystals of uraninite following U (VI) bioreduction (Madden et al. 2012; Bargar et al. 2013).

Successful pilot-scale studies have been demonstrated for in situ bioremediation of U (VI) in the field (Anderson et al. 2003; Williams et al. 2011). The success of in situ application is dependent on the stability of the final mineral phases so that there is less likelihood of reoxidation of U (IV) leading to its remobilization. The concentration of U (VI) existing in natural water like groundwater plays an important role as high concentrations of uranium might impose an inhibitory effect on the yield and growth rate of bacterial cells employed for bioreduction. Uranium concentrations at the Oak Ridge (DOE) site were found to range from 11 to 250 μM. Bioreduction experiments with enrichment cultures from the same site showed 50 % inhibition in their growth rate and yield at 100 μM of U(VI) (Nyman et al. 2007). Another factor contributing to the success of bioreduction is the choice of organic carbon source as electron donor. Acetate is the most preferred electron donor in laboratory and field experiments so far, followed by ethanol and lactate. The presence of nitrate as co-contaminant with uranium at nuclear sites seems to be a major impediment for in situ bioreduction as the former is known to be the preferred electron acceptor over U (VI) and Fe(III) (DiChristina 1992; Istok et al. 2004). Another challenge for bioreduction of uranium under oxic conditions is the presence of U(VI) as uranyl carbonate in natural water at $pH > 6.5$. Carbonate complexes of uranium are known to be very stable and are much less favourable electron acceptors (Brooks et al. 2003). However, abundant carbonate concentrations at Rifle site (US DOE) demonstrated rapid rates of in situ bioreduction of U (VI) (Williams et al. 2011). Field trials at Rifle site (where 90 % of U (VI) was present as recalcitrant uranyl–calcium–carbonate complexes), with a provision of acetate as electron donor, reduced the U concentrations from 1–1.5 μM to 0.05–0.1 μM which was maintained over 140 days (Williams et al. 2011).

It was observed that acetate amendment to contaminated groundwater facilitated uranium bioreduction over longer periods of time. Another interesting field trial for uranium bioreduction with a microbial consortium belonging to *Firmicutes* and *Proteobacteria* involved usage of emulsified vegetable oil as electron donor which reduced the uranium concentration in contaminated groundwater from 9.1 to 1 μM (Gihring et al. 2011). The denitrifying bacteria like *Rhodanobacter* dominated in acidic, nitrate-rich contaminated sediments or groundwater at Oak Ridge (Green et al. 2012). The known U(VI) reducers, *Desulfovibrio*, *Geobacter*, *Anaeromyxobacter*, *Desulfosporosinus* and *Acidovarar* species, were detected in abundance in the uranium-contaminated wells which were biostimulated or amended for over 2 years (Cardenas et al. 2008).

As mentioned earlier, one of the major challenges for successful application of bioreduction is reoxidation of poorly soluble U (IV) to U (VI), thereby remobilizing the latter in aqueous form. The reoxidation could be due to exposure to oxygen or to nitrate. The reoxidation due to oxygen was alleviated by exposing the contaminated sediments to natural oxic waters resulting in the predominance of the microbial community capable of oxidizing complex organic matter from dead biomass. The microbes used low levels of oxygen and prevented reoxidation of U (IV) by formation of uraninite nanoparticle aggregate which were found to be recalcitrant to reoxidation (Anderson et al. 2003; Senko et al. 2007). Introducing Fe (III)-reducing culture like *Clostridium* and sulphate-reducing culture like *Desulfovibrio* sp. prevented reoxidation of U(IV) by nitrate exposure in contaminated sediments (Boonchayaanant et al. 2009).

Field trials of bioreduction over long periods have shown promising potential for uranium bioremediation from contaminated groundwater and sediments.

7.2.4 Bioprecipitation

Microorganisms are capable of precipitating uranium by (a) activity of phosphatase enzyme which liberates inorganic phosphate from supplemented organic phosphate donor (Macaskie et al. 1992), (b) hydrolysis or degradation of intracellular polyphosphate granules resulting in phosphate release or efflux (Van Groenestijn et al. 1988), (c) localized alkalization at the cell surface (Van Roy et al. 1997) or (d) sulphides produced by sulphate-reducing bacteria.

The phosphatase activity (identified as acid phosphatase PhoN) was demonstrated in *Serratia* sp. which facilitated the precipitation of U (VI) as uranyl phosphate at the cell surface (Macaskie et al. 1992). As high as 9 g of uranium was precipitated per g of bacterial dry weight following 3 weeks of incubation period in a flow-through bioreactor with cells immobilized in polyacrylamide gel (Macaskie 1990). Isolates from contaminated sites like *Rahnella*, *Bacillus* and *Aeromonas* species have demonstrated biomineralization of uranium in the form of autunite through their phosphatase activity (Beazley et al. 2007). Uranium biomineralization through phosphatase activity was demonstrated using flow-through columns containing contaminated sediments and groundwater supplemented with glycerol 3 phosphate wherein 97 % of 200 μM uranium was removed at pH 5.5 and 7.0 (Beazley et al. 2011). One of the major limitations of uranium biomineralization through phosphatase activity is the use of organophosphate like glycerol phosphate which is not considered to be economically viable. Other substitutes of phosphate donors like phytic acid (from plant waste) (Paterson-Beedle et al. 2012) or tributyl phosphate (TBP, a solvent to extract actinides during nuclear fuel reprocessing,) (Thomas and Macaskie 1996) have been tested for uranium biomineralization.

Uranium biomineralization via controlled polyphosphate metabolism was demonstrated in *Pseudomonas aeruginosa* wherein degradation of intracellular polyphosphate led to the release of phosphate which, in turn, precipitated uranium as uranyl phosphate on the cell surface (Renninger et al. 2004).

Microbially induced precipitation of carbonates leading to formation of vaterite/calcite mineral in the microenvironment around the cells has been reported (Reeder et al. 2001). The co-precipitation of contaminant like uranium by bacterially induced calcite or other minerals in the subsurface environment seems to be a promising long-term biomineralization strategy for contaminated sediments or groundwater.

Sulphide precipitation by sulphate-reducing bacteria like *Desulfovibrio* and *Desulfotomaculum* has generated interest for in situ bioremediation. These bacteria are used to precipitate uranium in acidic leachates in contaminated soil. The solubility products of the metal sulphides are extremely low, and therefore bioprecipitation of actinides like uranium or other heavy metals is considered to be an important component of engineered wetland treatment.

7.3 General Considerations for Uranium Bioremediation

Although microorganisms have demonstrated promising potential for bioremediation of heavy metals in general, there are very few studies on uranium or other actinides. Generally, the contaminated sites host a combination of actinides like uranium, polonium, technetium, etc. along with various organic chemicals and toxic metals. Some of the factors determining the efficacy of the microbial processes for uranium bioremediation in such environments are discussed below.

7.3.1 Uranium Toxicity and Tolerance

One of the most important factors determining the successful implementation of bioremediation technologies at the contaminated sites is the basic understanding of uranium toxicity. Any toxic metal would result in increased lag phase or decreased growth rate which might affect the bioremediation process at the contaminated sites. The concentrations of uranium that inhibited growth of various organisms like *Deinococcus radiodurans*, *Escherichia coli*, *Pseudomonas putida*, *Caulobacter crescentus* and *Shewanella putrefaciens* ranged from 500 μM to 3 mM (Hu et al. 2005; Ruggiero et al. 2005; Wade and DiChristina 2000). The minimum inhibitory concentration (MIC) for bacterial (aerobic, chemoheterotrophic) isolates from subsurface soils of uranium-rich deposits was found to be 4 mM (Kumar et al. 2013a). The chemical toxicity studies for uranium have reported inhibition of growth of aquatic microflora including algae, cyanobacteria and other aquatic microorganisms at 4.2 μM in freshwater systems, whereas the bactericidal activity of this radionuclide is reported at a concentration of ~420 μM (Driver 1994). While some strains of *Acidithiobacillus ferrooxidans* isolated from greater depths of uranium mining waste piles have shown tolerance up to 9 mM of uranium, it was interesting to note that the others of the same organism from area closer to the surface tolerated only 2 mM of uranium (Merroun and Selenska-Pobell 2001). This observation revealed the fact that the strains of *A. ferrooxidans* (existing at greater depth) adapted to higher concentrations of uranium by natural selection. The growth inhibition observed in all cases was found to be chemical in nature. Uranium is a hard acid and hard metal ions are known to have high ratios of charge to ionic radius and tend to hydrolyze at low pH. The resulting hydroxides are insoluble in nature and exist in less available form. Therefore, the natural isolates tend to grow at a much higher concentration of uranium in contaminated sites that is reported to be toxic for them under laboratory conditions.

Another factor to be considered for uranium bioremediation in contaminated sites is the ability of microbes to tolerate high concentrations of co-contaminating metals, especially the transition metals. While *Deinococcus radiodurans* exhibited high tolerance to (a) gamma radiation (6 KGy) and (b) uranium concentration (2.7 mM), concentration as low as 1.8 μM of cadmium caused 70 % inhibition in its growth (Ruggiero et al. 2005). Generally, uranium-contaminated sites co-host several toxic metals. Therefore, metal-tolerant rather than radiation-tolerant strains would be ideal for bioremediation of such contaminated environments. The levels of radioactivity at most contaminated sites are relatively low, and the natural bacterial isolates are proposed to possess the desired radiation tolerance. The radiation dose for ^{238}U (100 % ^{238}U) is reported to be 0.0004 Gy/day if all the radiation emitted by ^{238}U is absorbed by bacterial culture at a given decay energy of 4.2 MeV (Ruggiero et al. 2005). The gamma radiation dose was found to be ~11 μGy/h at a sampling site of uranium-rich deposit harbouring 5 mM of uranium (~1,200 ppm) (Acharya et al. unpublished result). The strains isolated from such site showed high tolerance not only to uranium (4 mM) but also to various other heavy metals like copper, cadmium and lead (Kumar et al. 2013a).

7.3.2 Uranium Speciation

In oxidizing aqueous systems, uranium(VI) presents as the linear dioxo cation or the so-called uranyl ion $(O{=}U{=}O)^{2+}$ which dominates at pH ≤ 2. Depending on pH and uranium concentration, uranyl (UO_2^{2+}) entity undergoes a number of hydrolysis reactions. In the pH range above 3.5 till pH 14, uranium (VI) forms several mononuclear and polynuclear species: UO_2OH^+, $UO_2(OH)_2$(aq), $(UO_2)_2(OH)_2^{2+}$, $(UO_2)_3(OH)_5^+$, $(UO_2)_4(OH)_7^+$, $(UO_2)_3(OH)_7^-$ and $UO_2(OH)_3^-$ (Guillaumont et al. 2003). Under atmospheric conditions, the U(VI)/hydrolysis systems are in equilibrium with atmospheric CO_2 ($pCO_2 = 10^{-3.5}$ bar) and eventually various U (VI)/H_2O/CO_3^{2-} are formed. With rising carbonate concentrations, mono-, di- and tri-nuclear uranyl carbonate species, UO_2CO_3(aq), $UO_2(CO_3)_2^{2-}$ and $UO_2(CO_3)_3^{4-}$, become increasingly more important than hydrolytic species.

An important inorganic ligand is the phosphate, which plays a role not only in environmental systems but also in biological fluids (urine, saliva, blood) and can form several complex species with uranium (VI) in aquatic solutions (Scapolan et al. 1998). Depending on uranium and phosphate concentrations, the complexes $UO_2H_2PO_4^+$, $UO_2(H_2PO_4)_2$(aq), $UO_2(HPO_4)$(aq) and $UO_2PO_4^-$ are likely to be found in a pH range of 1–8 (Bonhoure et al. 2007; Brendler et al. 1996). Most of the work on the $UO_2^{2+}/PO_4^{3-}/H_2O$ systems has been done at relatively low concentrations to avoid precipitation of uranyl phosphate. Sandino and Bruno (1992) showed that UO_2^{2+} phosphate complexes [UO_2HPO_4'' (aq) and $UO_2PO_4^-$] could be important in aqueous systems with a pH between 6 and 9 when the total concentration ratio PO_4 (total)/CO_3 (total) is greater than 0.1. In acidic sulphate-rich water, such as seepage waters from uranium mines after in situ leaching by sulfuric acid, uranyl ion forms binary sulphate complexes: $UO_2SO_4(aq)$, $UO_2(SO_4)_2^{2-}$ and $UO_2(SO_4)_3^{4-}$ (Geipel et al. 1996). However, increase in pH and/or uranium concentration leads to several ternary sulphate complexes with hydroxo bridging such as $UO_2(OH)SO_4^-$, $UO_2(OH)(SO_4)_2^{3-}$, $(UO_2)_2(OH)_2(SO_4)_2^{2-}$ and $(UO_2)_3(OH)_4(SO_4)_3^{4-}$ (Moll et al. 2000).

Nitrate does not complex with uranium (VI) and has no role in the presence of other inorganic ligands in environmentally relevant waters. But at highly concentrated salt solutions ($I = 7.0$ M), the complexation becomes remarkable as the $UO_2NO_3^+$ is formed (Choppin and Du 1992). However, in a nitric acid medium, uranium (VI) coordinates with two nitrate groups to produce $UO_2(NO_3)_2{\cdot}xH_2O$ ($x = 2, 3, 6$); x value depends on acid concentration. The valuable property of this complex is its high solubility in both water and organic solvent, which is an important feature in the processing of nuclear waste (Cotton 2006).

Chloride, which is a common ion present in all natural waters, forms very weak complexes with uranium (VI) like UO_2Cl^+ and $UO_2Cl_{2\ (aq)}$ (Guillaumont et al. 2003). With increasing chloride concentration up to 9.0 M, higher complexes were observed: $UO_2Cl_3^-$, $UO_2Cl_4^{2-}$ and $UO_2Cl_5^{3-}$ (Hennig et al. 2005). However, their formation constants have not been reported in the literature so far.

In general, the relative strength of the uranyl ion (hard acid) complexation towards inorganic ligands (hard bases) decreases in the following order: $CO_3^{2-} > OH^- > F^-$, $HPO_4^{2-} > SO_4^{2-} > Cl^-$ and NO_3^- (Langmuir 1997).

The exact knowledge of uranium concentration in natural waters is a basic requirement for calculation and spectroscopic determination of its binding forms. Uranium is introduced to natural waters by weathering of U-bearing rocks and is considered as a common element that is always present in surface water and groundwater with extremely broad range of concentrations varying from few ng/L to several mg/L depending on the geological formation (Palmer and Edmond 1993). Highly extreme values of uranium (up to 12.4 mg/L) in Finnish groundwater and in other Nordic countries have been reported (Karpas et al. 2005; Kurttio et al. 2005; Frengstad et al. 2000; NCRP 1999). Effect from human activities (mining, phosphate fertilizers, energy production, etc.) can add more uranium to both surface water and groundwater and eventually drinking water supplies.

It is very important to assess the speciation of uranium in the contaminated water/sediment/soil so as to implement any microbial process for bioremediation.

7.3.3 Uranium Interactions

Majority of the bacterial community residing in the subsurface or sedimentary environments are associated with the particulate phases of uranium minerals. It is very important to address the microenvironments of the microorganisms, i.e. the nature of interface between the microbes and the minerals. This allows us to understand the role of microbes in influencing the fate of toxic metals or radionuclides in contaminated environments. Advances in imaging hydrated samples from contaminated sites have revealed that the cells lacked direct contact with mineral surfaces. It was therefore interesting to know the mechanism of electron transfer in metal-reducing bacteria which used oxidized minerals like Fe (III) on U (VI) minerals as terminal electron acceptors. A combination of transmission electron microscopy, X-ray absorption spectroscopy and molecular studies revealed the extracellular reduction of U(VI) to mononuclear tetravalent U(VI) in *Geobacter sulfurreducens* that occurred via its conductive pili (anchored to cell envelope) (Cologgi et al. 2011). These pili acted as electrical conduits and could potentially accept electrons from the cell envelope, thereby, maximizing the cell's catalytic surface for U (VI) reduction. Controlled laboratory experiments are required to understand the (a) degree of specificity of the microbe–mineral interactions and (b) phenotypic and genotypic changes in the cells resulting from such interactions. This information could be used to develop and test the conceptual and mathematical models (inclusive of biotic and abiotic processes at mineral surfaces) which in turn could be utilized for designing long-term stewardship strategies.

7.4 High-Throughput Approaches to Address Uranium Bioremediation

The science of bioremediation has demonstrated the potential of reinstating contaminated environments inexpensively. It has been shown that 99 % of the microbial communities existing in natural environments are non-cultivable. Culture-independent molecular tools have proven to be a boon for studying the composition and functioning of environmental microbial communities. High-throughput approaches like fingerprinting techniques (16S rRNA, TRFLP, DGGE, etc.), real-time PCR, microarray for genome and transcriptome analysis, proteomics, metaproteomics and metabolomics have been found to provide valuable information on novel catabolic and biodegradation activities in environmental microbes. Brodie et al. (2006) monitored the changes in bacterial populations during U (VI) reduction in uranium-contaminated sites using high-density 16S rRNA-based microarray. They inferred that the uranium-reducing populations were enhanced during U (VI) reduction and did not decline during U (IV) reoxidation phase. A 'GeoChip' microarray was designed for all functionally known geochemical, ecological and environmental processes including metal reduction and tolerance. When applied for tracking the activities of metal-reducing bacteria in groundwater system undergoing biostimulation for uranium reduction, the microarray revealed the abundance of *c*-type cytochrome genes from *Geobacter* type and *FeRB* and *dsrAB* (dissimilatory sulfite reductase) genes from *Desulfovibrio*-type SRB which correlated with uranium concentrations in groundwater (He et al. 2007). These two groups of microorganisms are known to be major groups capable of U (VI) reduction. In a recent study, it was found that 7 % of the clones from total bacterial community analysis from subsurface soil sample of uranium ore deposits using culture-independent approaches represented novel bacterial lineage (Kumar et al. 2013b). Although the genome-enabled approaches to bioremediation in general are in infancy, there is a strong requirement of exploiting these approaches to predict the microbially assisted attenuation of environmental contaminants to accelerate the bioremediation process.

7.5 Conclusions

Microorganism-based biotechnologies are receiving interest as potential treatment methods for a wide range of contaminated aquatic terrestrial habitats generated from anthropogenic activities. These methods are environmental friendly, cost-effective and superior alternatives over physico-chemical clean-up options. The progressive emergence of biophysical, molecular and analytical tools in environmental microbiology has been found to be instrumental for characterizing contaminated sites and monitor their bioremediation.

Field trials seem to be promising for uranium bioremediation. Uranium speciation and its toxicity are fundamental for developing appropriate remediation strategies. Numerous bacteria isolated from uranium-contaminated environment harbour the tolerance to uranium and radiation necessary for their growth at these sites. Uranium bioremediation to advance as a science requires a strong interdisciplinary approach for coupling models of microbial growth and metabolism in contaminated environments with geochemical and hydrological models to predict accurately the microbially assisted attenuation of uranium.

References

Acharya C, Apte SK (2013a) Insights into the interactions of cyanobacteria with uranium. Photosynth Res 118:83–94

Acharya C, Apte SK (2013b) Novel surface associated polyphosphate bodies sequester uranium in the filamentous, marine cyanobacterium, *Anabaena torulosa*. Metallomics 5:1595–1598

Acharya C, Joseph D, Apte SK (2009) Uranium sequestration by a marine cyanobacterium, *Synechococcus elongatus* strain BDU/75042. Bioresour Technol 100:2176–2181

Acharya C, Chandwadkar P, Apte SK (2012) Interaction of uranium with a filamentous, heterocystous, nitrogen-fixing cyanobacterium, *Anabaena torulosa*. Bioresour Technol 116:290–294

Anderson RT, Vrionis HA, Ortiz-Bernad I, Resch CT, Long PE, Dayvault R, Karp K, Marutzky S, Metzler DR, Peacock A, White DC, Lowe M, Lovley DR (2003) Stimulating the in situ activity of *Geobacter* species to remove uranium from the groundwater of a uranium-contaminated aquifer. Appl Environ Microbiol 69:5884–5891

Bargar JR, Williams KH, Campbell KM, Long PE, Stubbs JE, Suvorova EI, Lezama-Pacheco JS, Alessi DS, Stylo M, Webb SM, Davis JA, Giammar DE, Blue LY, Bernier-Latmani R (2013) Uranium redox transition pathways in acetate-amended sediments. Proc Natl Acad Sci U S A 110:4506–4511

Beazley MJ, Martinez RJ, Sobecky PA, Webb SM, Taillefert M (2007) Uranium biomineralization as a result of bacterial phosphatase activity: insights from bacterial isolates from a contaminated subsurface. Environ Sci Technol 41:5701–5707

Beazley MJ, Martinez RJ, Webb SM, Sobecky PA, Taillefert M (2011) The effect of pH and natural microbial phosphatase activity on the speciation of uranium in subsurface soils. Geochim Cosmochim Acta 75:5648–5663

Bernier-Latmani R, Veeramani H, Vecchia ED, Junier P, Lezama-Pacheco JS, Suvorova EI, Sharp JO, Wigginton NS, Bargar JR (2010) Non-uraninite products of microbial U(VI) reduction. Environ Sci Technol 44:9456–9462

Beveridge TJ, Fyfe WS (1985) Metal fixation by bacterial cell walls. Can J Earth Sci 22:1892–1898

Bonhoure I, Meca S, Marti V, De Pablo J, Cortina JL (2007) A new time-resolved laser-induced fluorescence spectrometry (TRLFS) data acquisition procedure applied to the uranyl-phosphate system. Radiochim Acta 95:165–172

Boonchayaanant B, Nayak D, Du X, Criddle CS (2009) Uranium reduction and resistance to reoxidation under iron-reducing and sulfate-reducing conditions. Water Res 43:4652–4664

Brendler V, Geipel G, Bernhard G, Nitsche H (1996) Complexation in the system $UO_2^{2+}/PO_4^{3-}/OH^-$(aq): potentiometric and spectroscopic investigations at very low ionic strengths. Radiochim Acta 74:75–80

Brodie EL, DeSantis TZ, Joyner DC, Baek SM, Larsen JT, Andersen GL (2006) Application of a high-density oligonucleotide microarray approach to study bacterial population dynamics during uranium reduction and reoxidation. Appl Environ Microbiol 72:6288–6298

Brooks SC, Fredrickson JK, Carroll SL, Kennedy DW, Zachara JM, Plymale AE, Kelly SD, Kemner KM, Fendorf S (2003) Inhibition of bacterial U (VI) reduction by calcium. Environ Sci Technol 37:1850–1858

Cardenas E, Wu WM, Leigh MB, Carley J, Carroll S, Gentry T, Luo J, Watson D, Gu B, Ginder-Vogel M, Kitanidis PK, Jardine PM, Zhou J, Criddle CS, Marsh TL, Tiedje JA (2008) Microbial communities in contaminated sediments, associated with bioremediation of uranium to submicromolar levels. Appl Environ Microbiol 74:3718–3729

Choppin GR, Du M (1992) f-Element complexation in brine solutions. Radiochim Acta 58–9:101

Choppin T, Buschmann AH, Troell M, Kautsky N, Neori A, Yarish C (2011) Integrating seaweeds into marine aquaculture systems: a key toward sustainability. J Phycol 37:975–986

Cologgi DL, Lampa-Pastirk S, Speers AM, Kelly SD, Reguera G (2011) Extracellular reduction of uranium via Geobacter conductive pili as a protective cellular mechanism. Proc Natl Acad Sci U S A 108:15248–15252

Cotton S (2006) Coordination chemistry of the actinides. In: Cotton S (ed) Lanthanide and actinide chemistry. Wiley, Chichester, pp 173–199

DiChristina TJ (1992) Effects of nitrate and nitrite on dissimilatory iron reduction by *Shewanella putrefaciens* 200. J Bacteriol 174:1891–1896

Driver CJ (1994) Ecotoxicity literature review of selected Hanford Site contaminants. US Department of Energy, Pacific Northwest Laboratory, PNL-9394, Richland, p 141

Fortin C, Denison FH, Garnier-Laplace J (2007) Metal phytoplankton interactions: modelling the effect of competing ions (H^+, Ca^{2+}, and Mg^{2+}) on uranium uptake. Environ Toxicol Chem 26:242–248

Fowle DA, Fein JB, Martin AM (2000) Experimental study of uranyl adsorption onto *Bacillus subtilis*. Environ Sci Technol 34:3737–3741

Frengstad B, Kjersti A, Skrede M, Banks D, Krog JR, Siewers U (2000) The chemistry of Norwegian groundwaters: III. The distribution of trace elements in 476 crystalline bedrock groundwaters, as analysed by ICP-MS techniques. Sci Total Environ 246:21–40

Gavrilescu M, Vasile L, Cretescu I (2009) Characterization and remediation of soils contaminated with uranium. J Hazard Mater 163:475–510

Geipel G, Brachmann A, Brendler V, Bernhard G, Nitsche H (1996) Uranium (VI) sulfate complexation studied by time resolved laser-induced fluorescence spectroscopy. Radiochim Acta 75:199–204

Gihring TM, Zhang GX, Brandt CC, Brooks SC, Campbell JH, Carroll S, Criddle CS, Green SJ, Jardine P, Kostka JE, Lowe K, Mehlhorn TL, Overholt W, Watson DB, Yang ZM, Wu WM, Schadt CW (2011) A limited microbial consortium is responsible for extended bioreduction of uranium in a contaminated aquifer. Appl Environ Microbiol 77:5955–5965

Green SJ, Prakash O, Jasrotia P, Overholt WA, Cardenas E, Hubbard D, Tiedje JM, Watson DB, Schadt CW, Brooks SC, Kostka JE (2012) Denitrifying bacteria from the genus *Rhodanobacter* dominate bacterial communities in the highly contaminated subsurface of a nuclear legacy waste site. Appl Environ Microbiol 78:1039–1047

Guillaumont R, Fanghaenel T, Neck V, Fuger J, Palmer DA, Grenthe I, Rand MH (2003) Chemical thermodynamics 5, Update on the chemical thermodynamics of uranium, neptunium, plutonium, americium and technetium. Elsevier, Amsterdam

He Z, Gentry TJ, Schadt CW, Wu L, Liebich J, Chong SC et al (2007) GeoChip: a comprehensive microarray for investigating biogeochemical, ecological and environmental processes. ISME J 1:67–77

Hennig C, Tutschku J, Rossberg A, Bernhard G, Scheinost AC (2005) Comparative EXAFS investigation of uranium(VI) and -(IV) aquo chloro complexes in solution using a newly developed spectroelectrochemical cell. Inorg Chem 44:6655–6661

Hsi CD, Langmuir D (1985) Adsorption of uranyl onto ferric oxyhydroxides. Application of the surface complexation site-binding model. Geochim Cosmochim Acta 49:1931–1941

Hu P, Brodie EL, Suzuki Y, McAdams HH, Andersen GL (2005) Whole-genome transcriptional analysis of heavy metal stresses in *Caulobacter crescentus*. J Bacteriol 187:8437–8844

Istok JD, Senko JM, Krumholz LR, Watson D, Bogle MA, Peacock A, Chang YJ, White DC (2004) In situ bioreduction of technetium and uranium in a nitrate contaminated aquifer. Environ Sci Technol 38:468–475

Karpas Z, Paz-Tal O, Lorber A, Salonen L, Komulainen H, Auvinen A, Saha H, Kurttio P (2005) Urine, hair, and nails as indicators for ingestion of uranium in drinking water. Health Phys 88:229–242

Kelly SD, Kemner KM, Fein JB, Fowle DA, Boyanov MI, Bunker BA, Yee N (2002) X-ray absorption fine structure determination of pH-dependent U-bacterial cell wall interactions. Geochim Cosmochim Acta 66:3855–3871

Koban A, Geipel G, Rossberg A, Bernhard G (2004) Uranium(VI) complexes with sugar phosphates in aqueous solution. Radiochim Acta 92:903–908

Kumar R, Nongkhlaw M, Acharya C, Joshi SR (2013a) Uranium (U)-tolerant bacterial diversity from U ore deposit of Domiasiat in North-East India and its prospective utilisation in bioremediation. Microbes Environ 28:33–41

Kumar R, Nongkhlaw M, Acharya C, Joshi SR (2013b) Soil bacterial metagenomic analysis from uranium ore deposit of Domiasiat in Northeast India. Curr Sci 105:495–498

Kurttio P, Komulainen H, Leino A, Salonen L, Auvinen A, Saha H (2005) Bone as a possible target of chemical toxicity of natural uranium in drinking water. Environ Health Perspect 113:68–72

Langmuir D (1997) Aqueous geochemistry of uranium. In: McConnin R (ed) Aqueous environmental chemistry. Prentice-Hall, Upper Saddle River, NJ, pp 494–512

Lovley DR, Phillips EJP (1988) Novel mode of microbial energy metabolism: organic carbon oxidation coupled to dissimilatory reduction of iron or manganese. Appl Environ Microbiol 54:1472–1480

Lovley DR, Phillips EJP (1992) Bioremediation of uranium contamination with enzymatic uranium reduction. Environ Sci Technol 26:2228–2234

Lovley DR, Phillips EJP, Gorby YA, Landa ER (1991) Microbial reduction of uranium. Nature 350:413–416

Lovley DR, Widman PK, Woodward JC, Phillips EJP (1993) Reduction of uranium by cytochrome-c3 of *Desulfovibrio vulgaris*. Appl Environ Microbiol 59:3572–3576

Macaskie LE (1990) An immobilized cell bioprocess for the removal of heavy metals from aqueous flows. J Chem Technol Biotechnol 49:357–379

Macaskie LE, Empson RM, Cheetham AK, Grey CP, Skarnulis AJ (1992) Uranium bioaccumulation by a *Citrobacter* sp. as a result of enzymically mediated growth of polycrystalline HUO_2PO_4. Science 257:782–784

Madden AS, Swindle AL, Beazley MJ, Moon JW, Ravel B, Phelps TJ (2012) Long term solid-phase fate of co-precipitated U(VI)–Fe(III) following biological iron reduction by *Thermoanaerobacter*. Am Miner 97:1641–1652

NCRP (National Council on Radiation Protection and Measurements) (1999) Biological effects and exposure limits for "hot particles". NCRP report no. 130, Bethesda, MD

Merroun ML, Selenska-Pobell S (2001) Interactions of three eco-types of *Acidithiobacillus ferrooxidans* strains with (VI). Biometals 14:171–179

Merroun ML, Raff J, Rossberg A, Hennig C, Hennig C, Reich T, Selenska-Pobell S (2005) Complexation of uranium by cells and S-layer sheets of *Bacillus sphaericus* JG-A12. Appl Environ Microbiol 71:5532–5543

Moll H, Reich T, Hennig C, Rossberg A, Szabo Z, Grenthe I (2000) Solution coordination chemistry of uranium in the binary UO^{2+}–$SO_4{}^{2-}$ and the ternary $UO_2{}^{2+}$ –$SO_4{}^{2-}$–OH^- system. Radiochim Acta 88:559–565

Myers CR, Nealson KH (1988) Bacterial manganese reduction and growth with manganese oxide as the sole electron acceptor. Science 240:1319–1321

Nyman JL, Wu HI, Gentile ME, Kitanidis PK, Criddle CS (2007) Inhibition of a U(VI)-and sulfate-reducing consortia by U(VI). Environ Sci Technol 41:6528–6533

Palmer MR, Edmond JM (1993) Uranium in river water. Geochim Cosmochim Acta 57:4947–4955

Paterson-Beedle M, Jeong BC, Lee CH, Jee KY, Kim WH, Renshaw JC, Macaskie LE (2012) Radiotolerance of phosphatases of a *Serratia* sp.: potential for the use of this organism in the biomineralization of wastes containing radionuclides. Biotechnol Bioeng 109:1937–1946

Priest ND (2001) Toxicity of depleted uranium. Lancet 357:244–246

Reeder RJ, Nugent M, Tait CD, Morris DE, Heald SM, Beck KM, Hess WP, Lanzirotti A (2001) Coprecipitation of uranium(VI) with calcite: XAFS, micro-XAS, and luminescence characterization. Geochim Cosmoschim Acta 65:3491–3503

Renninger N, Knopp R, Nitsche H, Clark D, Keasling J (2004) Uranyl precipitation by *Pseudomonas aeruginosa* via controlled polyphosphate metabolism. Appl Environ Microbiol 70:7404–7412

Ruggiero CE, Boukhalfa H, Forsythe JH, Lack JG, Hersman LE, Neu MP (2005) Actinide and metal toxicity to prospective bioremediation bacteria. Environ Microbiol 7:88–97

Sandino A, Bruno J (1992) The solubility of $(UO_2)_3(PO_4)_2{\cdot}4H_2O(S)$ and the formation of U (VI) phosphate complexes: their influence speciation in natural waters. Geochim Cosmochim Acta 56:4135–4145

Scapolan S, Ansoborlo E, Moulin C, Madic C (1998) Uranium (VI)–transferrin system studied by time resolved laser-induced fluorescence. Radiat Prot Dosim 79:505–508

Schiewer S, Volesky B (2000) Biosorption by marine algae. In: Valdes JJ (ed) Remediation. Kluwer, Dordrecht, pp 139–169

Senko JM, Kelly SD, Dohnalkova AC, McDonough JT, Kemner KM, Burgos WD (2007) The effect of U(VI) bioreduction kinetics on subsequent reoxidation of biogenic U(IV). Geochim Cosmochim Acta 71:4644–4654

Suzuki Y, Banfield JF (1999) Geomicrobiology of uranium. In: Redfern SAT, Carpenter MA (eds) Transformation processes in minerals, Reviews in mineralogy and geochemistry. Mineralogical Society of America, Washington, DC, pp 393–432

Suzuki Y, Banfield JF (2004) Resistance to, and accumulation of, uranium by bacteria from a uranium-contaminated site. J Geomicrobiol 21:113–121

Thomas RAP, Macaskie LE (1996) Biodegradation of tributyl phosphate by naturally occurring microbial isolates and coupling to the removal of uranium from aqueous solution. Environ Sci Technol 30:2371–2375

VanEngelen MR, Field EK, Gerlach R, Lee BD, Apel WA, Peyton BM (2010) UO_2^{2+} speciation determines uranium toxicity and bioaccumulation in an environmental *Pseudomonas* sp. isolate. Environ Toxicol Chem 29:763–769

Van Groenestijn JW, Vlekke GJFM, Anink DME, Deinema MH, Zehnder AJB (1988) Role of cations in accumulation and release of phosphate by *Acinetobacter* strain 210A. Appl Environ Microbiol 54:2894–2901

Van Roy S, Peys K, Dresselaers T, Diels L (1997) The use of an *Alcaligenes eutrophus* biofilm in a membrane bioreactor for heavy metals recovery. Res Microbiol 148:526–528

Wade R, DiChristina TJ (2000) Isolation of U (VI) reduction-deficient mutants of *Shewanella putrefaciens*. FEMS Microbiol Lett 184:143–148

Williams KH, Long PE, Davis JA, Wilkins MJ, N'Guessan AL, Steefel CI, Yang L, Newcomer D, Spane FA, Kerkhof LJ, McGuinness L, Dayvault R, Lovley DR (2011) Acetate availability and its influence on sustainable bioremediation of uranium-contaminated groundwater. Geomicrobiol J 28:519–539

Chapter 8
Environmental-Microbial Biotechnology Inside Mining Operations from an Engineering Viewpoint Based on LCA

Alex Godoy-Faúndez, Douglas Aitken, Lorenzo Reyes-Bozo, and Diego Rivera

8.1 Environmental-Microbial Biotechnologies Applied to Sustainable Mining Operations

The concept of "sustainable development" has been most commonly used since 1987 in accordance to the Brundtland report—Our Common Future—which defined sustainable development as "development that meets the needs of the present without compromising the ability of future generations to meet their own needs" (Brundtland Commission 1987). This concept can be adopted by companies as part of their operational strategy when their activities are based on renewable resources; however, in intensive industrial activities based on nonrenewable resources, this concept needs better understanding due to the environmental impacts caused by extraction of finite raw materials. Today, the terms "sustainable development" and "sustainability" have been used synonymously; according to an engineering viewpoint, the sustainability concept can be defined as "a continual improving the quality of human life while living within the carrying capacity of supporting ecosystems, throughout strategies or course of actions to achieve a balance among social, economic and environmental issues." In this sense, the concept of sustainability is understood as a task in progress or "journey" and therefore requires political decision-making set as common goals and values.

A. Godoy-Faúndez (✉) • D. Aitken
iSustainability Lab, Centro para la Sustentabilidad y Gestión Estratégica de Recursos, Facultad de Ingeniería, Universidad del Desarrollo, Concepción, Chile
e-mail: alexgodoy@ingenieros.udd.cl

L. Reyes-Bozo
Departamento de Ciencias de la Ingeniería, Facultad de Ingeniería, Universidad Andres Bello, Santiago, Chile

D. Rivera
Laboratorio de Políticas Comparadas en Gestión de Recursos Hídricos, Universidad de Concepción, Concepción, Chile

L.B. Sukla et al. (eds.), *Environmental Microbial Biotechnology*, Soil Biology 45,
DOI 10.1007/978-3-319-19018-1_8

Therefore, with regard to the mining industry, the idea of sustainable mining needs to reflect the philosophy of sustainability as a specific framework, which companies and governing institutions can embrace (Azapagic 2014). A useful operational approach to facilitate incorporation of sustainability into mining companies with a long vision could be declared as "the need to satisfy human needs within economically viable operations including environmentally friendly mineral processing to produce as an outcome improved social welfare." Currently, the main challenges facing the mining industry are social and environmental conflicts due to the impacts of ore extraction linked to high water and energy consumption, the main driving forces on legislation today (Emtairah et al. 2002). A further conflict area is the huge amount of hazardous materials disposed on landfills and tailing dams near local communities that need to be resolved and included in mine closure planning to avoid impacts upon future generations.

In addition to these localized impacts, mining activities face hard environmental decisions related to global warming impacts such as water scarcity and loss of biodiversity near mining areas. These impacts result in strong economic repercussions to local communities such as increasing health costs and reduced benefits and welfare. These problems are large social concerns for governments because they will eventually impact areas such as employment, health, and human rights.

The controversial application of the concept of sustainability into mining activities is related to the extraction of nonrenewable resources as commodities with high environmental impacts (Mudd 2010). These paradoxes provide the arguments to nongovernmental organizations (NGOs) to disclaim the opinion that "mining is inherently unsustainable" (Young and Septoff 2002). In opposite to this argument, the International Council on Mining and Metals (ICMM) claims that mining can be "undertaken in such a way that the activity itself and the products produced provide a net positive long-term contribution to human and ecosystem well-being" (ICMM 2012; MAC 2014). According to both disclosures, an alternative vision is necessary to put focus beyond extraction and towards mineral recycling, reducing its impacts using "triple bottom line thinking" within the context of a Circular Economy (European Commission 2014). Conflicts, however, exist due to the fact that there is no agreement on how to assess the sustainability in mining operations.

Until today, an approach to assess the sustainability has been reported (Fonseca et al. 2013) that can serve to make the impacts of mining operations more transparent and include innovation within mineral processing by incorporating environmental-microbial biotechnologies. Regarding reporting, the Global Compendium of Sustainability Indicators Initiatives includes at least 20 records of frameworks to assess the sustainability into mining operations, as well as the Global Reporting Initiative (GRI 2012) for mining and metals sector supplement (MMSS)—the most widely adopted in the mining sector—indicators to assess the impacts of replacing old processes with new green technologies based on environmental biotechnologies are, however, not included. Finally, the rules regarding the relationships between stakeholders within the mining industry, the driver of government legislation, have been towards the promotion of corporate social responsibility (Dummet 2006; Emtairah et al. 2002) developing public policies to encourage the compliance of companies to achieve higher standards with respect to sustainability issues. Within this framework, environmental-microbial biotechnologies can become useful tools to be deployed into mining operations to contribute towards the abatement of environmental impacts.

In the twenty-first century, emerging biotechnologies are providing environmental solutions to economic, social, and environmental challenges (Rawlings 2002; Dua et al. 2002) replacing old technologies and reducing the consumption of raw materials and the production of emissions. Today, bioprocessing of mineral ores and concentrates applied to extract base (e.g., copper, cobalt, and nickel) and precious metals (e.g., gold and silver) is already used to achieve a number of goals although it remains a niche technology as you seen in Table 8.1 (Johnson 2013). Today, the use of biotechnologies has been on the rise for the last 20 years. One example about this was documented by a statistics Canada survey applied to companies on 17 different industrial sectors. According to this survey, three classes of biotechnologies had been applied: Genetic Engineering, Environmental Biotechnology, and Bioculture/Industrial process biotechnology. However, biotechnologies have been deployed mainly into firms concerned with resource-based extraction and manufacturing at rate of 92 % per year. The barriers and troubles to deploy these technologies are related to the development of expertise and reporting regarding costs savings to control pollution (Arundel and Rose 1999).

According to current projections in the mining industry, which show a greater reliance upon low-grade primary metal ores due to increasing demand of metals into the economy, biotechnologies can assist in improving the operational

Table 8.1 Application of biotechnology processes

Biotechnology process	Goals	Applications
Bioreactors	Enclosed containers in which microorganisms are maintained under controlled conditions for the purpose of creating or destroying specific compounds	Bioleaching
Biological gas cleaning	Use of microorganism to break down or degrade hazardous substances in a gas stream into less hazardous or nontoxic substances	Emissions from foundries
Bioremediation and biodegradation	The use of naturally or genetically modified microorganism to break down or degrade hazardous substances to less hazardous or nontoxic substances. This process increases the efficiency of the microbial population to concentrate or accumulate specific compounds by adding nutrients, oxygen, or water	Hazardous reagents used in froth flotation, oil spills, and mineral ore recovery in tailing dams
Phytoremediation	Use of vegetative species for the purpose of site remediation	Tailing dams, green covers, and mine closure process

sustainability. Biotechnology allows the reprocessing of mining wastes, tailing dams, and applying in situ leaching for material recovery to convert bioprocessing into a conventional approach (e.g., pyrometallurgy) used in biomining; these techniques are likely to feature heavily in the future of mining. Biomining will be the gate to promoting further innovations beyond old technologies related to extraction and ore processing, achieving opportunities in mine closure processing including securing mine wastes (rocks and tailings) sensoring, monitoring, and process controls (Johnson 2014).

In this framework, environmental biotechnologies are key in controlling mine-related pollution through the introduction of "ecological engineering" approaches and also to remediate and recover metals from wastewaters, such as acid mine drainage. Indeed, with depletion of ore/minerals and implementation of stricter environmental rules, environmental biotechnologies will increasingly be useful for metal recovery in solid industrial wastes, allowing the technique to be environmentally friendly, economical, and cost-effective (Mishra and Ha Rhee 2014). Furthermore, great advances have been achieved in the understanding and development of bioleaching of copper from primary copper sulfide minerals (e.g., Chalcopyrite) with a focus on low oxidation–reduction potential by thermophilic bacteria (mainly Archaea). The heap bioleaching of Nickel has been a successful commercial process used on a global scale to enhance precious metal recovery; this process has also successfully been applied to gold through biooxidation-heap process for pretreatment of sulfuric-refractory gold ore (Brierley and Brierley 2013). A further example is in response to continual toxic chemicals disposed on landfills—hazardous wastes; green engineering techniques using experimental bioremediation at in situ and ex situ mining areas have proven successful (Balba et al. 1998).

Beginning in the late 1970s, under governmental pressure in the United States, environmental-microbial biotechnologies were supported by environmental regulatory agencies through funds for environmental cleanups (e.g., Environmental Protection Agency, EPA). Since the 1960s, the mining industry had resolved the issue of hazardous waste landfilling by the use of mono-landfill including clean production strategies complementarily to end-of-pipe technologies. The remediation processes involved a wide range of strategies and technological tools such as biological, physical, and chemical systems, used for treating water and soils by composting or in-vessel methodologies using microbial communities or genetically modified microorganism (Fig. 8.1).

Since the 1990s, these treatment technologies were focused on intensive treatment of waste excluding the analysis of their environmental impacts to air, water, land, and ecosystems, as well as issues related to water and energy consumption. The bias is due to our assumption that biotechnologies are beneficial by itself. However, these technologies need an analytical processing of information about environmental and health impacts, in order to replace old processes with new biotechnologies that will produce new kinds of wastes and energy and water consumptions.

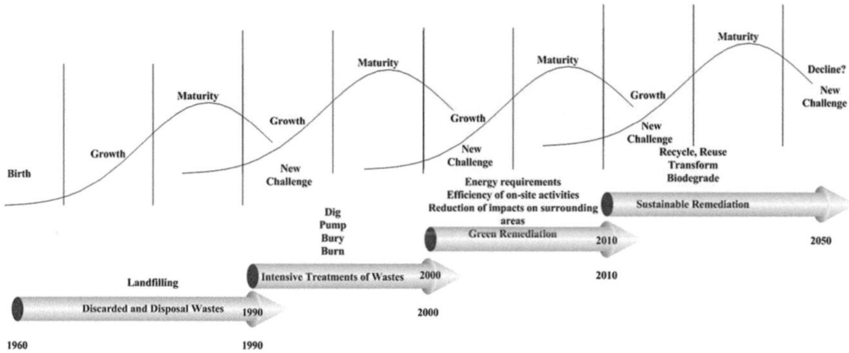

Fig. 8.1 Evolution of remediation wastes and cleanups

8.1.1 Bioremediation as an Environmental-Friendly Process

The bioremediation of polluted environments is a highly promising environmentally friendly technology feasible to be applied by several kinds of industries around the world. The advantages of this technology are its feasibility to be applied to a wide range of environments and contaminants. The disadvantage of this technology is the requirement for preliminary and feasibility studies to be approved for a posteriori escalation to industrial sized processes.

A wide range of available technologies have been applied to contaminated soil, sediment, bedrock, and sludge categorized as in situ or ex situ treatments in reference to the location of the decontamination process applied (on-site vs. off-site) (CLU-IN 2014). According to the literature, the in situ treatments demonstrate a high variability in success related to the optimization of the decontamination end point with hard control process and operational verification. It can, however, be applied on site and without additional costs related to containment and confinement in comparison with ex situ treatments considered highly successful (require shorter times for decontamination) and easier optimization of operation (see Table 8.2). However, both kinds of treatments depend on metabolism present in microbial communities by the addition of external agents used as enhancers and the existence of environmental conditions (aerobic and anaerobic environments) depending upon the target pollutants. Setting parameters such as oxygen content, nutrient availability, and inoculum type allows the development of autochthonous or allochthonous microorganisms and facilitates the bioavailability of pollutants to enzymatic actions.

For contaminated ground water, surface water, and leachate (see Table 8.3), the bioremediation process has been successfully applied to different geographical contexts showing a lower structure of costs associated to remediation treatments compared with physical and chemical treatments that depend on the use of energy, water, and land and ecosystem resources.

Table 8.2 Treatment technologies applied to soil, sediment, bedrock, and sludge

		Relative overall cost and performance						
Development status	Treatment train	Operation and maintenance intensive	Capital	System reliability and maintainability	Relative costs	Time	Availability	
In situ biological treatment								
Bioventing	AA	AA	AA	AA	AA	AA	A	AA
Enhanced bioremediation	AA	AA	BA	A	A	AA	A	AA
Phytoremediation	AA	AA	AA	AA	BA	AA	BA	A
In situ physical/chemical treatment								
Chemical oxidation	AA	AA	BA	A	A	A	AA	AA
Electrokinetic separation	AA	BA	BA	A	A	BA	A	A
Fracturing	AA	A	A	BA	A	A	A	AA
Soil flushing	AA	AA	BA	A	A	A	A	AA
Soil vapor extraction	AA	BA	BA	A	AA	AA	A	AA
Solidification/stabilization	AA	AA	A	BA	AA	AA	AA	AA
In situ thermal treatment								
Thermal treatment	AA	BA	BA	BA	AA	A	AA	AA
Ex situ biological treatment (assuming excavation)								
Biopiles	AA	AA	AA	AA	AA	AA	A	AA
Composting	AA	AA	AA	AA	AA	AA	A	AA
Land farming	AA	AA	AA	AA	AA	AA	A	AA
Slurry-phase biological treatment	AA	BA	BA	BA	A	A	A	AA
Ex situ physical/chemical treatment (assuming excavation)								
Chemical extraction	AA	BA	BA	BA	A	A	A	AA
Chemical reduction/oxidation	AA	A	A	BA	AA	A	AA	AA
Dehalogenation	AA	A	BA	BA	BA	BA	A	A

Separation	AA	A	BA	A	AA	A	AA	AA
Soil washing	AA	BA	BA	BA	AA	A	AA	AA
Solidification/stabilization	AA	AA	A	BA	AA	AA	AA	AA
Ex situ thermal treatment (assuming excavation)								
Hot gas decontamination	BA	AA	BA	BA	AA	AA	AA	A
Incineration	AA	AA	BA	BA	A	BA	AA	AA
Open burn/open detonation	AA	AA	BA	BA	AA	AA	AA	AA
Pyrolysis	AA	AA	BA	BA	BA	BA	AA	AA
Thermal desorption	AA	AA	BA	BA	A	A	AA	AA
Containment								
Landfill cap	AA	AA	A	BA	AA	AA	BA	AA
Enhancements/alternatives	AA	AA	A	BA	AA	AA	BA	AA
Other treatment								
Excavation, retrieval, off-site disposal	AA	AA	AA	AA	AA	?	AA	AA

AA, above average; A, average; BA, below average; N/A, not applicable; I/D, insufficient data; ?, level of effectiveness highly dependent of specific contaminant and its application

Table 8.3 Treatment technologies applied to groundwater, surface water, and leachate

			Relative overall cost and performance					
Development status	Treatment train	Operation and maintenance intensive	Capital	System reliability and maintainability	Relative costs	Time	Availability	
In situ biological treatment								
Enhanced bioremediation	AA	AA	BA	A	A	AA	?	AA
Monitored natural attenuation	AA	AA	BA	A	A	AA	?	AA
Phytoremediation	AA	AA	AA	AA	BA	AA	BA	A
In situ physical/chemical treatment								
Air sparging	AA	AA	AA	AA	AA	AA	AA	AA
Bioslurping	AA	A	AA	AA	A	AA	A	AA
Chemical oxidation	AA	AA	BA	A	A	A	AA	AA
Directional wells	AA	AA	A	BA	A	A	A	AA
Dual phase extraction	AA	BA	BA	BA	A	A	A	AA
Thermal treatment	AA	BA	BA	BA	A	A	AA	AA
Hydrofracturing	AA	A	AA	AA	AA	A	A	AA
In-well air stripping	AA	A	A	BA	A	A	BA	AA
Passive/reactive treatment walls	AA	AA	A	BA	AA	A	BA	AA
Ex situ biological treatment								
Bioreactors	AA	AA	A	BA	AA	AA	BA	AA
Constructed wetlands	AA	AA	AA	AA	A	AA	BA	AA
Ex situ physical/chemical treatment (assuming pumping)								
Adsorption/absorption	AA	AA	BA	A	A	BA	BA	AA
Advanced oxidation processes	AA	A	BA	BA	A	A	BA	AA
Air stripping	AA	A	BA	A	AA	AA	BA	AA
Granulated activated carbon/liquid phase carbon adsorption	AA	A	BA	A	AA	A	BA	AA

Groundwater pumping/pump and treat	AA	A	BA	BA	AA	BA	BA	AA
Ion exchange	AA	A	BA	BA	AA	A	BA	AA
Precipitation/coagulation/ flocculation	AA	A	A	BA	AA	A	BA	AA
Separation	AA	A	BA	BA	AA	BA	AA	AA
Sprinkler irrigation	AA	AA	AA	AA	AA	BA	BA	AA
Containment								
Physical barriers	AA	AA	A	BA	AA	AA	BA	AA
Deep well injection	AA	AA	AA	AA	A	AA	BA	AA
Air emissions/off-gas treatment								
Biofiltration	AA	N/A	AA	AA	?	AA	AA	BA
High-energy destruction	BA	N/A	I/D	I/D	BA	A	I/D	A
Membrane separation	BA	N/A	I/D	I/D	BA	A	I/D	A
Oxidation	AA	N/A	AA	AA	AA	AA	I/D	AA
Scrubbers	AA	N/A	A	BA	AA	AA	I/D	AA
Vapor-phase carbon adsorption	AA	N/A	AA	AA	AA	AA	I/D	AA

AA, above average; A, average; BA, below average; N/A, not applicable; I/D, insufficient data; ?, level of effectiveness highly dependent of specific contaminant and its application

Until today, the decision-making analysis to decide between environmental biotechnologies and old technologies has included variables such as operational costs and time necessary to achieve a cleanup according to governmental rules for contaminated sites. Economical variables such as investment or the use of other wastes as amended to create an industrial ecology between companies amortizing costs in raw materials are also being used (Reyes-Bozo et al. 2014). The final decision, however, will depend mainly upon engineering costs related to type of pollutant(s) present, extensiveness of the oil spill, the type of polluted soils (wastes), the water and energy requirements, and the potential environmental and human risks from a toxicological viewpoint (USEPA 2002). Nevertheless, from a green engineering viewpoint, the application of bioremediation as environmental-friendly technology does not depend much on economic analysis. As a result, it needs to be based on holistic tools such as life cycle assessment (LCA) to compare different scenarios on environmental and health risks according to the whole process.

Regarding pollutant removal, bioremediation treatments such as bioventing are enhanced not only by adding microorganisms but also by nutrient abundance, water, air/oxygen, and surfactants to achieve complete removal of contaminants (Dejonghe et al. 2001; USEPA 2002) with a consideration to the full life cycle based on cradle-to-cradle thinking. Indeed, the phytoremediation process has been used as complementary approach to bioremediation process on contaminated soil, sediment, bedrock, and sludge successfully. However, phytoremediation, which is defined as the process by which plants remove contaminants by accumulating, degrading, or rendering harmless environmental pollutants (Cunningham et al. 1996), is an attractive alternative to avoid the disposal of waste to landfill or the stabilization of tailing dams (Macek et al. 2000; Kechavarzi et al. 2007). The water consumption of deploying such techniques for remediation in arid mining areas has, however, received no attention. Biopiles, landfarming, and composting bioremediation have been widely used as ex situ approaches to treat oil spills, which involve soil excavation (and the consumption of fuel oils) to relocate in a remote site the polluted materials (energy and fuel oil consumption). These treatments also require irrigation to achieve appropriate moisture content creating the necessary soil–water interface for pollutant removal and to promote microbial communities and the uptake of pollutants with periodic aeration (Guerin 1999, 2000).

Despite the numerous different alternatives to apply green remediation processes by physical, chemical, and biological methods, bioremediation has the potential to provide real sustainable remediation. Sustainable bioremediation can use innovative reagents from nontraditional sources, i.e., biodegradable or bio-surfactants that can significantly reduce consumption of chemical reagents which can sometimes be more recalcitrant than pollutants. Indeed, bioremediation mediated by microbial communities can replace the use of natural resources used as an amendment by waste products in such approaches as bioremediation via composting (Al-Daher et al. 2001; Cai et al. 2007; Canet et al. 2001).

Bioremediation by composting could provide an efficient approach to contaminated soils through the use of green wastes or sludge as amendments

(Antizar-Ladislao et al. 2004) and balance the indirect environmental impacts (e.g., water and energy consumption) in oil spills. In both cases, the rate of biodegradation would depend on biological and physicochemical factors present in site (Antizar-Ladislao et al. 2006) and on several physicochemical properties related to the contaminant (i.e., hydrophobicity, volatility, polarity) (Mrayyan and Battikhi 2005), solid matrix (i.e., organic matter, porosity) (Mader et al. 1997), and its bioavailability (Semple et al. 2003) but not considering external factors as in biopiles or landfarming. In all cases, we need to consider environmental issues related to impacts and use of resources in regard to bioremediation as an environmental-friendly pathway to green remediation. Green remediation can be defined as the cleanup of soils or hazardous wastes improving and protecting human health and the local environment while maximizing global environmental benefits by reducing risks in relation to climate change. Such considerations need to include energy and water requirements and the reduction of impacts on surrounding areas as well as the expansion of life cycle analysis practice within mining operations. In this framework, the US Environmental Protection Agency through its website Clean-Up Information (CLU-IN 2014) has enounced about Best Management Practice (BMP) incorporating five core elements (USEPA 2010) which provide inputs to LCA.

- Energy requirements of the treatment system: "BMPs focus on opportunities to improve energy efficiency of on-site equipment and buildings, meet electricity demands of treatment systems through renewable energy instead of fossil-fuel-based energy, and reduce consumption of petroleum-based fuels for routine vehicles and heavy machinery."
- Air emissions: "BMPs focus on opportunities to further reduce emission of greenhouse gas and criteria pollutants (ozone, particulate matter, carbon monoxide, nitrogen dioxide, sulfur dioxide, and lead)."
- Water requirements and impacts on water resources: "BMPs focus on opportunities to reduce quantities of water consumed (intensity) during cleanup, re-use treated water, store or divert storm water for beneficial use, and preserve natural hydraulic conditions."
- Land and ecosystem impacts: "BMPs focus on opportunities to preserve natural land features, maintain open space, sequester carbon, enhance biodiversity, increase wildlife habitat, and minimize surface and subsurface disturbance."
- Material consumption and waste generation: "BMPs focus on opportunities to reduce waste generation, recycle spent products, reuse materials, salvage items for donation or resale, beneficially use industrial byproducts, and purchase environmentally preferred products."

According to these recommendations, environmental biotechnology treatments can focus on achieving high-energy and water efficiencies with a reduction in resource consumption through the application of LCA. To minimize environmental impacts, we need to move incorporate systemic thinking towards sustainable environmental biotechnology to:

Table 8.4 Targets of a sustainable remediation by LCA

Sustainable remediation	Indicator[a]
Energy requirements of the treatment system	Siting
Air emissions	Energy efficiency
Water requirements and impacts on water resources	Water efficiency
Land and ecosystem impacts	Materials efficiency
Material consumption and waste generation	Occupant health and safety
Long-term stewardship actions	Building operations and maintenance
	Waste reduction

Source: [a]Taken from USEPA (2008)

1. Minimize or eliminate energy consumption or the consumption of other natural resources
2. Minimize or eliminate natural resources consumption or replacement of them
3. Reduce or eliminate dispersion of pollutant as well as releases to the environment such as emissions to the air
4. Harness or mimic a natural process and promote the use of biodegradable reagents
5. Reuse or recycling of land or otherwise undesirable materials
6. Encourage the use of remedial technologies that permanently destroy contaminants

In a global framework considering climate change, sustainable remediation could provide a new approach to cleaning up impacted environments when adopted in current and future practices. In addition, sustainability concepts have the potential to minimize the deleterious environmental and human health beyond conventional end-of-pipe strategies based on targets and indicators throughout ore processing (see Table 8.4).

The use of innovative reagents from non-traditional sources can significantly reduce consumption of chemical reagents, sometimes more recalcitrant than pollutants. Indeed, these new materials can replace the use of virgin natural resources or by using waste products, a network between wastes and processing can be created. This is the baseline for promoting good industrial ecology and sustainable bioremediation (Reyes-Bozo et al. 2014).

8.2 Environmental Impacts of Mineral Processing: The Chilean Case

Chile has a per capita income of approximately US$24,000, corrected by purchasing power parity (PPP), similar to figures for developed countries (reference, last report IMF adjusted by CENSUS 2012). According to this data, Chile is considered a developing country with a strong mining economy with several other key industries such as agriculture, fishing, or stockbreeding. Chile's wealth results mainly from the extraction of natural resources, mainly ore extraction and predominantly copper ores (Klaus Schmidt-Hebbel 2012). Today, Chilean copper production represents 40 % of commercial transactions and between 20 and 35 % of the gross national product dependent upon international copper prices and stocks (COCHILCO 2011), thanks to Codelco (Corporación Nacional del Cobre de Chile or, in English, the National Copper Corporation of Chile) and continuous private investments within the mining industry (i.e., BHP-Billiton; Minera Escondida 2007). Furthermore, copper represents up to 60 % of total exports (Meller 2013). According to the volume of copper exports, Chile produces one third of global copper production; in 2013, the country achieved a total production of 58 million tons.

Since the end of the eighteenth century, the mining industry has been sustaining the Chilean economy but at the same time causing numerous environmental impacts due to continuous material extraction, disposal of hazardous wastes, production of tailing dams and acid mine drainage, as well as associated higher consumption of water and energy (mainly electricity) causing habitat modifications with the consequent perturbation of native flora and fauna. Indeed, after mine closures with large tailing dams on the ground near local communities, confrontations are ranging from the type of exploitation (e.g., open pit mine) to cultural disturbances (Vallejos 1994). Despite these impacts, the mining industry in Chile has been widely accepted by local communities as an important economic activity that has contributed to the rapid development of the country's economy (Oyarzún and Oyarzún 2011). Today, the most important ores produced in Chile are cathodes representing 80 % of total exports (Siliverstovs and Herzer 2005), both sulfide and oxide copper ores, but sulfide ores are the more predominant (Bulatovic 2007). According to mineral processing, sulfide ores are traditionally processed by means of pyrometallurgical and hydrometallurgical techniques (Biswas and Davenport 2002) with the pyrometallurgical process (Bulatovic 2007) being the main method, as seen in Fig. 8.2.

The environmental impacts of the pyrometallurgical processes include five basic stages:

1. Comminution, aimed to reduce ore grain size; producing noise and dust emissions with disposal of material in tailing dams
2. Froth flotation, aimed to obtain copper ore concentrates with production of hazardous materials and slags
3. Smelting, aimed to separate the metal from their minerals, producing hazardous emissions, foundry slags, and mineral residues with ashes due to combustion of fuel oils
4. Converting, aimed to obtain blister copper
5. Electro-refining, aimed to purify the resulting final product (Biswas and Davenport 2002; Memary et al. 2012) producing acid frog mainly

However, every stage of copper processing requires huge quantity of raw materials and expensive reagents to increase its concentration in a cathode copper as well

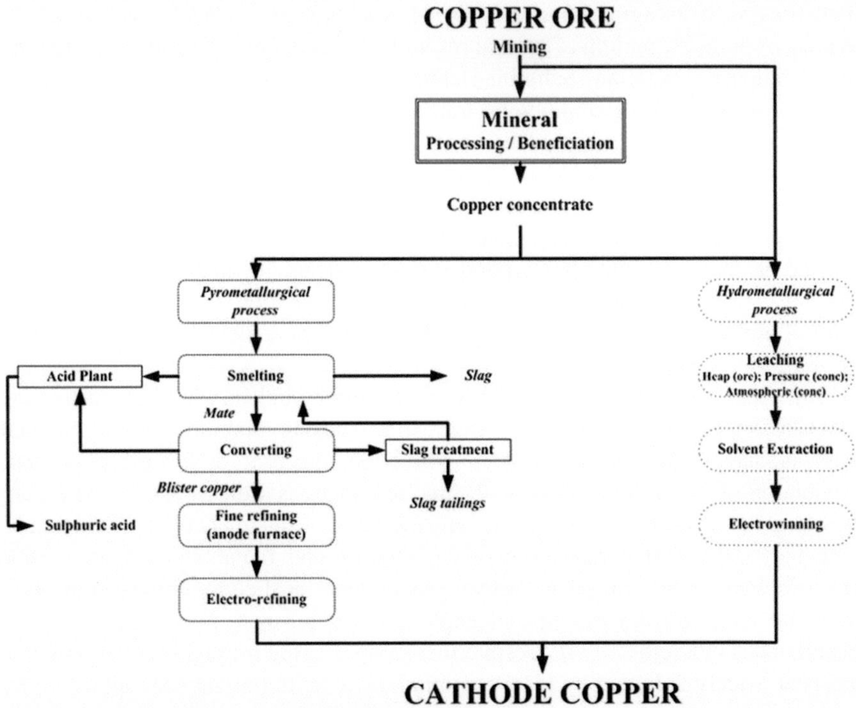

Fig. 8.2 Copper ore pathway processing

as solid wastes (slags and tailings in general) disposition as tailing dams. There exists unseen indirect impacts related to the water requirements for each stage which in turn produce direct impacts on watersheds through use and decreasing the water quality by wastewater discharges containing persistent chemicals and acid mine drainage into rivers. Other impacts are due to consumption of energy by direct combustion of fuel oils and electricity with the consequent air emissions (CO_2, SO_2/H_2SO_4).

The main metallic and nonmetallic ore deposits are located in the northern regions (especially the Region of Antofagasta), where the climate is arid and forest and hydrological resources are scarce. Good examples of impact-generating processes in this region are the comminution process which requires high amounts of energy to reduce ore grain size; the froth flotation stage which produces voluminous tailings and concentrates (25 % copper) and uses chemical reagents as collector, frothers, and modifiers; and the smelting process which requires that concentrates are fed into smelter to produce blister copper (98 % copper) to separate the metal from their minerals producing hazardous emissions, foundry slags, and mineral residues with ashes due to combustion of fuel oils and slags; then, blister copper is refined into copper with high purity (99.99 % copper). These last steps are high-energy-consuming processes and produce hazardous materials (Memary et al. 2012; McLellan and Corder 2012; Moors et al. 2005) where electro-refining, aimed to purify the resulting final product (Biswas and Davenport 2002; Memary

Table 8.5 Pyrometallurgical and hydrometallurgical impacts

Parameter	Pyrometallurgy	Hydrometallurgy
Gas emission	High, greenhouse gas emissions and volatile organic carbon	Low and easy to treat
Dioxin potential	High	No dioxins
Dust potential	High, during material handling and transport	Low, can be dissolved in solution or taken care of by pollution control equipment
Energy	High (up 1,200 °C)	Low, room temperature
Recovering rate	Low (only fraction of metals), useful nonmetals are incinerated and impossible to recover	High recovery, clean separation of material types enable individual effective recovery
Final residue	High (slag and dusts), potential metal trapped to reduce recovery effectiveness	Low only mixed plastics, which could be recycled into engineered plastic
Social acceptance	Low, due to high environmental liabilities	High, cleaner environment with highly effective and mature pollution control methodology

et al. 2012), produces acid mainly acid fog. From an LCA viewpoint, the hydrometallurgical processes (leaching) could be cleaner than pyrometallurgical process due to operational conditions being conducted at room temperature and through the use of pollution control equipment capable of reducing emissions and discharges (Table 8.5). This can lead to a higher recovery rate due to the possibility of cascading and recirculating solid waste achieving a high recovery rate with chemical precipitation of electrowinning. Regarding ore processing, direct fuel consumption is almost negligible as compared with the pyrometallurgical method. Indeed, recently the hydrometallurgical method has become more popular due to lower energy costs and reduced water and air consumption. However, despite the fact that leaching could be cleaner than other techniques, the scale of mineral ores processed, resultant water, and electricity consumption could present great challenges to be controlled, especially on arid mining places.

Today, strategies to improve the environmental performance of conventional leaching promote the use of bioleaching by using genetically modified organism to extract copper from oxide ore and avoid the use of hazardous reagents into mining operations. In Chile, the hydrometallurgical method is incorporated into about 23 % of total copper production. The proportion of bioleaching of the total leached copper production stands at around 42 %, and the projections are towards increase in its application to mineral recovery ores from low-grade primary metal ores and tailing dams (Gentina and Acevedo 2013). Furthermore, the environmental concerns of the Chilean government related to climate change impacts resulting from (Ministry of Environment 2012) mining operations have increased the restrictions in accordance with the environmental impact assessment (EIA) system (Ministry of Environment 1994). The goal has been to introduce specific regulations on closed and abandoned mine sites and on the production and management of hazardous wastes with the inclusion of cradle-to-grave thinking. It is therefore clear that the Chilean mining industry will require the development and implementation of new protocols on environmental biotechnologies.

8.3 Bioremediation Process and Mining Wastes: Life Cycle Assessment as Next Step

A non-evident impact of the mining industry is the production of hazardous wastes resulting from the daily operation of heavy machinery. Such operations cause repetitive spills of fuel oils during reparation and maintenance of the machinery, as well as casual accidents. Using soils and sawdust as cheap and readily available sorbent materials helps the cleanup which must then be contained in Hazardous Wastes Landfills (HWL) whereby large amounts of hazardous wastes accumulate over time. According to Current Chilean legislation, bioremediation technologies can be tested at a lab scale as a valid alternative practice to environmental decontamination process (Ministry of Health 2004; Balba et al. 1998).

In 2002, Minera Escondida Ltd, the most important private investment by BHP-Billiton, located in the Atacama Desert (north of Chile; 170 km SE of Antofagasta; 3,100 m above sea level) and responsible for the production of 8 % of the world's copper and represents 20 % of total copper production, started a new program to confine and dispose hazardous material by mono landfilling. Common materials to be disposed of are soils and sawdust polluted with fuel oils from oil spills due to accidents that have been used as cheap adsorbent material. However, due to high costs of final disposal in a mine closure, feasibility analysis of bioremediation was necessary; 30 laboratory-scale cylindrical-aerated composting reactors at five ratios of contaminated soil to contaminated sawdust—S:SD 1:0, 3:1, 1:1, 1:3, and 0:1 (2,000 g total composting mixture, wet weight)—were operated continuously for 56 days at controlled temperatures; moisture content was close to 50 %. Atmospheric air warmed to 60 °C was introduced to each reactor and circulated through an internal perforated piping to ensure sufficient oxygen concentration in the reactors (Godoy et al. 2008). Direct extraction and purification of DNA from soils was conducted by minor modifications of previously described methods using MoBio kits. Terminal-restriction fragment length polymorphism and Biolog Ecoplate™ tests were conducted to find Operational Taxonomic Units through detection of restriction fragments, richness (S), and diversity expressed as the Shannon–Wiener index (H′). The average well color development and profiles of utilization of C-source were also determined. Sampling was conducted in triplicate after 0, 14, 28, 42, and 56 days and analyzed following standard methods (USEPA 2005).

Following 56 days of treatment, different biodegradation curves were obtained in all reactors. Biodegradation curves indicated a similar trend with smooth

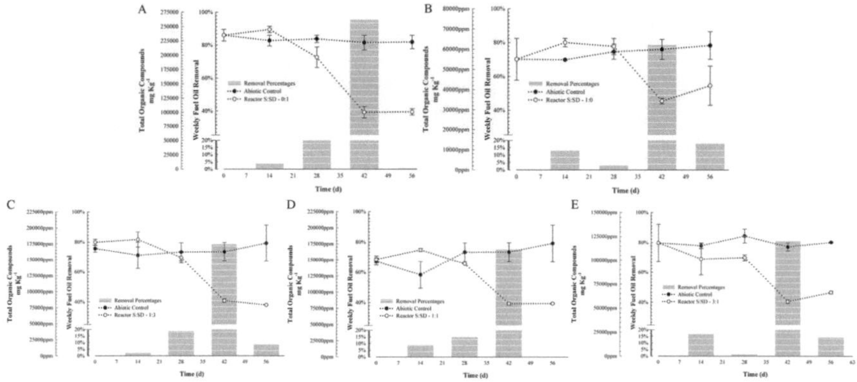

Fig. 8.3 Temporal removal curves of fuel oil-contaminated desert soil: (**a**) reactor with 100 % sawdust, (**b**) reactor with 100 % soil, (**c**) reactor with ratio S:SD—1:3, (**d**) reactor with ratio S: SD—1:1, (**e**) reactor with ratio S:SD—3:1

decreases reaching steady-state and contaminant removal between 35 and 60 %. Reactors with higher levels of sawdust presented higher removal rates (Fig. 8.3).

The results indicate that bioremediation of an aged and weathered contaminated mixture of desert mining soil and sawdust with fuel oil is feasible at S:SD ratio of 1:3 and a correct nutrient balance in order to achieve a maximum overall hydrocarbon removal of fuel oil in the weathered and aged contaminated wastes (Atlas and Bartha 1998; Godoy-Faundez et al. 2008). According to microbial communities, direct counts and percentage of viable bacteria indicated that count numbers were high at the beginning of the treatment and that no significant changes were observed during the course of the treatment (over 10^6 cells/g) with over 60 % of total number of cells. Cutting-edge biomonitoring tools based on molecular and metabolic profiles showed contradictory results, indicating that the changes in number of phylotypes were not in agreement with the changes in the metabolic activity. Metabolic profiles presented correlations with temporal removal curves, indicating that the microbial activity in composting reactors increased with the duration of the treatment and the removal rates, and thus higher microbial activities resulted in higher fuel removal rates. It is suggested that the metabolic profiles and not the molecular results can be an important monitoring tool since they provide information of the metabolic activity based of differential consumption of organic sources (Godoy-Faúndez et al. 2008).

However, despite bioremediation of aged fuel oil-contaminated desert mining soils and sawdust being feasible via aerated in-vessel composting, the water and energy consumption of the technique remains uncertain. Due to the necessary high moisture content for vessel composting (50 % w/w), this is potentially a limiting resource in mine sites located in arid zones. The issues of water availability can be an environmental risk in LCA through overextraction and consumption. On a pilot scale, rotating drums are the preferred vessel to treat large amount of hazardous material that require high-energy inputs to provide aeration inside reactor and start

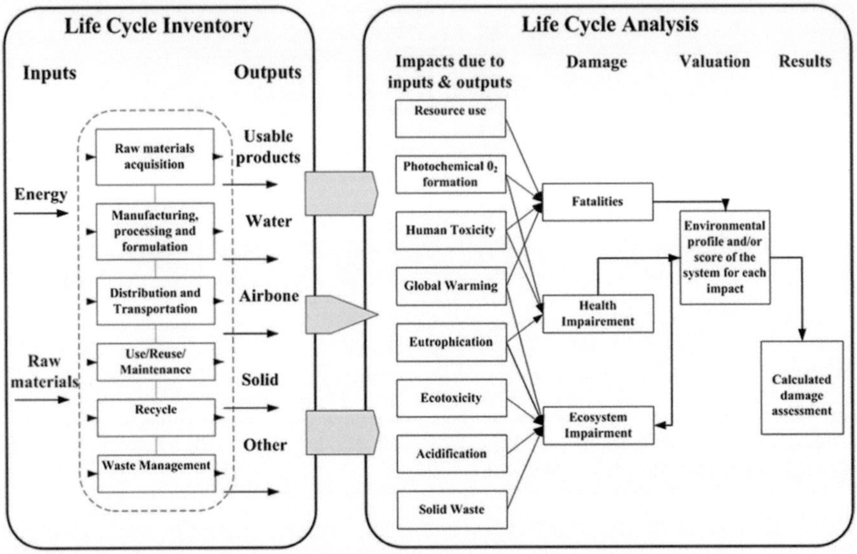

Fig. 8.4 Life cycle assessment: linking life cycle inventory and life cycle analysis

the aerobic degradation of organic compounds. These unseen requirements require analysis conducted using life cycle analysis (Fig. 8.4).

Life cycle assessment is a tool that allows the quantification of the environmental impacts of different processes within a larger system to determine the environmental damage or benefit of a product or system (Baumann and Tillman 2004; Murray et al. 2008). LCA considers all stages of the systems life cycle and examines all the inputs and outputs associated with a particular unit, the functional unit (Lofgren et al. 2011; Northey et al. 2013), and can be used to consider a wide variety of interactions and environmental impacts (energy use, global warming potential, acidification potential) (Baumann and Tillman 2004; ISO 2006; Norgate and Haque 2010) (Fig. 8.4).

Several LCA studies have been conducted within the copper mining industry (Memary et al. 2012; Northey et al. 2013), but few studies have been applied to environmental-microbial biotechnologies such as the bioremediation process in mining operations. Cadotte et al. (2007) applied LCA to select different scenarios comparing in situ and ex situ treatments to manage treatment of diesel-contaminated sites. Their result demonstrated that in situ treatment impacts required longer time periods than ex situ scenarios considering different approaches (e.g., composting, biopiles, landfarming). Indeed, when combining chemical with biological processes, the primary (local toxicity impacts of the contamination) versus secondary (stemming from the remedial actions) impacts were different when the time variable was considered. Then, in situ remediation can generate more secondary impacts than ex situ treatments. Furthermore, Cadotte and coauthors demonstrated that primary impacts should not be neglected because they can represent up to 92 % of global impacts. This study shows the big difference between the primary impacts (ecotoxicity and human toxicity categories) and secondary environmental impacts as a function of the duration of treatment and the achievement of regulatory criteria.

Toffoletto et al. (2005) studied the remediation of diesel-contaminated soil using biopiles applied to soils impacted with an average of 6,145 mg of diesel fuel/kg soil over a 2-year period, comparing two scenarios, for a single-use treatment facility on site or a permanent treatment center—ex situ—accepting 25,000 m^3 soil per year. According to their results, a permanent treatment center requires additional operational conditions that increase the environmental impacts (site preparation). In this case, ex situ treatments contribute to secondary impacts related to site preparation and transportation that can be alleviated by LCA recommendations (Table 8.6), while in situ treatment contributes with primary impacts. Therefore, environmental biotechnologies will provide different impacts depending upon whether the technologies are applied in situ or ex situ. Furthermore, the monitoring of residual soil contamination as well as spatial and temporal data need to be collected and integrated to make characterization models in a whole picture into mining operations (Lemming et al. 2010; Blanc et al. 2004).

Finally, Table 8.7 displays 12 LCA studies reported in literature of remediation technologies and contaminant types. The environmental remediation biotechnologies show a high diversity of impacts related to environmental and resource consumption over the perception that these are cleaner than traditional clean technologies.

The bias behind these results are based on the fact that life cycle assessments have been mainly conducted for in situ remediation technologies where the majority of the existing literature focuses on ex situ remediation of contaminated soil. On the other hand, the LCAs use, as the functional unit, the volume of contaminated soil to be treated disregarding the rest of the mineral processing operations and their

Table 8.6 Recommendations applied to bioremediation processes based on LCA

Element	Requirement
Siting	Use engineered barriers to reduce air emissions from on-site construction demolition equipment and from trucking waste materials
	Impose idling restrictions on construction equipment
	Sequence work to minimize double handling of materials
Energy efficiency	Incorporate renewable energy sources into treatment systems
	Use energy efficient systems and office equipment
Materials efficiency	Identify salvage options for materials from existing structures
	Consider reuse options for existing structures
	Consider structural reuse of slabs or foundations
	Crush existing structures to optimize scrap recovery and produce fill materials
	Use of waste wood, food, and other organics for on-site use
Waste reduction	Identify recycling options for waste and debris, such as metal and food wastes

Taken from Illinois (2008)

Table 8.7 Environmental impacts by LCA from environmental remediation biotechnologies

Reference	Beinat et al. (1997)	Diamond et al. (1999)	Page et al. (1999)	Volkwein et al. (1999)	ScanRail Consult et al. (2000)	Ribbenhed et al. (2002)	Godin et al. (2004); Toffoletto et al. (2005)	Bayer and Finkel (2006)	Cadotte et al. (2007)
LCIA method					EDIP97	USES-LCA[a]	EDIP97		TRACI
Environmental impacts									
Global warming		Q	x	x	x	x	x	x	x
Ozone depletion		Q			x		x		x
Photochemical ozone formation		Q		x	x	x	x	x	x
Acidification		Q		x	x	x	x	x	x
Nutrient enrichment		Q			x	x	x	x	x
Eco toxicity		Q[b]			x	x[h]	x[h]		x[h]
Human toxicity		Q[b]	x	x	x	x[h]	x[h]	x	x[h]
Waste	x		x		x		x		
Surface water pollution	x								
Air pollution	x								
Land use	x[c]	Q[d]		x[c]					
Odor				x	Q				
Noise		Q		x	Q				
Other site-related impacts		Q[e]		x[f]					
Positive impacts	x[g]				x[h]				

Resource consumption									
Fossil energy	x	Q	x	x	x	x		x	
Scarce metals					x				
Clean groundwater	x	Q	x	x					
Clean soil/ sand/graved	x	Q	x		x				

Q categories that are used only for a qualitative impact assessment
[a]For eco and human toxicity
[b]Includes both process related (secondary) and site related (primary toxic impacts)
[c]Land use due to remediation
[d]Land use due to landfilling
[e]Soil quality disturbance, heat damage, habitat attenuation, effects on soil moisture, interrupted drainage, land stagnation, and human social disturbances
[f]Residual human toxicity burlen
[g]Amount of cleared soil and groundwater
[h]Prevented toxic impacts due to remediation calculated in a separate environmental benefit module

geographical location, which will affect the environmental impacts on the ecosystem.

Another bias to be resolved relates to the establishment of methodologies beyond the conventional set of impact categories. Ecotoxicity and human toxicity are the impact categories varying the most between these methodologies and pollutants. Therefore, in a holistic approach to be applied to environmental-microbial biotechnologies such as bioremediation process, more studies are necessary to address the importance of evaluating both primary and secondary impacts. This concerns primarily impacts related to local impacts regarding residual contamination, which varies between different remediation technologies and secondary impacts related to resource use and emissions arising in other stages of the life cycle of the biotechnological projects.

8.4 Final Remarks

Life cycle assessment is becoming an increasingly widespread tool within support systems for environmental decision-making and is becoming increasingly important within environmental-microbial biotechnological projects such as bioremediation of soils. Despite the perception that biotechnology, such as bioleaching or bioremediation, is cleaner than old operational processing materials, use of bioreactors for cultures can be linked to unseen primary and second impacts that can be analyzed using LCA where all processes are considered. In this review, the use of LCA to compare the environmental impacts of different remediation technologies shows that despite local environmental problems being reduced, at the same time, the remediation activities may cause negative environmental impacts on the local, regional, and global scale. Therefore, LCA can be used to evaluate the inherent trade-off and to compare scenarios in terms of their associated environmental burden independent of the type of biotechnology to be applied.

Acknowledgement The authors would like to thank Water Research Center for Agriculture and Mining (WARCAM) supported by CONICYT/Chile in the framework of FONDAP 2013 (Fifth National Competition for Research Centers in Priorities Areas)—CRHIAM/CONICYT/FONDAP 15130015.

References

Al-Daher R, Al-Awadhi N, Yateem A, Balba MT, ElNawawy A (2001) Compost soil piles for treatment of oil-contaminated soil. Soil Sediment Contam 10:197–209

Antizar-Ladislao B, Lopez-Real JM, Beck AJ (2004) Bioremediation of polycyclic aromatic hydrocarbon (PAH)-contaminated waste using composting approaches. Crit Rev Environ Sci Technol 34:249–289

Antizar-Ladislao B, Lopez-Real J, Beck AJ (2006) Degradation of polycyclic aromatic hydrocarbons (PAHs) in an aged coal-tar contaminated soil under in-vessel composting conditions. Environ Pollut 141:459–468

Arundel A, Rose A (1999) The diffusion of environmental biotechnology in Canada: adoption strategies and cost offsets. Technovation 19:551–560

Atlas R, Bartha R (1998) "Ecología microbiana y microbiología ambiental" 4ª Edición Cap13: Interacciones microbianas con contaminantes xenobióticos e inorgánicos Cap 14: Ensayos de biodegradabilidad y seguimiento de la bioremediación de contaminantes xenobióticos

Azapagic A (2014) Sustainability considerations for integrated biorefineries. Trends Biotechnol 32 (1):1–4

Balba MT, Al-Daher R, Al-Awadhi N, Chino H, Tsuji H (1998) Bioremediation of oil-contaminated desert soil: the Kuwaiti experience. Environ Int 24:163–173

Baumann H, Tillman AM (2004) The Hitch Hiker's guide to LCA: an orientation in life cycle assessment methodology and applications. Studentliettaratur AB, Lund

Bayer P, Finkel M (2006) Life cycle assessment of active and passive groundwater remediation technologies. J Contam Hydrol 83(3):171–199

Beinat E, van Drunen MA, Janssen R, Nijboer MH, Koolenbrander JGM, Okx JP, Schütte AR (1997) REC: a methodology for comparing soil remedial alternatives based on the criteria of risk reduction, environmental merit and costs. CUR/Nobis report 95-10-3, Gouda, the Netherlands

Biswas AK, Davenport WG (2002) Extractive metallurgy of copper, 4th edn. Pergamon, Oxford

Blanc A, Métivier-Pignon H, Gourdon R, Rousseaux P (2004) Life cycle assessment as a tool for controlling the development of technical activities: application to the remediation of a site contaminated by sulfur. Adv Environ Res 8:613–627

Brierley C, Brierley J (2013) Progress in bioleaching: part B: applications of microbial processes by the minerals industries. Appl Microbiol Biotechnol 97:7543–7552

Brundtland GH, United Nations WCED (1987) Our common future. Oxford University Press, Oxford

Bulatovic M (2007) Handbook of flotation reagents: chemistry, theory and practice: flotation of sulphides ores. Elsevier, Amsterdam, 446 p

Cadotte M, Deschênes L, Samson R (2007) Selection of a remediation scenario for a diesel-contaminated site using LCA. Int J Life Cycle Assess 12(4):239–251

Cai QY, Mo CH, Wu QT, Zeng QZ, Katsoyiannis A, Ferard JF (2007) Bioremediation of polycyclic aromatic hydrocarbons (PAHs)-contaminated sewage sludge by different composting processes. J Hazard Mater 142:535–542

Canet R, Birnstingl JG, Malcolm DG, Lopez-Real JM, Beck AJ (2001) Biodegradation of polycyclic aromatic hydrocarbons (PAHs) by native microflora and combinations of white-rot fungi in a coal-tar contaminated soil. Bioresour Technol 76:113–117

Cochilco, Corporación Chilena del Cobre (2011) Yearbook: copper and other mineral statistics 1992–2011. Santiago, 172 p

Cunningham SD, Andersson TA, Schwab AP, Hsu FC (1996) Phytoremediation of soils contaminated with organic pollutants. Adv Agron 56:55–113

Dejonghe W, Boon N, Seghers D, Top EM, Verstraete W (2001) Bioaugmentation of soils by increasing microbial richness: missing links. Environ Microbiol 3(10):649–657

Diamond ML, Page CA, Campbell M, McKenna S, Lall R (1999) Life-cycle framework for assessment of site remediation options: method and generic survey. Environ Toxicol Chem 18(4):788–800

Dua M, Singh A, Sethunathan N, Johri A (2002) Biotechnology and bioremediation: successes and limitations. Appl Microbiol Biotechnol 59:143–152

Dummet K (2006) Drivers for corporate environmental responsibility (CER). Environ Dev Sustain 8:375–389

Emtairah T, Jacobsson N, Kogg K, Lissinger L, Mont O (2002) Who creates the market for green products? An analysis of the role of different actors in relation to supply and demand of green products. Swedish Environmental Protection Agency, Stockholm

European Commissions (2014) Communication from the Commission to the European Parliament, The Council, The European Economic and Social Committee and the Committee of the regions towards a circular economy: a zero waste programme for Europe. /* COM/2014/0398 final */. http://eur-lex.europa.eu/legal-content/EN/TXT/?uri=CELEX:52014DC0398

Fonseca A, McAllister ML, Fitzpatrick P (2013) Sustainability reporting among mining corporations: a constructive critique of the GRI approach. J Clean Prod 46–47:180–186

Gentina J, Acevedo F (2013) Application of bioleaching to copper mining in Chile. Electron J Biotechnol 16(3):1–16. doi:10.2225/vol16-issue3-fulltext-12

Godin J, Menard JF, Hains S, Deschenes L, Samson R (2004) Combined use of life cycle assessment and groundwater transport modeling to support contaminated site management. Hum Ecol Risk Assess 10(6):1099–1116

Godoy-Faundez A, Antizar-Ladislao B, Reyes-Bozo L, Camaño A, Sáez-Navarrete C (2008) Bioremediation of contaminated mixtures of desert mining soil and sawdust with fuel oil by aerated in-vessel composting in the Atacama region (Chile). J Hazard Mater 151:649–657

Godoy-Faúndez A, Reyez-Bozo L, Sáez-Navarrete C, Antizar-Ladislao B (2008) Differential metabolic changes in microbial communities, explain the bioremediation of organic contaminants in the Atacama Desert (Chile). In: 4th European bioremediation conference, 3–6 Sept 2008, Chania, Crete, Greece

GRI (2012) GRI sustainability disclosure database global reporting initiative (GRI), Amsterdam

Guerin TF (1999) Bioremediation of phenols and polycyclic aromatic hydrocarbons in creosote contaminated soil using ex-situ land treatment. J Hazard Mater 1999(65):305–315

Guerin TF (2000) The differential removal of aged polycyclic aromatic hydrocarbons from soil during bioremediation. Environ Sci Pollut Res 7:19–26

ICMM (2012) Our work: sustainable development framework international council on mining and metals, London

Illinois EPA (2008) Illinois' greener cleanups matrix: how to maximize the environmental benefits of site remediation, Illinois Environmental Protection Agency. Available from http://www.epastateilus/land/greener-cleanups/matrixpdf. Accessed 7 July 2010

ISO (2006) Environmental management—life cycle assessment—requirements and guidelines. International Organization for Standardization, Geneva

Johnson DB (2013) Development and application of biotechnologies in the metal mining industry. Environ Sci Pollut Res 20:7768–7776

Johnson DB (2014) Biomining—biotechnologies for extracting and recovering metals from ores and waste materials. Curr Opin Biotechnol 30:24–31

Kechavarzi C, Pettersson K, Leeds-Harrison P, Ritchie L, Ledin S (2007) Root establishment of perennial ryegrass (L-perenne) in diesel contaminated subsurface soil layers. Environ Pollut 145:68–74

Lemming G, Hauschild M, Bjerg P (2010) Life cycle assessment of soil and groundwater remediation technologies. Int J Life Cycle Assess 15:115–127

Lofgren B, Tillman A, Rinde B (2011) Manufacturing actor's LCA. J Clean Prod 19 (17–18):2025–2033

MAC Mining Association of Canada (2014) Towards sustainable mining. Available at http://www.mining.ca

Macek T, Mackova M, Kas J (2000) Exploitation of plants for the removal of organics in environmental remediation. Biotechnol Adv 18:23–34

Mader BT, UweGoss K, Eisenreich SJ (1997) Sorption of nonionic, hydrophobic organic chemicals to mineral surfaces. Environ Sci Technol 31:1079–1086

McLellan BC, Corder GD (2012) Risk reduction through early assessment and integration of sustainability in design in the minerals industry. J Clean Prod. doi:10.1016/jjclepro201202014

Meller P (2013) Analysis of Chilean economic and development performance from the copper perspective. In: Copper conference, Santiago, 1–4 Dec 2013
Memary R, Giurco D, Mudd G, Mason L (2012) Life cycle assessment: a time-series analysis of copper. J Clean Prod 33:97–108
Minera Escondida (2007) Minera Escondida, certified with ISO 14001 Atacama. www.escondidacl/Escondida/ingles/indexhtml
Ministry of Environment (1994) General baseline of environmental La N°19,300, Government of Chile
Ministry of Environment (2012) National climate change action plan 2008–2012, Government of Chile. http://www.mmagobcl/1304/w3-article-54783html
Ministry of Health (2004) Decreto Supremo Nª148, Sanitary normative on handling hazardous wastes, Government of Chile
Mishra D, Ha Rhee Y (2014) Microbial leaching of metals from solid industrial wastes. J Microbiol 52(1):1–7. doi:10.1007/s12275-014-3532-3
Moors EHM, Mulder KF, Vergragt PJ (2005) Towards cleaner production: barriers and strategies in the base metals producing industry. J Clean Prod 13:657–668
Mrayyan B, Battikhi MN (2005) Biodegradation of total organic carbons (TOC) in Jordanian petroleum sludge. J Hazard Mater 120:127–134
Mudd GM (2010) The ultimate sustainability of mining: linking key mega-trends with 21st century challenges. In: Sustainable mining 2010 conference, AusIMM, Kalgoorlie, pp 351–373, August 2010
Murray A, Horvath A, Nelson KL (2008) Hybrid life-cycle environmental and cost inventory of sewage sludge treatment and end-use scenarios: a case study from China. Environ Sci Technol 42:3163–3169
Norgate T, Haque N (2010) Energy and greenhouse gas impacts of mining and mineral processing operations. J Clean Prod 18:266–274
Northey S, Haque N, Mudd G (2013) Using sustainability reporting to assess the environmental footprint of copper mining. J Clean Prod 40:118–128
Oyarzún J, Oyarzún R (2011) Sustainable development threats, inter-sector conflicts and environmental policy requirements in the arid, mining rich, northern Chile territory. Sustain Dev 19 (4):263–274
Page CA, Diamond ML, Campbell M, McKenna S (1999) Life-cycle framework for assessment of site remediation options: case study. Environ Toxicol Chem 18(4):801–810
Rawlings D (2002) Heavy metal mining using microbes. Annu Rev Microbiol 56:65–91
Reyes-Bozo L, Godoy-Faúndez A, Herrea-Urbina R, Higueras P, Salazar J, Valdés-Gonzalez H, Vhymeister E, Antizar-Ladislao B (2014) Greening Chilean copper mining operations through industrial ecology strategies. J Clean Prod 84:671–679
Ribbenhed M, Wolf-Watz C, Almemark M, Palm A, Sternbeck J (2002) Livscykelanalys av marksaneringstekniker för förorenad jord och sediment (Life cycle assessment of remediation technologies for contaminated soil and sediment). IVL Rapport/report B1476 IVL. Swedish Environmental Research Institute, Stockholm
ScanRail Consult, HOH Water Technology A/S, NIRAS Consulting Engineers and Planners A/S, Revisorsamvirket/Pannell Kerr Forster (2000) Environmental/economic evaluation and optimising of contaminated sites remediation EU LIFE Project no 96ENV/DK/0016, Copenhagen
Schmidt-Hebbel K (2012) Fiscal policy for commodity exporting countries: Chile's experience, Documentos de Trabajo 415, Instituto de Economia Pontificia, Universidad Católica de Chile
Semple KT, Morriss AWJ, Paton GI (2003) Bioavailability of hydrophobic organic contaminants in soils: fundamental concepts and techniques for analysis. Eur J Soil Sci 564:1–10
Siliverstovs B, Herzer D (2005) Manufacturing exports, mining exports and growth: cointegration and causality analysis for Chile (1960–2001). Appl Econ 39(2):153–167
Toffoletto L, Deschênes L, Samson R (2005) LCA of ex-situ bioremediation of diesel-contaminated soil. Int J LCA 10(6):406–416

USEPA (US Environmental Protection Agency) (2002) Application, performance, and costs of biotreatment technologies for contaminated soils, EPA/600/R-03/037. USEPA, Washington, DC

USEPA (US Environmental Protection Agency) (2005) SW846 Methods for evaluating solid wastes physical/chemical methods 8000 series: method 8260 B-method 8270 in agency, USEP (Ed)

USEPA (US Environmental Protection Agency) (2008) Green remediation: incorporating sustainable environmental practices into remediation of contaminated sites. United States Environmental Protection Agency [online]. Available from http://www.clu-in.org/download/remed/green-remediation-primer.pdf. Accessed 7 July 2010

USEPA (US Environmental Protection Agency) (2010) Principles for greener cleanups [online]. Available from http://www.epa.gov/oswer/greencleanups/principles.html. Accessed 7 July 2010

USEPA (US Environmental Protection Agency) (2014) Clean up-information (CLU-IN), US Environmental Protection Agency. http://clu-in.org/contaminantfocus/defaultfocus/sec/Fractured_Rock/cat/Overview/

Vallejos JP (1994) Historia y minería en Chile: estudios y fuentes. América Latina en la Historia Económica 1:65–88

Volkwein S, Hurtig HW, Klöpffer W (1999) LCA concepts and methods—life cycle assessment of contaminated sites remediation. Int J Life Cycle Assess 4(5):263–273

Young J, Septoff A (2002) Digging for change: towards a responsible minerals future. An NGO and community perspective, Mineral Policy Centre, Consejo Minero de Chile

Chapter 9
Neutrophilic Bacteria in Iron Mineral Transformation and Their Applications

Jacintha Esther and Lala Behari Sukla

9.1 Introduction

Iron is ubiquitous in the earth's crust, atmosphere, and oceans which is being recycled within these reservoirs via different fluxes such as aeolian dust, atmospheric precipitation, rivers, and hydrothermal activity. The activity of flowing rivers and hydrothermal vents results in delivery of iron to ocean basins and sediments. Fe is a predominant redox-active metal in such oxic–anoxic interfacial sedimentary environments, which exists in +2 and +3 oxidation states. Iron exists either as solid-phase ferric [Fe (III)]/ferrous [Fe (II)] minerals or as dissolved ions depending on the geochemical parameters. Some of the iron-based minerals are tabulated in Table 9.1. The redox transformations of iron mineral are a result of chemical processes and to a major extent of microbial respiration (Kappler and Straub 2005). Redox cycling of Fe regulates the carbon and energy flow and the speciation of various aqueous and solid-phase inorganic components in sedimentary interfacial environments.

Microorganisms play a key role in the iron redox cycle in the redox transition zones of these environments (Straub et al. 2001; Blöthe and Roden 2009). The aqueous Fe (II) at circumneutral pH is oxidized to bioavailable Fe (III) mineral forms by anaerobic/aerobic iron-oxidizing bacteria (IOB) which in turn serve as the terminal electron acceptor for anaerobic dissimilatory iron-reducing bacteria (IRB). The biogenic Fe (II) minerals which are the metabolic product of IRB are utilized as electron donors by the IOB. This ongoing microbe-mediated Fe redox cycle in such

J. Esther (✉) • L.B. Sukla
Bioresources Engineering Department, CSIR – Institute of Minerals and Materials Technology, Bhubaneswar 751013, Odisha, India

Academy of Scientific and Innovative Research (AcSIR), CSIR – Institute of Minerals and Materials Technology, Bhubaneswar, 751013 Odisha, India
e-mail: jacesther@gmail.com

L.B. Sukla et al. (eds.), *Environmental Microbial Biotechnology*, Soil Biology 45,
DOI 10.1007/978-3-319-19018-1_9

Table 9.1 List of iron-containing minerals

Mineral redox state	Mineral class	Mineral name	Mineral formula
Fe (III)	Oxides	Hematite	α-Fe_2O_3
		Maghemite	γ-Fe_2O_3
	Oxyhydroxides	Ferrihydrite	$Fe_4HO_8 \cdot 4H_2O$
		Lepidocrocite	γ-FeOOH
		Goethite	α-FeOOH
		Amorphous	$Fe_2O_3 \cdot nH_2O$
	Oxyhydroxyl-sulfate	Schwertmannite	$Fe_3O_8(OH)_{1-1.8} \cdot H_2O$
	Silicate	Illite	$(K, H_3O)\ (Al, Mg, Fe)_2\ (Si, AL)_4\ O_{10}\ (OH)_2\ (H_2O)$
Fe (II)	Sulfides	Pyrite	FeS_2
		Mackinawite	FeS
	Carbonates	Siderite	$FeCO_3$
	Phosphates	Vivianite	$Fe_3(PO)_4$
Mixed Fe(II)–Fe(III)	Oxide	Magnetite	Fe_3O_4
	Sulfide	Greigite	Fe_3S_4
	Hydroxide	Green rusts	$Fe_x^{III}Fe_y^{II}(OH)_{3x+2y-z(A-)z}$; $A^- = Cl^-$; . SO_4^{2-}
	Silicate	Smectite	$(Ca, Na)\ (Al, Mg, Fe)_2\ (Si, Al)_4\ O_{10}\ (OH)_2 \cdot xH_2O$
		Chlorite	$(Mg, Fe, Mg)_6\ (Si, Al)_4\ O_{10}\ (OH)_8$

environments develop into self-sustained microbial community playing a pivotal role in iron mineral transformations (Blöthe and Roden 2009) as illustrated in Fig. 9.1.

Iron is also a significant component of living cells and essential micronutrient for the physiological functions of all eukaryotes and most prokaryotes. However, the low solubility of Fe (III) ions under neutral and aerobic conditions (abiotic oxidation by molecular oxygen) restricts their availability to microorganisms compelling them to grow best under microaerobic conditions by evolving strategies for iron acquisition. Siderophores are chelating agents (ligands) produced by these microbes that help in dissolution of Fe (III) to meet their nutritional demands. Siderophores have higher affinity for Fe (III) than Fe (II). Most microorganisms utilize iron as nutrition source, while some archaea and bacteria utilize iron as energy source contributing significantly to the Fe redox cycle. The role of microorganisms in the Fe redox cycle of neutrophilic environments under different favorable conditions is discussed in this chapter.

9.2 Neutrophilic Iron-Oxidizing Bacteria

Microorganism-mediated Fe (II) oxidation dominates chemical oxidation in acidic pH and aerobic conditions due to the stability of Fe (II) which is a significant feature of acid mine drainage process. However, at circumneutral pH and aerobic

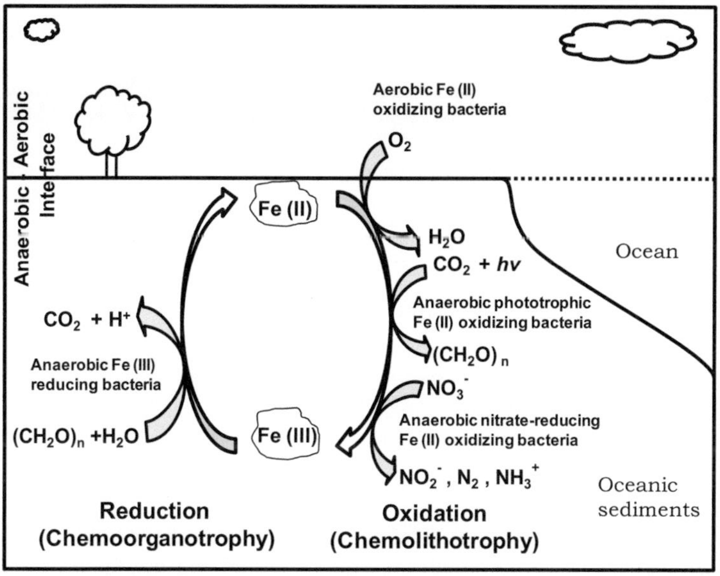

Fig. 9.1 Iron redox cycle at aerobic/anaerobic niches mediated by neutrophilic microorganism

conditions, microorganism-mediated Fe (II) oxidation competes with rapid chemical oxidation (by molecular oxygen) due to the higher stability of Fe (III) over Fe (II) (Emerson and Moyer 1997). Hence, most neutrophilic iron-oxidizing microorganisms thrive under microaerobic or near-surface conditions.

The biogenic Fe (III) minerals produced by NIOB are mostly amorphous or poorly crystalline such as green rust or ferrihydrite which on further stabilization develop into crystalline Fe (III) minerals of lepidocrocite, goethite, or hematite. Neutrophilic iron-oxidizing bacteria contribute to the formation of biogenic Fe (III) oxides by (1) anaerobic photosynthesis coupled to Fe (II) oxidation, (2) anaerobic nitrate reduction coupled to Fe (II) oxidation, and (3) aerobic, lithotrophic oxidation of Fe (II), whose detailed description is provided below. Detailed illustration of the biochemical molecular mechanisms of Fe (II) oxidation by these NIOB has been provided by Dubinina and Sorokina (2014).

9.2.1 *Anaerobic Phototrophic Fe (II)-Oxidizing Bacteria*

Anaerobic phototrophic Fe (II)-oxidizing bacteria were discovered two decades ago capable of growing in the presence of light using CO_2 as electron acceptor along with Fe (II) as their sole electron donor (Widdel et al. 1993). Anaerobic phototrophic Fe (II)-oxidizing bacteria were found to grow optimally in the neutral pH range of 6.5–7. They can oxidize dissolved Fe (II) according to the stoichiometric equation [Eq. (9.1)] established by Ehrenreich and Widdel (1994) and relatively soluble minerals; FeS (ferrous monosulfide) and $FeCO_3$ (siderite), as

Table 9.2 Anaerobic phototrophic Fe (II)-oxidizing bacteria

Phylogenetic lineage	Anaerobic phototrophic NIOB	Habitat	Reference
Purple non-sulfur bacteria	*Rhodomicrobium vannielii* BS-1	Freshwater	Widdel et al. (1993)
	Rhodobacter ferrooxidans SW2	Freshwater ditch	Ehrenreich and Widdel (1994)
	Rhodovulum iodosum N1	Marine sediment	Straub et al. (1999)
	Rhodovulum robiginosum N2	Marine sediment	Straub et al. (1999)
	Rhodopseudomonas palustris TIE-1	Iron-rich freshwater mat	Jiao et al. (2005)
Purple sulfur bacteria	*Thiodictyon* sp. F4	Freshwater marsh	Croal et al. (2004)
Green sulfur bacteria	*Chlorobium ferrooxidans* KoFox	Freshwater ditch	Heising et al. (1999)

per the stoichiometric equation ([Eq. (9.2)] established by Kappler and Straub (2005). However, they were found incapable of oxidizing insoluble FeS_2 (pyrite) and Fe_3O_4 (magnetite) minerals:

$$CO_2 + 4\,Fe(II) + 11\,H_2O + h\nu \rightarrow \langle CH_2O\rangle + 4\,Fe(OH)_3 + 8\,H^+ \tag{9.1}$$

$$FeCO_s + 7\,H_2O + h\nu \rightarrow \langle CH_2O\rangle + 4\,Fe(OH)_3 + 3\,CO_2 \tag{9.2}$$

Thus far, seven anaerobic phototrophic Fe (II)-oxidizing bacteria belonging to three phylogenetic lineages have been discovered which are summarized in Table 9.2.

The generation process of banded iron formations (BIFs) (colossal iron mineral deposits) had remained a mystery due to the abundance of Fe (II) with hardly any oxygen during the Precambrian age. Photochemical oxidation of Fe (II) was believed to play a major role (Cairns-Smith 1978; Francois 1986) in these formations till the discovery of chemical oxidation of Fe (II) by oxygen produced by photosynthesis of cyanobacteria (Konhauser et al. 2002). The discovery of anaerobic phototrophic Fe (II)-oxidizing bacteria has traced back their existence to the Precambrian age suggesting their significant role in the generation of banded iron formations (BIFs) (Kappler et al. 2005a; Trouwborst et al. 2007). The findings and suggestions of the role of various microbial processes in BIF generation have been critically reviewed by Posth et al. (2013).

9.2.2 Anaerobic Nitrate-Reducing Fe (II)-Oxidizing Bacteria

Anaerobic nitrate-reducing Fe (II)-oxidizing bacteria were first identified by Hafenbradl et al. (1996) and Straub et al. (1996) which depend on nitrate as their electron acceptor to provide energy for their growth. They thrive in environments

Table 9.3 Anaerobic nitrate-reducing Fe (II)-oxidizing bacteria

Anaerobic nitrate-reducing IOB	Nutrition	Habitat	Reference
Acidovorax sp. strain BrG1	Mixotroph	Freshwater mud	Straub et al. (1996)
Aquabacterium sp. strain BrG2	Mixotroph	Freshwater mud	Straub et al. (1996)
Thermomonas sp. strain BrG3	Mixotroph	Freshwater mud	Straub et al. (1996)
Acidovorax strain BoFeN1	Mixotroph	Freshwater lake sediment	Kappler et al. (2005b)
A. oryzae strain PS	Heterotrophic	Swine waste lagoon	Chaudhuri et al. (2001)
Pseudogulbenkiania strain 2002	Autotroph	Freshwater lake sediment	Weber et al. (2009)
Paracoccus ferrooxidans	Facultative autotroph	Denitrifying bioreactor	Kumaraswamy et al. (2006)
Ferroglobus placidus	Autotroph and heterotroph	Submarine hydro-thermal vent	Hafenbradl et al. (1996)

having neutrophilic (or higher) pH conditions where the redox potential of Fe (III)/Fe (II) pair is much lesser than in acidic conditions. Moreover, the redox potential of nitrate redox pairs in the nitrate reduction pathway is much higher than the redox pair Fe (III)/Fe (II). Anaerobic nitrate-reducing Fe (II)-oxidizing bacteria are found to flourish in freshwater, brackish, and anaerobic sediments (Hedrich et al. 2011). They oxidize soluble Fe (II) coupled to reduction of nitrate resulting in the formation of Fe (III) mineral (ferrihydrite) and release of nitrogen gas as explained in the following Eq. (9.3):

$$10\,Fe^{2+} + 2\,NO_3^- + 24\,H_2O \rightarrow 10\,Fe(OH)_3 + N_2 + 18\,H^+ \qquad (9.3)$$

The nutrition and habitat of some common anaerobic nitrate-reducing Fe (II)-oxidizing bacteria are summarized in Table 9.3.

9.2.3 Aerobic, Lithotrophic Fe (II)-Oxidizing Bacteria

Aerobic, lithotrophic Fe (II)-oxidizing bacteria were recognized as early as the nineteenth century to utilize oxygen as their electron acceptor (Vatter and Wolfe 1956). They need to compete with the abiotic Fe (II) oxidation in order to oxidize Fe (II) for their survival and hence grow in neutral or microaerobic conditions where abiotic Fe (II) oxidation process is slowed down by low-oxygen concentrations (Sobolev and Roden 2001). The species of genera *Gallionella* and *Leptothrix* belonging to the domain β-proteobacteria are the significant neutrophilic, aerobic Fe (II)-oxidizing bacteria. *Mariprofundus ferrooxydans* is the first Fe (II)-oxidizing

Table 9.4 Aerobic lithotrophic Fe (II)-oxidizing bacteria

A-IOB	Nutrition	Characteristic growth	Reference
Gallionella ferruginea	Autotroph or mixotroph	Bean-shaped cells with twisted stalk	Hallbeck and Pedersen (1990)
Leptothrix discophora SS-1	Obligate heterotroph	Sheath-forming cells which lost the ability to form sheaths in 18 months though were active in Fe (II) oxidation	Ghiorse and Chapnick (1983)
Sphaerotilus natans	Heterotroph	Unicellular rods with unbranched sheath formation	Pellegrin and Juretschko (1999)
Strain TW2	Autotroph and mixotroph	Slightly curved rod	Sobolev and Roden (2004)
Sideroxydans sp. ES-1	Autotroph	Unicellular rods without stalk or sheath formation	Blöthe and Roden (2009)
Sideroxydans paludicola	Autotroph	Bean-shaped cells without stalk or sheath formation	Weiss et al. (2007)
Ferritrophicum radicicola	Autotroph	Bean-shaped cells without stalk or sheath formation	Weiss et al. (2007)
Sideroxydans lithotrophicus	Autotroph	Bean-shaped cells without stalk or sheath formation	Weiss et al. (2007)
M. ferrooxydans	Autotroph	Bean-shaped cells with filamentous stalk	Emerson et al. (2007)

bacteria belonging to the domain ς-proteobacteria isolated from hydrothermal vent site (Emerson et al. 2007). Oxidation of Fe (II) by these bacteria results in the formation of Fe (III) ions which precipitate immediately as Fe (III) oxides and oxyhydroxides on the stalks and sheaths of the microorganisms away from the living cell:

$$4\,Fe^{2+} + O_2 + 4\,H^+ \rightarrow 4\,Fe^{3+} + 2\,H_2O \tag{9.4}$$

$$Fe^{3+} + 3\,H_2O \rightarrow Fe(OH)_3 + 3\,H^+ \tag{9.5}$$

The nutrition, habitat, and characteristic growth to prevent mineral encrustation of some aerobic, lithotrophic Fe (II) oxidizing bacteria are summarized in Table 9.4.

9.2.4 Distribution of NIOB

Neutrophilic iron-oxidizing bacteria (NIOB) inhabit low-oxygen, Fe (II)-rich environments having circumneutral pH (Emerson and Weiss 2004). They are commonly observed in wetland surface sediments (Emerson et al. 1999) and groundwater seeps and springs (James and Ferris 2004) where atmospheric oxygen comes in contact with anoxic, Fe (II)-rich groundwater (Emerson and Weiss 2004). They are also found in habitats like rhizosphere of wetlands, cave walls, irrigation ditches, drinking water wells and pipes, deep-ocean basalt (Edwards et al. 2003), mines, and ocean floor hydrothermal vents (Toner et al. 2009; McAllister et al. 2011; Sun et al. 2013). The NIOB can thrive at the hydrothermal vents due to the large flux of Fe (II) (about 3×10^{11} mol per year) and production of sheer redox gradients at the interface of cold seawater and hydrothermal vent fluids (Glazer and Rouxel 2009; McAllister et al. 2011).

Freshwater neutrophilic iron-oxidizing bacteria are commonly encountered in iron plaque on root systems of most wetland and submerged aquatic plants together with ferric iron-reducing bacteria (Weiss et al. 2007).

9.3 Neutrophilic Iron-Reducing Bacteria

Iron-reducing bacteria (IRB) complete the iron redox biogeochemical cycle by dissimilative reduction of Fe (III) to Fe (II) in neutrophilic environments under anaerobic conditions. IRB respire Fe (III) as their terminal electron acceptor coupled to oxidation of organic carbon or hydrogen as their electron donor conserving energy for their growth (Zegeye et al. 2010). Some IRB also use oxygen, sulfur, humic acids, and nitrate as electron acceptors. A phylogenetically diverse range of microorganisms are capable of reducing Fe (III) in the natural ecosystem. However, strict anaerobes belonging to *Geobacter* genus of the δ-proteobacteria domain and facultative anaerobes belonging to *Shewanella* genus of the γ-proteobacteria domain are the best characterized and extensively studied IRB. The nutrition and habitat distribution of various IRB have been reviewed by Esther et al. (2014).

The amorphous and poorly crystalline Fe (III) oxides such as ferrihydrite formed biotically or abiotically are the more bioavailable form for iron reduction by neutrophilic iron-reducing bacteria (NIRB) than crystalline Fe (III) oxides such as goethite or hematite (Zachara et al. 2002). *Shewanella* spp. gain energy from complete oxidation of lactate coupled to reduction of ferrihydrite at pH 7:

$$\begin{aligned} &4\,Fe(OH)_3 + CH_3CH(OH)COO^- + 7\,H^+ \\ &= 4\,Fe^{2+} + CH_3COO^- + HCO_3{}^- + 10\,H_2O \end{aligned} \tag{9.6}$$

Geobacter spp. gain energy from oxidation of acetate coupled to reduction of ferrihydrite to magnetite at pH 7:

$$24\,Fe(OH)_3 + CH_3COO^- = 8\,Fe_3O_4 + 2\,HCO_3{}^- + H^+ + 36\,H_2O \tag{9.7}$$

The strict anaerobe *Geobacter metallireducens* has also been reported to oxidize Fe (II) by utilizing nitrate as its electron acceptor (Finneran et al. 2002). Similarly,

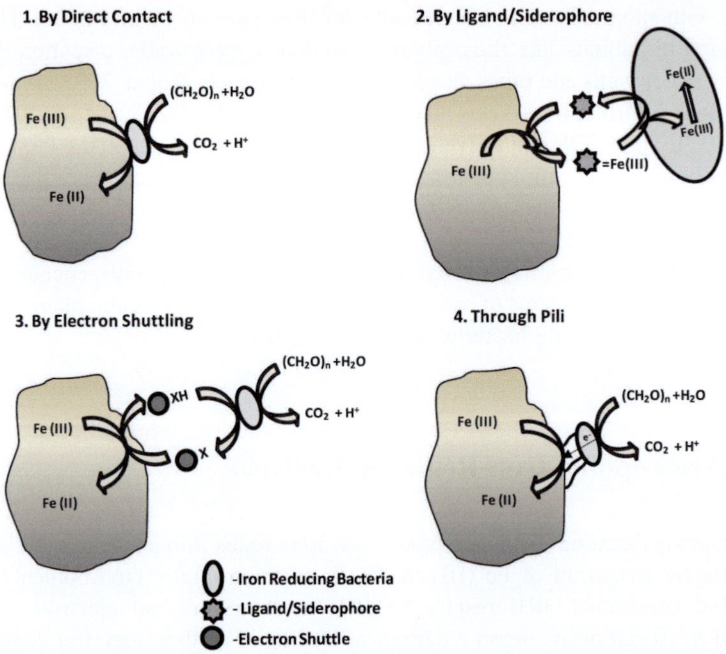

Fig. 9.2 Strategies of solid-phase Fe (III) uptake as electron acceptor by IRB (1) by direct contact, (2) by ligand/siderophore, (3) by electron shuttling, and (4) through pili

Desulfitobacterium frappieri has also been found capable of reducing Fe (III) with H2 as its terminal electron donor and oxidizing Fe (II) with nitrate as its terminal electron acceptor (Shelobolina et al. 2003). These two single bacterial strains have been exceptionally discovered to utilize both Fe (II) and Fe (III) minerals for their energy and growth.

IRB can competently reduce the mineral-bound Fe (III) to Fe (II) resulting in reductive dissolution of iron minerals (Zachara et al. 2002; Crowe et al. 2007; Esther et al. 2013). IRB obtain Fe (III) from solid phase by four different strategies: (1) via bacteria–mineral direct contact, (2) via redox-active electron shuttles, (3) via ligands or chelating agents, and (4) via pili. Extensive research has gone into deciphering the exact molecular mechanism involved in Fe (III) reduction (Esther et al. 2014; Gralnick and Newman 2007) as depicted in Fig. 9.2.

9.3.1 Distribution of NIRB

Neutrophilic iron-reducing bacteria fill the Fe (III)-rich, anaerobic niches of wetlands and aquatic sediments, contaminated aquifers, groundwater contaminated with landfill leachate and organics, mangrove sediments , rice paddy field soils (Zhu et al. 2009), and methanogenic digesters (Morita et al. 2011). Some IRB have also been reported to inhabit the hot environments of hydrothermal vents. IRB are capable to thrive in heavily contaminated sites containing benzene, toluene, methyl tert-butyl ether (MTBE), and azo and anthraquinone dyes by degrading these to nontoxic substances.

9.4 Biomineralization-Induced Iron Mineral Transformation

A significant proportion of the sedimentary iron minerals are formed and transformed biologically, mediated by several biotic and abiotic mechanisms which are well known as biomineralization. Biomineralization is a natural phenomenon wherein inorganic mineral phases are formed by microorganisms due to several cellular-mediated physiological processes. Around 60 widespread biominerals have been identified (Bazylinski et al. 2007). Mineral formation occurs in two different modes depending on the degree of bacterial involvement (Lowenstam 1981). "Biologically induced mineralization" (BIM) is a common process of biomineralization that occurs due to close association between the bacterial activity and their surrounding milieu. It is a passive process in which the biominerals are generated without any regulatory control. However, "biologically controlled mineralization" (BCM) is a regulatory process in which bacteria precipitate biominerals within a preformed framework that serves physiological and structural roles.

Bacterial biomineralization may be extracellular or intracellular depending on the site of mineral deposits. The formation and deposit of biogenic minerals on the outer cell wall or in the immediate surrounding tissues refer to the extracellular type of biomineralization usually occurring as a result of BIM. Neutrophilic iron-oxidizing and iron-reducing bacteria display an extracellular type of biomineralization (Fortin and Langley 2005). Extracellular biomineralization takes place in two steps: (1) electrostatic interaction of metals with the anionic cell wall surface or immediate organic polymers forming nucleation sites for crystal growth and (2) preferential binding of iron to these organic sites followed by inorganic binding of available counterions forming specific iron mineral types accordingly (Konhauser 1997). Hence, the mineralogy and chemistry of Fe (III) minerals depend on the environmental conditions and available inorganic ions (Miot et al. 2009).

BCM usually results in the formation of biogenic minerals in a preformed vesicle or matrix within the cell which is referred to as intracellular biomineralization. Magnetotactic bacteria are the common example of both BIM and BCM wherein single magnetite crystals align themselves in chains within the cell (Frankel et al. 1997).

If the biogenic minerals formed by a characteristic bacterial group have unique texture, mineralogy, and/or morphology, then biomineralization by various microbes can serve as biosignatures. Iron biomineral serves as an iron store to meet future metabolic needs of the microorganisms. They are also useful to the microorganisms in terms of hardness, density, and magnetism (Frankel 1991).

9.4.1 Biomineralization by NIOB

The metabolic product of NIOB, i.e., Fe (III), is insoluble at circumneutral pH and hence tends to precipitate immediately either in the periplasm or on cell surface or in the cell environs which have been captured through SEM by various researchers (Miot et al. 2009; Schädler et al. 2009). This Fe (III) precipitation leads to the formation of mineral-encrusted or non-encrusted cells according to the metabolic need and survival of the microorganism.

Accordingly, different NIOB strains have been reported to possess different biomineralization mechanisms toward formation of Fe (III) precipitates (Kappler et al. 2005b). *Acidovorax* sp. strain BoFeN1, a nitrate-reducing iron-oxidizing strain, encrusts itself with biogenic Fe (III) precipitates (Kappler et al. 2005b), while the photosynthetic IOB, *Rhodobacter* sp. strain SW2, does not encrust itself (Fig. 9.3) but localizes the Fe (III) precipitates at a distance along the polymer fibers emerging from the cells (Miot et al. 2009; Chan et al. 2004). The existence of low-pH microenvironment around the microbial cell may be a possible strategy of photoferrotrophs to avoid cell encrustation, thereby facilitating the entry of light and other nutrients for their metabolic activity (Hegler et al. 2010). However, *Rhodomicrobium vannielii* cultures are an exception which has been reported to encrust themselves with Fe (III) precipitates resulting in restricted light access followed by incomplete Fe (II) oxidation (Heising and Schink 1998).

Gallionella sp. and *Mariprofundus* (sp.) *ferrooxydans* of aerobic lithotrophic iron-oxidizing bacterial group produce organic extracellular, ribbonlike, twisted stalks, whereas the *Leptothrix* spp. form tubular sheaths (Fig. 9.4) that capture and precipitate Fe (III) leaving the cell free of mineral encrustation. Studies have reported that the near-neutral surface charge and mineral-repelling surface properties of the aerobic IOB equip them to avoid cell encrustation sequestering the Fe (III) onto their organic filaments (Saini and Chan 2013). This activity of aerobic lithotrophic IOB to avoid encrustation of bacterial cells prevents the impairment of substrate uptake and metabolite release which would otherwise lead to cell death (Hallberg and Ferris 2004).

The iron-encrusted filaments produced by neutrophilic iron-oxidizing bacteria (NIOB) serve as biosignatures in both the present and ancient geological deposits. The filaments may be twisted stalks mostly found on ocean floor hydrothermal vents (Fleming et al. 2013) or sheaths noticed in flooded underground tunnels (Chan et al. 2004).

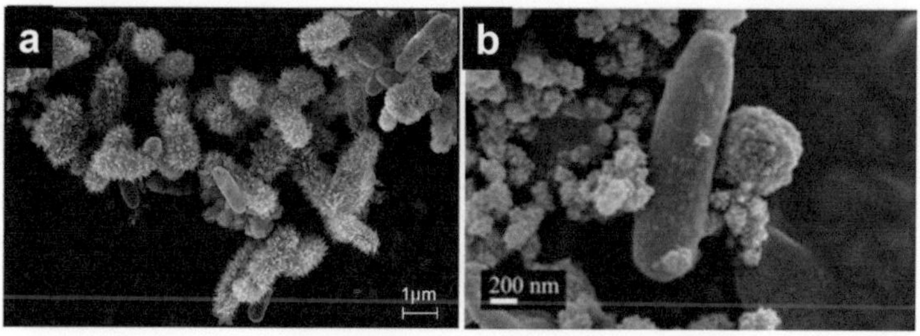

Fig. 9.3 Scanning electron micrograph of (**a**) biogenic Fe (III)-encrusted cells of nitrate-reducing Fe (II)-oxidizing bacterium *Acidovorax* sp. strain BoFeN1 [reprinted with permission from Hohmann et al. (2010). Copyright © 2010, American Chemical Society] and (**b**) non-encrusted cells of phototrophic Fe (II)-oxidizing *Rhodobacter ferrooxidans* strain SW2 [reprinted with permission from Hegler et al. (2008). Copyright © 2008 Federation of European Microbiological Societies. Published by Blackwell Publishing Ltd.]

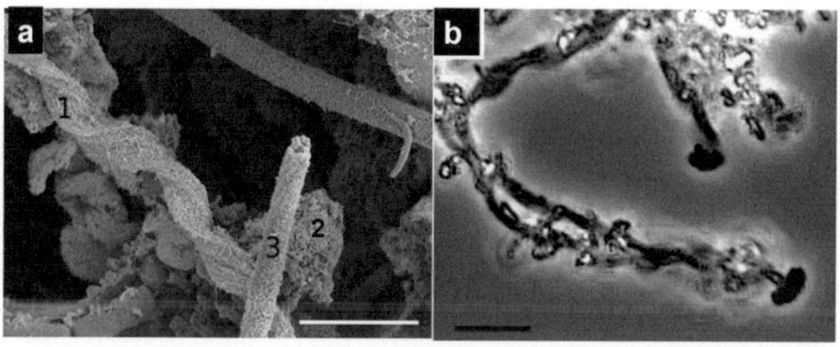

Fig. 9.4 (**a**) Scanning electron micrograph of iron-oxidizing microbial mat with (1) twisted stalks of *Gallionella*, (2) iron oxyhydroxide particulates, and (3) tubular sheaths of *Leptothrix*. *Scale bar* = 30 μm [reprinted with permission from Fleming et al. (2014). Copyright, Nature Publishing Group]. (**b**) Phase contrast micrograph of twisted stalks growing out the iron-oxidizing, bean-shaped cells. *Scale bar* = 5 μm [reprinted with permission from Krepski et al. (2012). Copyright © 2011 Society for Applied Microbiology and Blackwell Publishing Ltd.]

9.4.2 Biomineralization by NIRB

Biomineralization by IRB is mostly extracellular via BIM mode. The metabolic product, i.e., Fe (II), forms extracellular fine-grained minerals mostly found embedded in the extrapolysaccharide (EPS) matrix. Unlike in NIOB, biomineralization has not yet found to influence the cell or their structure. However, the net negative surface charge of gram-negative NIRB allows the attachment of positive metal ions and serves as nucleation sites toward biogenic mineral formation. Moreover, these cells have high affinity to Fe minerals (goethite, hematite) encompassing the complete cell (Glasauer et al. 2001) (Fig. 9.5a).

Predominantly, biomineralization of Fe (II) results in the formation of magnetite (Fe_3O_4). Magnetites produced by the proteobacteria domain of NIRB are epicellular crystals or granules and are not aligned in chains (Bazylinski

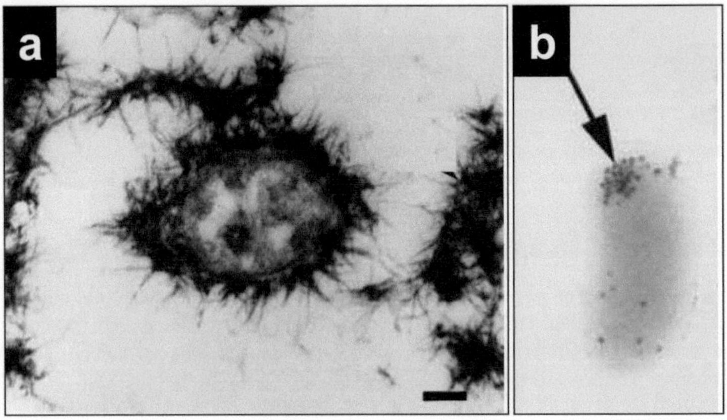

Fig. 9.5 Transmission electron micrograph of (**a**) crystalline, acicular goethite grains encompassing the bacteria. *Scale bar* = 50 nm [reprinted with permission from Konhauser (1997). Copyright © 2006, John Wiley and Sons]. (**b**) Intracellular magnetite crystals in *Shewanella putrefaciens* CN32 on transformation of HFO; *arrows* indicate intracellular granules. *Scale bar* = 1 μm [reprinted with permission from Glasauer et al. (2003). Copyright © 2003 Elsevier Science Ltd]

et al. 2007). However, intracellular magnetite crystals were found in *Shewanella putrefaciens* CN32 (Glasauer et al. 2003) on reduction of hydrous ferric oxide (HFO) mineral which is unusual in prokaryotes (Fig. 9.5b). The significant activity of the IRB can be traced back to the Precambrian age due to the presence of highly negative δ^{56} Fe values in magnetite-rich BIF (Bazylinski et al. 2007).

The aqueous Fe (II) released from Fe (III) reduction by IRB forms different biogenic Fe (II) minerals depending on the available anion. Some of the solid-phase biogenic Fe (II) minerals formed are vivianite ($Fe_3(PO_4)_2 \cdot 8H_2O$) with phosphate, siderite ($FeCO_3$) with carbonate, magnetite (Fe_3O_4) with oxygen, and a complex mineral green rust ($[Fe^{II}_{(6-x)}Fe^{III}_{(x)}(OH)_{12}]^{x+}[(A^{2-}_{x/2} \cdot yH_2O]^{x-}$). These biogenic minerals may either enhance bioreduction via lowering the thermodynamic activity of by-products or impede the bioreduction activity by passivating the Fe (III) or IRB surface (Zachara et al. 2002).

9.4.3 Magnetotactic Bacteria and Iron Mineralization

Magnetotactic bacteria as the name suggests are gram-negative bacteria that swim along the geomagnetic field lines of the earth. They form intracellular chains of magnetite crystals via both BCM and BIM mode of biomineralization (Blakemore and Blakemore 1991).

Magnetotactic bacteria also exist under microaerophilic conditions in the anaerobic/aerobic interfacial niches and utilize oxygen, nitrate, nitrous oxide, Fe (III), or sulfate as their terminal electron acceptor (Konhauser 1997). Magnetotactic bacteria produce magnetite in the oxic zone and greigite (Fe_3S_4) in the anoxic zone which is regulated by local oxygen and hydrogen sulfide availability (Bazylinski et al. 1995). Unicellular magnetotactic bacteria mineralize magnetite, but multicellular magnetotactic bacteria have been reported to mineralize mostly greigite as well as magnetite in bullet-shaped crystals (Lins et al. 2007).

NIRB have been reported to generate 5,000 times more magnetite than magnetotactic bacteria as magnetite is the end product of their energy-generating metabolism.

9.5 Application of Biogenic Iron Minerals

9.5.1 Iron Oxides as Bio-sorbents

Biogenic iron oxides formed by IOB serve as excellent bio-sorbents and find their application in filtration materials. Most biogenic iron oxides carry a net positive surface charge at neutral pH and hence serve as sorbents of negatively charged metal ions such as phosphate, arsenate, arsenite, chromate, etc. Biogenic iron oxides (BIOs) are found to have high capacity for phosphorus removal from wastewater effluents compared to other synthetic iron oxides and bio-sorbents (Rentz et al. 2009). BIO is an efficient bio-sorbent of heavy metals and radionuclides (Katsoyiannis 2007) and can serve as scavengers of inorganic pollutants (Seder-Colomina et al. 2014).

In addition to biomineralized iron oxides, the bacterial cell surface, sheaths, and filaments, extrapolysaccharides (EPSs) of the IOB can also directly adsorb the pollutants paving their way into the field of bioremediation to remove contaminants and heavy metals from polluted environments (Seder-Colomina et al. 2014). The high affinity of IRB cells to iron can serve as bio-sorbents of Fe from iron-contaminated sites and water (Glasauer et al. 2001) (Table 9.5).

9.5.2 Sequestration of Heavy Metals in Biogenic Iron Oxides

Both Fe (II) oxidation and Fe (III) reduction have been reported to sequester heavy metals via coprecipitation of biogenic Fe (III) oxyhydroxide minerals (Hohmann et al. 2010) leading to immobilization of toxic metal ions. Anaerobic Fe (II) oxidation by *D. suillum* was found capable of immobilizing radionuclide and heavy metals like uranium and cobalt, respectively (Lack et al. 2002). The chief heavy metal contaminant of water, i.e., arsenic, can be strategically immobilized by Fe (II)-oxidizing bacteria on its coprecipitation with the Fe (III) mineral, thus finding its application toward purification of water (Hohmann et al. 2010).

Table 9.5 List of Fe (II)-oxidizing, Fe (III)-reducing bacteria as metal bio-sorbents

Bacteria	Sorbent	Metal	References
Leptothrix and/or *Gallionella*	Biogenic iron oxide	Arsenic	Keim (2011)
		Strontium	Langley et al. (2009)
		Cesium, lead, strontium, uranium	Ferris et al. (2000)
		Iodide	Kennedy et al. 2011
Gallionella	Biogenic iron oxide	Cadmium	Martinez and Ferris (2005)
		Uranium	Katsoyiannis (2007)
Shewanella algae	Iron oxide-coated cells	Strontium	Small et al. (2001)
Sphaerotilus natans	Magnetite-coated cells	Lead, chromium	Zhao et al. (2007)
	Bacterial cell	Chromium (III)	Caravelli et al. (2008)
		Iron	Shopska et al. (2013)

Fe (III) reduction of goethite by *Shewanella putrefaciens* 200 led to immobilization of zinc by its incorporation into siderite or more crystalline goethite. Similar immobilization of Zn was also observed during reduction of lepidocrocite to magnetite (Cooper et al. 2000). Strontium was found immobilized in the reduced Fe (II) mineral siderite on reduction of hydrous ferric oxide by *Shewanella putrefaciens* CN32 (Roden et al. 2002).

9.5.3 Reduction of Contaminants

The continuous redox cycling of iron produces a constant flow of Fe (II) from its aqueous state to solid state which majorly gets adsorbed to the Fe (III) minerals. Surface-bound Fe (II) has been illustrated to be highly reactive than in the aqueous state. The oxidation of the surface-bound Fe (II) was found to transform numerous reducible organic contaminants such as nitroaromatic and chlorinated compounds, oxamyl and carbamate pesticides, and polyhalogenated methanes (Hofstetter et al. 1999; Kenneke and Weber 2003; Strathmann and Stone 2009) and inorganic contaminants like U (VI), Cr (VI), or Tc (VI) (Buerge and Hug 1999; Cui and Eriksen 1996; Liger et al. 1999). Continuous regeneration of the surface-bound Fe (II) by means of microbial Fe (III) reduction can contribute to sustained reductive transformation of these contaminants in polluted environments.

In addition to reactive surface-bound Fe (II), the reduced palladium nanoparticles adsorbed on the cell surface of *Shewanella oneidensis* were found capable of reductive dechlorination of polychlorinated biphenyls (De Windt et al. 2005).

9.5.4 *Extraction of Metals via Reductive Dissolution of Iron Oxides*

Microbial Fe (III) reduction of amorphous and crystalline iron oxides by IRB may release the metal cations which were coprecipitated or incorporated structurally into the environment (Zachara et al. 2001). Hence, microbial reductive dissolution of metal-containing iron oxides was found to effectively extract metals of value such as gold (Kashefi et al. 2001), nickel–cobalt (Esther et al. 2013), etc., finding their application in the field of biomineral processing (Sukla et al. 2014). IRB find their profitable use in beneficiation of kaolin (Guo et al. 2010) and bauxite (Papassiopi et al. 2010) for removal of iron impurities via reductive dissolution of the Fe (III).

9.5.5 *Putrefaction of Organic Compounds*

Various aromatic organic compounds are significant pollutants of the environment. These organic compounds are easily degraded on oxidation by the IRB coupled to Fe (III) reduction. IRB can degrade the numerous organic pollutants such as phenol, benzene, toluene, *p*-cresol, etc. (Jahn et al. 2005; Lovley and Anderson 2000).

9.6 Conclusion

Diverse neutrophilic IOB and IRB belonging to various domains are found to play a closely associated role in iron mineral transformations of interfacial environments having circumneutral pH. In addition, the iron mineral-transforming potential of neutrophilic bacteria has made significant contributions in various fields such as bioremediation, biosorption, biomining, etc. However, extensive research toward exploitation of prospective neutrophilic bacteria and the mechanisms involved therein for application in distinct fields is an urgent need of the day to overcome the mounting rate of pollution caused by anthropogenic activities.

References

Bazylinski DA, Frankel RB, Heywood BR, Mann S, King JW, Donaghay PL, Hanson AK (1995) Controlled biomineralization of magnetite and greigite in a magnetotactic bacterium. Appl Environ Microbiol 61:3232–3239

Bazylinski DA, Frankel RB, Konhauser KO (2007) Modes of Biomineralization of Magnetite by Microbes. Geomicrobiol J 24:465–475

Blakemore RP, Blakemore NA (1991) Magnetotactic magnetogens. In: Frankel RB, Blakemore RP (eds) Iron biominerals. Plenum, New York, pp 51–67

Blöthe M, Roden EE (2009) Microbial iron redox cycling in a circumneutral-pH groundwater seep. Appl Environ Microbiol 75:468–473

Buerge IJ, Hug SJ (1999) Influence of mineral surfaces on chromium (VI) reduction by iron (II). Environ Sci Technol 33:4285–4291

Cairns-Smith AG (1978) Precambrian solution photochemistry, inverse segregation, and banded iron formation. Nature 276:807–808

Caravelli AH, Giannuzzi L, Zaritzky E (2008) Reduction of hexavalent chromium by *Sphaerotilus natans* a filamentous micro-organism present in activated sludges. J Hazard Mater 156:214–222

Chan CS, De Stasio G, Welch SA, Girasole M, Frazer BH, Nesterova MV, Fakra S, Banfield JF (2004) Microbial polysaccharides template assembly of nanocrystal fibers. Science 303:1656–1658

Chaudhuri SK, Lack JG, Coates JD (2001) Biogenic magnetite formation through anaerobic biooxidation of Fe (II). Appl Environ Microbiol 67:2844–2848

Cooper DC, Picardal F, Rivera J, Talbot C (2000) Zinc immobilization and magnetite formation via ferric oxide reduction by Shewanella putrefaciens 200. Environ Sci Technol 34:100–106

Croal LR, Gralnick JA, Malasarn D, Newman DK (2004) The genetics of geochemistry. Ann Rev Genet 38:175–202

Crowe SA, Roberts JA, Weisener CG, Fowle DA (2007) Alteration of iron-rich lacustrine sediments by dissimilatory iron-reducing bacteria. Geobiology 5:63–73

Cui D, Eriksen TE (1996) Reduction of pertechnetate by ferrous iron in solution: influence of sorbed and precipitated Fe(II). Environ Sci Technol 30:2259–2262

De Windt W, Aelterman P, Verstraete W (2005) Bioreductive deposition of palladium (0) nanoparticles on Shewanella oneidensis with catalytic activity towards reductive dechlorination of polychlorinated biphenyls. Environ Microbiol 7:314–325

Dubinina GA, Sorokina AY (2014) Neutrophilic lithotrophic iron oxidizing prokaryotes and their role in the biogeochemical processes of the iron cycle. Microbiology 8:1–14

Edwards KJ, Rogers DR, Wirsen CO, McCollom TM (2003) Isolation and characterization of novel psychrophilic, neutrophilic, Fe-oxidizing, chemolithoautotrophic α- and γ-*Proteobacteria* from the deep sea. Appl Environ Microbiol 69:2906–2913

Ehrenreich A, Widdel F (1994) Anaerobic oxidation of ferrous iron by purple bacteria, a new type of phototrophic metabolism. Appl Environ Microbiol 60:4517–4526

Emerson D, Moyer C (1997) Isolation and characterization of novel iron-oxidizing bacteria that grow at circumneutral pH. Appl Environ Microbiol 63:4784–4792

Emerson D, Weiss JV (2004) Bacterial iron oxidation in circumneutral freshwater habitats: findings from the field and the laboratory. Geomicrobiol J 21:405–414

Emerson D, Weiss JV, Megonigal JP (1999) Iron-oxidizing bacteria are associated with ferric hydroxide precipitates (Fe-plaque) on the roots of wetland plants. Appl Environ Microbiol 65:2758–2761

Emerson D, Rentz JA, Lilburn TG, Davis RE, Aldrich H, Chan C, Moyer CL (2007) A novel lineage of proteobacteria involved in formation of marine Fe-oxidizing microbial mat communities. PLoS One 2, e667

Esther J, Panda S, Behera SK, Sukla LB, Pradhan N, Mishra BK (2013) Effect of dissimilatory Fe (III) reducers on bio-reduction and nickel–cobalt recovery from Sukinda chromite-overburden. Bioresour Technol 146:762–766

Esther J, Sukla LB, Pradhan N, Panda S (2014) Fe (III) reduction strategies of dissimilatory iron reducing bacteria: a review. Korean J Chem Eng. doi:10.1007/s11814-014-0286-x

Ferris F, Hallberg R, Lyvén B, Pedersen K (2000) Retention of strontium, cesium, lead and uranium by bacterial iron oxides from a subterranean environment. Appl Geochem 15:1035–1042

Finneran K, Anderson R, Nevin K, Lovley DR (2002) Potential for bioremediation of uranium-contaminated aquifers with microbial U(VI) reduction. Soil Sediment Contam 11:339–357

Fleming EJ, Davis RE, McAllister SM, Chan CS, Moyer CL, Tebo BM, Emerson D (2013) Hidden in plain sight: discovery of sheath-forming, iron-oxidizing *Zetaproteobacteria* at Loihi Seamount, Hawaii, USA. FEMS Microbiol Ecol 85:116–127

Fleming EJ, Cetinic I, Chan CS, King DW, Emerson D (2014) Ecological succession among iron-oxidizing Bacteria. ISME J 8:804–815

Fortin D, Langley S (2005) Formation and occurrence of biogenic iron-rich minerals. Earth Sci Rev 72:1–19

Francois LM (1986) Extensive deposition of banded iron formations was possible without photosynthesis. Nature 320:352–354

Frankel RB (1991) Iron biominerals: an overview. In: Blakemore R, Frankel R (eds) Iron biominerals. Plenum, New York, pp 1–6

Frankel RB, Bazylinski DA, Johnson MS, Taylor BL (1997) Magneto-aerotaxis in marine coccoid bacteria. Biophys J 73:994–1000

Ghiorse WC, Chapnick SD (1983) Metal-depositing bacteria and the distribution of manganese and iron in swamp waters. Environ Biogeochem Ecol Bull (Stockh) 35:367–376

Glasauer S, Langley S, Beveridge TJ (2001) Sorption of Fe (hydr)oxides to the surface of *Shewanella putrefaciens*: cell-bound fine-grained minerals are not always formed de novo. Appl Environ Microbiol 67:5544–5550

Glasauer S, Weidler PG, Langley S, Beveridge TJ (2003) Controls on Fe reduction and mineral formation by a subsurface bacterium. Geochim Cosmochim Acta 67:1277–1288

Glazer BT, Rouxel OJ (2009) Redox speciation and distribution within diverse iron-dominated microbial habitats at Loihi Seamount. Geomicrobiol J 26:606–622

Gralnick JA, Newman DK (2007) Extracellular respiration. Mol Microbiol 65:1–11

Guo M, He Q, Li Y, Lu X, Chen Z (2010) Removal of Fe from kaolin using dissimilatory Fe (III)-reducing bacteria. Clays Clay Miner 58:515–521

Hafenbradl D, Keller M, Dirmeier R, Rachel R, Rosnagel P, Burggraf S, Huber H, Stetter KO (1996) *Ferroglobus placidus* gen nov., sp. nov., a novel hyperthermophilic archaeum that oxidizes Fe^{2+} at neutral pH under anoxic conditions. Arch Microbiol 166:308–314

Hallbeck L, Pedersen K (1990) Culture parameters regulating stalk formation and growth-rate of *Gallionella ferruginea*. J Gen Microbiol 136:1675–1680

Hallberg R, Ferris FG (2004) Biomineralization by Gallionella. Geomicrobiol J 21:325–330

Hedrich S, Schlömann M, Johnson DB (2011) The iron-oxidizing proteobacteria. Microbiology 157:1551–1564

Hegler F, Posth NR, Jiang J, Kappler A (2008) Physiology of phototrophic iron (II)-oxidizing bacteria: implications for modern and ancient environments. FEMS Microbiol Ecol 66:250–260

Hegler F, Schmidt C, Schwarz H, Kappler A (2010) Does a low-pH microenvironment around phototrophic FeII oxidizing bacteria prevent cell encrustation by FeIII minerals? FEMS Microbiol Ecol 74:592–600

Heising S, Schink B (1998) Phototrophic oxidation of ferrous iron by a *Rhodomicrobium vannielii* strain. Microbiology 144:2263–2269

Heising S, Richter L, Ludwig W, Schink B (1999) *Chlorobium ferrooxidans* sp. nov., a phototrophic green sulfur bacterium that oxidizes ferrous iron in coculture with a '*Geospirillum*' sp. strain. Arch Microbiol 172:116–124

Hofstetter T, Heijman CG, Haderlein SB, Holliger C, Schwarzenbach RP (1999) Complete reduction of TNT and other (poly)nitroaromatic compounds under iron-reducing subsurface conditions. Environ Sci Technol 33:1479–1487

Hohmann C, Winkler E, Morin G, Kappler A (2010) Anaerobic Fe(II)-oxidizing bacteria show as resistance and immobilize As during Fe(III) mineral precipitation. Environ Sci Technol 44:94–101

Jahn MK, Haderlein SB, Meckenstock RU (2005) Anaerobic degradation of benzene, toluene, ethylbenzene, and o-xylene in sediment-free iron-reducing enrichment cultures. Appl Environ Microbiol 71:3355–3358

James RE, Ferris FG (2004) Evidence for microbial-mediated iron oxidation at a neutrophilic groundwater spring. Chem Geol 212:301–311

Jiao Y, Kappler A, Croal L, Newman DK (2005) Isolation and characterization of a genetically tractable photoautotrophic Fe(II)-oxidizing bacterium *Rhodopseudomonas palustris* strain TIE1. Appl Environ Microbiol 71:4487–4496

Kappler A, Straub KL (2005) Geomicrobiological cycling of iron. Rev Miner Geochem 59:85–108

Kappler A, Pasquero C, Konhauser KO, Newman DK (2005a) Deposition of banded iron formations by anoxygenic phototrophic Fe(II)-oxidizing bacteria. Geology 33:865–868

Kappler A, Schink B, Newman DK (2005b) Fe (III) mineral formation and cell encrustation by the nitrate-dependent Fe (II)-oxidizer strain BoFeN1. Geobiology 3:235–245

Kashefi K, Tor JM, Nevin KP, Lovley DR (2001) Reductive precipitation of gold by dissimilatory Fe (III)-reducing *bacteria* and *archaea*. Appl Environ Microbiol 67:3275–3279

Katsoyiannis IA (2007) Carbonate effects and pH-dependence of uranium sorption onto bacteriogenic iron oxides: kinetic and equilibrium studies. J Hazard Mater 139:31–37

Keim CN (2011) Arsenic in biogenic iron minerals from a contaminated environment. Geomicrobiol J 28:242–251

Kennedy CB, Gault AG, Fortin D, Clark ID, Ferris FG (2011) Retention of iodide by bacteriogenic iron oxides. Geomicrobiol J 28:387–395

Kenneke JF, Weber EJ (2003) Reductive dehalogenation of halomethanes in iron- and sulfate-reducing sediments. 1. Reactivity pattern analysis. Environ Sci Technol 37:713–720

Konhauser KO (1997) Bacterial Iron biomineralization in nature. FEMS Microbiol Rev 20:315–326

Konhauser KO, Hamade T, Raiswell R, Morris RC, Ferris FG, Southam G, Canfield DE (2002) Could bacteria have formed the Precambrian banded iron formations? Geology 30:1079–1082

Krepski ST, Hanson TE, Chan CS (2012) Isolation and characterization of a novel biomineral stalk-forming iron-oxidizing bacterium from a circumneutral groundwater seep. Environ Microbiol 14:1671–1680

Kumaraswamy R, Sjollema K, Kuenen G, van Loosdrecht M, Muyzer G (2006) Nitrate-dependent [Fe(II)EDTA]2- oxidation by Paracoccus ferrooxidans sp. nov., isolated from a denitrifying bioreactor. Syst Appl Microbiol 29:276–286

Lack JG, Chaudhuri SK, Kelly SD, Kemner KM, O'Connor SM, Coates JD (2002) Immobilization of radionuclides and heavy metals through anaerobic bio-oxidation of Fe (II). Appl Environ Microbiol 68:2704–2710

Langley S, Gault AG, Ibrahim A, Takahashi Y, Renaud R, Fortin D, Clark ID, Ferris FG (2009) Sorption of strontium onto bacteriogenic iron oxides. Environ Sci Technol 43:1008–1014

Liger E, Charlet L, Van Capellen P (1999) Surface catalysis of uranium (VI) reduction by iron (II). Geochim Cosmochim Acta 63:2939–2955

Lins U, Keim CN, Evans FF, Farina M, Buseck PR (2007) Magnetite (Fe_3O_4) and greigite (Fe_3S_4) crystals in multicellular magnetotactic prokaryotes. Geomicrobiol J 24:43–50

Lovley DR, Anderson RT (2000) The influence of dissimilatory metal reduction on the fate of organic and metal contaminants in the subsurface. J Hydrol 238:77–88

Lowenstam HA (1981) Minerals formed by organisms. Science 211:1126–1131

Martinez RE, Ferris FG (2005) Review of the surface chemical heterogeneity of bacteriogenic iron oxides: proton and cadmium sorption. Am J Sci 305:854–871

McAllister SM, Davis RE, McBeth JM, Tebo BM, Emerson D, Moyer CL (2011) Biodiversity and emerging biogeography of the neutrophilic iron-oxidizing Zetaproteobacteria. Appl Environ Microbiol 77:5445

Miot J, Benzerara K, Obst M, Kappler A, Hegler F, Schädler S, Bouchez C, Guyot F, Morin G (2009) Extracellular iron biomineralization by photoautotrophic iron-oxidizing bacteria. Appl Environ Microbiol 75:5586–5591

Morita M, Malvankar NS, Franks AE, Summers ZM, Giloteaux L, Rotaru AE, Rotaru C, Lovley DR (2011) Potential for direct interspecies electron transfer in methanogenic wastewater digester aggregates. mBio 2:e00159-11

Papassiopi N, Vaxevanidou K, Paspaliaris I (2010) Effectiveness of iron reducing bacteria for the removal of iron from bauxite ores. Miner Eng 23:25–31

Pellegrin V, Juretschko S (1999) Morphological and biochemical properties of a Sphaerotilus sp. isolated from paper mill slimes. Appl Environ Microbiol 65:156–162

Posth NR, Konhauser KO, Kappler A (2013) Microbiological processes in banded iron formation deposition. Sedimentology 60:1733–1754

Rentz JA, Turner IP, Ullman JL (2009) Removal of phosphorus from solution using biogenic iron oxides. Water Res 43:2029–2035

Roden EE, Leonardo MR, Ferris FG (2002) Immobilization of strontium during iron biomineralization coupled to dissimilatory hydrous ferric oxide reduction. Geochim Cosmochim Acta 66:2823–2839

Saini G, Chan CS (2013) Near-neutral surface charge and hydrophilicity prevent mineral encrustation of Fe-oxidizing micro-organisms. Geobiology 11:191–200

Schädler S, Burkhardt C, Hegler F, Straub KL, Miot J, Benzerara K, Kappler A (2009) Formation of cell-iron-mineral aggregates by phototrophic and nitrate-reducing anaerobic Fe(II)-oxidizing bacteria. Geomicrobiol J 26:93–103

Seder-Colomina M, Morin G, Benzerara K, Ona-Nguema G, Pernelle J, Esposito G, Van Hullebusch ED (2014) Sphaerotilus natans, a neutrophilic iron-related sheath-forming bacterium: perspectives for metal remediation strategies. Geomicrobiol J 31:64–75

Shelobolina ES, Vanpraagh CG, Lovley DR (2003) Use of ferric and ferrous iron containing minerals for respiration by *Desulfitobacterium frappieri*. Geomicrobiol J 20:143–156

Shopska M, Cherkezova-Zheleva ZP, Paneva DG, Iliev M, Kadinov GB, Mitov IG, Groudeva VI (2013) Biogenic iron compounds: XRD, Mossbauer and FTIR study. Cent Eur J Chem 11:215–227

Small TD, Warren LA, Ferris FG (2001) Influence of ionic strength on strontium sorption to bacteria, Fe (III) oxide, and composite bacteria-Fe (III) oxide surfaces. Appl Geochem 16:939–946

Sobolev D, Roden EE (2001) Suboxic deposition of ferric iron by bacteria in opposing gradients of Fe(II) and oxygen at circumneutral pH. Appl Environ Microbiol 67:1328–1334

Sobolev D, Roden EE (2004) Characterization of a neutrophilic, chemolithoautotrophic Fe(II)-oxidizing β-proteobacterium from freshwater wetland sediments. Geomicrobiol J 21:1–10

Strathmann TJ, Stone AT (2009) In: 219th ACS national meeting, vol 40, Division of Environmental Chemistry, American Chemical Society, Washington, DC, pp 141–144

Straub KL, Benz M, Schink B, Widdel F (1996) Anaerobic, nitrate-dependent microbial oxidation of ferrous iron. Appl Environ Microbiol 62:1458–1460

Straub KL, Rainey FA, Widdel F (1999) *Rhodovulum iodosum* sp. nov. and *Rhodovulum robiginosum* sp. nov., two new marine phototrophic ferrous-iron-oxidizing purple bacteria. Int J Syst Bacteriol 49:729–735

Straub KL, Benz M, Schink B (2001) Iron metabolism in anoxic environments at near neutral pH. FEMS Microbiol Ecol 34:181–186

Sukla LB, Esther J, Pradhan N, Panda S (2014) Biomineral processing: a valid eco-friendly alternative for metal extraction. Res Rev J Microbiol Biotechnol 3:1–10

Sun Z, Zhou H, Glasby GP, Sun Z, Yang Q, Yin X, Li J (2013) Mineralogical characterization and formation of Fe-Si oxyhydroxide deposits from modern seafloor hydrothermal vents. Am Mineral 98:85–97

Toner BM, Santelli CM, Marcus MA, Wirth R, Chan CS, McCollom T, Bach W, Edwards KJ (2009) Biogenic iron oxyhydroxide formation at mid-ocean ridge hydrothermal vents: Juan de Fuca Ridge. Geochim Cosmochim Acta 73:388–403

Trouwborst RE, Johnston A, Koch G, Luther GW III, Pierson BK (2007) Biogeochemistry of Fe (II) oxidation in a photosynthetic microbial mat: implications for Precambrian Fe(II) oxidation. Geochim Cosmochim Acta 71:4629–4643

Vatter AE, Wolfe RS (1956) Electron microscopy of *Gallionella ferruginea*. J Bacteriol 72:248–252

Weber KA, Hedrick DB, Peacock AD, Thrash JC, White DC, Achenbach LA, Coates JD (2009) Physiological and taxonomic description of the novel autotrophic, metal oxidizing bacterium, Pseudogulbenkiania sp. strain 2002. Appl Microbiol Biotechnol 83:555–565

Weiss JV, Rentz JA, Plaia T, Neubauer SC, Merrill-Floyd M, Lilburn T, Bradburne C, Megonigal JP, Emerson D (2007) Characterization of neutrophilic Fe (II)-oxidizing bacteria isolated from the rhizosphere of wetland plants and description of *Ferritrophicum radicicola* gen. nov. sp. nov., and Sideroxydans paludicola sp. nov. Geomicrobiol J 24:559–570

Widdel F, Schnell S, Heising S, Ehrenreich A, Assmus B, Schink B (1993) Ferrous iron oxidation by anoxygenic phototrophic bacteria. Nature 362:834–836

Zachara JM, Fredrickson JK, Smith SC, Gassman PL (2001) Solubilization of Fe (III) oxide-bound trace metals by a dissimilatory Fe (III) reducing bacterium. Geochim Cosmochim Acta 65:75–93

Zachara JM, Kukkadapu RK, Fredrickson JK, Gorby YA, Smith SC (2002) Biomineralization of poorly crystalline Fe(III) oxides by dissimilatory metal reducing bacteria (DMRB). Geomicrobiol J 19:179–207

Zegeye A, Mustin C, Jorand F (2010) Bacterial and iron oxide aggregates mediate secondary iron mineral formation: green rust versus magnetite. Geobiology 8:209–222

Zhao J, Guan X, Unuma H (2007) Preparation of Fe_3O_4 nanoparticles and their application to composite biosorbent. J Ceram Soc Jpn 115:475–478

Zhu YG, Wang XJ, Yang J, Chen XP, Sun GX (2009) Phylogenetic diversity of dissimilatory ferric iron reducers in paddy soil of Hunan, South China. J Soils Sediment 9:568–577

Chapter 10
Anaerobic Bioleaching by Acidophilic Bacterial Strains

Sradha Singh and Swaranjit Singh Cameotra

10.1 Introduction

Use of microbially based processes for the recovery of metals from ores, concentrates and wastes is a well-known and established process. Unlike other processes, it has several applications in mining industries. However, it comprises two major areas; chemistry and microbiology, and was put into a board subject area called biohydrometallurgy. Nevertheless, the exact mechanism governing the entire process is still under debate. Speculations, assumptions and theories have been proposed by many workers from time to time, but coincidently the question still remains unanswered.

Nonetheless, it is well understood that the leaching reaction is primarily accountable by a chemical process involving the action of ferric ions and protons. This also depends on the nature and composition of the ore mineralogy crucial for the process to occur. Very often it relies in the realm of chemistry rather than microbiology. This does not imply a negative or less role of microbiology rather required to generate or regenerate the ions such as Fe^{3+} and H^{+} ions for continuity of the process. Therefore, both the processes, i.e. chemical and biological, are interrelated and dependable. Hence, it is worthwhile to consider both the process for an efficient metal recovery from the ore matrix.

The most important role played by the microorganisms in solubilisation of metals in the process is their ability to oxidise iron- and sulphur-containing minerals. In the last decade, the role of several bacteria and archaea in mineral oxidation process has become clear. Microbial identification and their ecology study will bring out the core information needed to understand their nature and behaviour. Not only this, it certainly answered the unsolved puzzle regarding how and why they do so?

S. Singh • S.S. Cameotra (✉)
Institute of Microbial Technology-CSIR, Sector 39A, Chandigarh 160036, India
e-mail: ssc@imtech.res.in

L.B. Sukla et al. (eds.), *Environmental Microbial Biotechnology*, Soil Biology 45,
DOI 10.1007/978-3-319-19018-1_10

Since man's arrival on earth, he has been continuously exploiting the natural resources for his own benefit and well-being. Excessive population growth and developmental activities are the two major causes responsible for this exploitation. Due to uncontrolled use, the nonrenewable sources are on the verge of exhaustion, and several environmental problems are becoming increasingly serious inviting global effect to develop cleaner techniques based on renewable natural resources. Future sustainable development requires measures to reduce the dependence on nonrenewable raw materials and the demand for primary resources. New resources for metals must be developed with the aid of novel technologies. Biotechnology can help in finding a solution to this problem. Like many technologies, biological-based science took a long time to cross the boundary into the mineral processing industry and did not gain wide acceptance until the later part of the twentieth century. In context of **mineral biotechnology**, **biohydrometallurgy** is a promising approach towards replacing the conventional methods and covers two broad areas of microbial activity—(1) biosorption and (2) redox reactions. It involves metal–microbe interaction commonly referred to as **bioleaching** in which naturally occurring microorganisms bring about metal extraction from ores and concentrates in the presence of air and nutrients.

Biohydrometallurgy has been developed as a discipline and an alternative process for extraction of metals like copper, uranium, gold, nickel, cobalt, etc., and offers many innovative and cost-effective solutions in particular to the extraction of low-grade ore (Tang and Valix). Biological activity related to mineral dissolution has long been known to occur in nature and in fact is primarily responsible for acid mine drainage (AMD). The earliest exploitation of biological systems by the metallurgical industry was in the leaching of uranium in the 1950s. Natural populations of acidophilic iron- and sulphur-oxidising bacteria that are readily isolated from mine waters aid these processes. Until late 1980s this process was exploited for the treatment of refractory gold ores and in past few years, it has been commercialized for laeching of base metals such as copper, zinc, nickel, cobalt etc.

In view of the present scenario, biohydrometallurgy is considered as one of the most promising and revolutionary solutions for the treatment of wastes, overburdens and low-grade ores. The importance of biohydrometallurgy is increasing day by day due to the following two factors:

(a) Depletion of high-grade ore
(b) Enforcement of strict antipollution laws

Almost all microorganisms interact with metals either directly or indirectly. Bioleaching is the first and primary component of biogeotechnology. Leaching can also be explained as the solubilisation of one or more components of a complex solid by contact with a liquid phase. The potential benefits of bioleaching process for treating ores and concentrates can be summarised as follows:

Advantages of bioleaching process

(a) Low capital investment.

(b) A reduced amount of energy consumption.

- The relative absence of air, water, land pollution.

(c) The process can be applied for low/complex ores.

(i) Degradation of a variety of mineral forms
(ii) Selective leaching possible, one mineral solubilised while the other remains insoluble
(iii) Low energy requirement and operating costs as compared to conventional process for recovery of metals from low-grade ores

(d) Zero discharge, i.e. recycling of effluents.
(e) The process is site specific.
(f) The microorganisms are available indigenously.
(g) A reduced amount of technical sophistication.

- Simple technology which is easy to operate and maintain in heap, dump or bioreactors (single or multiple).

(h) The process is safe and conforms to nature.

(i) Free from gaseous and dust emission. Can handle a variety of simple and complex materials of low grade as well as concentrates.
(ii) The processes are more environmentally friendly than traditional extraction methods; for the company, this can translate into profit, since necessary limiting of sulphur dioxide emissions during smelting is expensive.
(iii) Application of most useful is desulphurisation of coal for burning is free of sulphur and further pollution problem avoidable.

(i) A reduced amount of process control.
(j) Operate at ambient temperature and normal atmospheric pressure.
(k) Suitable for less-developed countries as it eliminates the need for some costly imported heavy mining equipment.

There are also some disadvantages of the process:

1. The bacterial leaching process is very slow compared to smelting.
2. This brings less profit as well as introducing a significant delay in case flow for new plants.
3. Toxic chemicals are sometimes produced in this process.
4. In situ application still under development.
5. Control difficult.
6. More possible for acid-producing minerals and not acid-consuming minerals.

10.1.1 Pyrometallurgy Versus Hydrometallurgy

There are various commercial methods involved in the processing of ores and concentrates from the ore bodies often containing more than one valuable metal.

10.1.1.1 Pyrometallurgy

Pyrometallurgy, or the use of heat for the treatment, includes smelting and roasting.

It involves heating in a blast furnace at temperatures above 1,500 °C to convert waste to a form that can be refined. The oxide waste is heated with a reducing agent, such as carbon in the form of coke or coal; the oxygen of the metal combines with the carbon and is removed in carbon dioxide gas. The waste material in e-waste (nonmetallic parts) is called gangue; it is removed by means of a substance called a flux which, when heated, combines with it to form a molten mass called slag. Being lighter than the metal, the slag floats on it and can be skimmed or drawn off.

10.1.1.2 Hydrometallurgy

Hydrometallurgy is sometimes called leaching.

It involves the selective dissolution of metals from their waste. It involves the use of aqueous chemicals and much lower temperatures to separate metal. Metal is recovered by electrolysis of the solution. If metal obtained from waste still contains impurities, special refining processes are required.

10.1.1.3 Electrometallurgy

It involves metallurgical processes that take place in some form of electrolytic cell. The most common types of electrometallurgical processes are electrowinning and electrorefining. Electrowinning is an electrolysis process used to recover metals in aqueous solution, usually as the result of an ore having undergone one or more hydrometallurgical processes. The metal of interest is plated onto the cathode, while the anode is an inert electrical conductor. Electrorefining is used to dissolve an impure metallic anode (typically from a smelting process) and produce a high purity cathode. Fused salt electrolysis is another electrometallurgical process whereby the valuable metal has been dissolved into a molten salt which acts as the electrolyte, and the valuable metal collects on the cathode of the cell. The fused salt electrolysis process is conducted at temperatures sufficient to keep both the electrolyte and the metal being produced in the molten state. The scope of electrometallurgy has significant overlap with the areas of hydrometallurgy and (in the case of fused salt electrolysis) pyrometallurgy. Additionally, electrochemical phenomena play a considerable role in many mineral processing and hydrometallurgical processes.

The decision whether to use hydrometallurgy or pyrometallurgy can be seen from various concerns including environment and economy. Low waste technology solutions provide a real answer to the increasing requirements of environmental legislation. Instead of waste treatment and waste disposal, new technologies for waste avoidance are a challenge today. Such technologies should meet the demand for economical use of raw materials and an energy-poor future and contribute to efficiency in industry. Environmental regulations will be more restrictive in the future. The expenditure for waste disposal and water use as well as energy, materials and labour will increase, and internal recycling processes will become indispensable tools for successful industrial operations. A low waste technology solution should always be preferred as it minimises the threat of environmental legislation.

Low waste technology solutions can be directed towards material recycling and heat recovery which contributes to economy. Chemical treatments improve on minimising impurity buildup, and toxic compounds may be substitutable for less harmful constituents.

10.1.2 Environment Impact

10.1.2.1 Gas Emission

Pyrometallurgy—The basic operation is the direct introduction of e-waste into a furnace mixed with a reducer and smelting agent. This operation is accompanied with strong gas emissions including:

- CO_2–CO coming from oxidation of carbon used as the reducer
- Dust of scrap metals and other components
- Green house effect gases like SO_2, Cl_2, HCl and NOx
- Organic volatile compounds
- Dioxins

The burning of waste in the presence of oxygen, especially waste with plastic and other organic material content, generates toxic gases such as furans and dioxins. Such gases are carcinogenic and increase the risk of contracting respiratory disease. Incinerators have been found to be the largest producers of dioxins and furans. Incineration produces ash with concentrated amounts of heavy metals, such as lead, arsenic and cadmium. These chemicals are well known to cause birth defects, cancer, respiratory ailments and reproductive dysfunction among people who live near incineration plants. Beside this, incomplete combustion may generate carbon monoxide and also volatile compounds, including formaldehyde and acetaldehyde. The treatment of these gases involves large capital investments in advanced technologies and equipment.

The amount of sulphur dioxide released depends:

- On the characteristics of the type of waste—complex ones may contain lead, zinc and nickel
- Whether facilities are in place for capturing and converting the sulphur dioxide

SO_2 emissions may range from less than 4 kilograms per metric ton (kg/t) of copper to 2,000 kg/t of copper. Particulate emissions can range from 0.1 kg/t of copper to as high as 20 kg/t of copper. Fugitive emissions occur at furnace openings and from launders, casting moulds and ladle carrying molten materials. Additional fugitive particulate emissions occur from materials handling and transport of ores and concentrates. The smelting furnace will generate process gas streams with SO_2 concentrations ranging from 0.5 to 80 %, depending on the process used. Vapours of arsenic and mercury are also present at high gas temperatures, which required additional scrubbing for removal. Modern plants using good industrial practices should set as targets total dust releases of 0.5–1.0 kg/t of copper and SO_2 discharges of 25 kg/t of copper.

Hydrometallurgy generates some hazardous gases such as chlorine, noxious and hydrogen cyanide gases which is possible to be treated by a simple 1–3 stage scrubber system with a chemical scrubbing solution. In contrast to a furnace process, chemical process also generates wastewater. However, as the treatment of these gases and wastewater utilises common established technology, its efficiency can be justified with much lower capital investment. No gases escape and solvents are fully trapped at room temperature, where it is not in position to produce dioxins or other greenhouse effects. Hydrometallurgy is more environmentally friendly also as sulphur is presented as either a stable sulphate or elemental sulphur rather than sulphur dioxide emissions. There are global environmental concerns to smelting activities, which spew extremely harmful pollutants into the atmosphere. Smelting with pollution control equipment is extremely expensive, which contributes directly to the high cost.

10.1.2.2 Solid Waste Generation

In pyrometallurgy, almost all waste content is burnt to ashes or carbon and leaving behind also a mixture of heavy metals. Useful materials such as plastics, which might otherwise be further recycled into re-engineering plastic are also being burnt (this is in the event that feed materials did not go through initial mechanical separation stage). Other lesser important content such as paper, ceramics, glass and fibres which could also be reused as filler or flux in certain products is also non-recoverable. It is estimated that for every three tonnes of waste that is incinerated, one ton of ash is generated. The main portion of the solid waste is discarded slag from the smelter. This ash is very toxic, containing concentrated amounts of heavy metals and dioxins which, when buried, will eventually leach into the soil, potentially polluting groundwater.

The end product of ashes or carbon eventually ends up in landfill, while the mixture of heavy metals undergoes further segregation and refining via chemical process or smelting process. The heavy metals mixture recovered may be covered or trapped within carbon residue, which makes it more complicated for downstream refining. Slag requires special treatment, e.g. slow cooling, grinding, and flotation or treatment in an electric furnace, to recover its copper content which represents a heavy recycling load.

By using hydrometallurgy, almost all waste components (not only heavy metals) could be segregated and recovered for further recycling or reuse. Each component refining stage could be accomplished in one process, without the need for diversion to another process. Leaching processes produce residues, while effluent treatment results in sludges which can be sent for metals recovery.

10.1.2.3 Wastewater Discharge

Wastewater from primary copper production contains dissolved and suspended solids that may include concentrations of copper, lead, cadmium, zinc, arsenic and mercury and residues from mould release agents (lime or aluminium oxides). Fluoride may also be present, and the effluent may have a low pH. Normally there is no liquid effluent from the smelter other than cooling water; wastewaters do originate in scrubbers (if used), wet electrostatic precipitators, cooling of copper cathodes and so on.

In the electrolytic refining process, by-products such as gold and silver are collected as slimes that are subsequently recovered. Sources of wastewater include spent electrolytic baths, slime recovery, spent acid from hydrometallurgy processes, cooling water, air scrubbers, washdowns, stormwater and sludges from wastewater treatment processes that require reuse/recovery or appropriate disposal.

10.1.3 Mass Balance and Energy Consumption

In terms of mass balance, smelting leads to higher loss of metals as compared to hydrometallurgy. The main factors are:

- Loss of metals into the slags
- Loss of metals on the refractories
- Loss of dust and high volatility products

Large amounts of fuel are required for the melting process. The route for production of cathode copper requires large amounts of energy per ton of copper: 30–40 million British thermal units (Btu) per ton cathode copper. In terms of energy use, there is no doubt that smelting is equal to high-energy consumption. Hydrometallurgy leads to a higher recovery rate due to relative ease in leaching of product and the possibility of cascading—recirculating solid waste to the next step and achieving a high recovery rate with chemical precipitation of electrowinning.

Compared to pyrometallurgy, direct fuel consumption of hydrometallurgy is almost negligible. On the other hand, it should be noted that to produce a ton of copper cathode, this process requires around 3,400 kWh. This means that, if the present fuel mix in electricity generation remains constant, indirect emissions due to hydrometallurgy production in the year 2001 would probably attain 3.6×10^6 ton of CO_2, considering unit emissions of 0.68 kg CO_2/kWh and a production of 1.56 million tons of cathodes.

10.1.4 Importance of Bioleaching

Bioleaching has gained increased interest as an alternative process over other techniques. It involves the use of iron- and sulphur-oxidising microorganisms to catalyse the dissolution of valuable metals species. It was not until 1947 that these phenomena were attributed to bacteria. Once identified, however, rapid steps were taken to commercialise the process. Commercial application of bacterial leaching began in the late 1950s at the Kennecott Utah Copper Company's Bingham Canyon Mine near Salt Lake City, Utah, where it was observed that blue copper-containing solutions were running out of waste piles that contained copper sulphide minerals—something that should not have happened in the absence of powerful oxidising agents and acid.

10.2 Microorganisms Involved

The mineral sulphide-oxidising microorganisms are acidophilic prokaryotes as their optimal growth varies between pH 2 and 4. They are autotrophic in nature as they use inorganic carbon (CO_2) as carbon source. They are strictly chemolithotrophic, i.e. they derive energy for growth from oxidation of reduced sulphur compounds and metal sulphides and some species also derive energy through oxidation of ferrous iron, while some species also can derive energy by oxidation of hydrogen. They are classified into three groups such as mesophiles (20–40 °C), moderate thermopiles (40–60 °C) and thermopiles (60–80 °C), based on the temperature requirements for optimal growth. The mesophiles actively involved in bio-oxidation and bioleaching are *Acidithiobacillus ferrooxidans*, *Acidithiobacillus thiooxidans*, *Acidithiobacillus caldus*, *Leptospirillum ferrooxidans*, *Leptospirillum ferrodiazotrophum*, *Leptospirillum thermoferrooxidans* and *Leptospirillum ferriphilum*; the moderate thermophiles are *Acidimicrobium ferrooxidans*, *Acidithiobacillus caldus* and *Sulfobacillus thermosulfidooxidans*; while the thermophiles are *Sulfolobus metallicus*,

Table 10.1 List of some commercially used acidophilic prokaryotic

Mineral-degrading acidophiles	Thermal classification
Iron oxidisers	
Leptospirillum ferrooxidans	Mesophile
L. ferriphilum	Mesophile
L. thermoferrooxidans	Moderate thermophile
Ferrimicrobium acidiphilum	Mesophile
Ferroplasma acidiphilum	Mesophile
Fp. acidarmanus	Mesophile/thermotolerant
Sulphur oxidisers	
Acidithiobacillus thiooxidans	Mesophile
At. caldus	Moderate thermophile
Metallosphaera sp.	Extreme thermophile
Sulfolobus sp.	Extreme thermophile
Iron and sulphur oxidisers	
Acidithiobacillus ferrooxidans	Mesophile
Acidianus sp.	Extreme thermophile
Sulfolobus metallicus	Extreme thermophile
Iron reducers	
Acidiphilium sp.	Mesophile
Iron oxidisers/reducers	
Acidimicrobium ferrooxidans	Moderate thermophile
Iron oxidisers/reducers and sulphur oxidisers	
Sulfobacillus sp.	Mesophiles and
	Moderate thermophiles

Mesophiles, $T_{optimum} < 40$ °C; moderate thermophiles, $T_{optimum}$ 40–60 °C; extreme thermophiles, $T_{optimum} > 60$ °C

Sulfobacillus spp. and *Metallosphaera sedula*. Characteristic features of some of the important microorganisms are given in Table 10.1.

10.2.1 Mechanism of Bacterial Oxidation

The chemistry of the bioleaching process is relatively straightforward. The bioleach microorganisms catalyse the oxidation of ferrous iron and sulphur, to produce ferric iron and sulphuric acid, according to

$$2Fe^{2+} + \tfrac{1}{2}O_2 + 2H^+ \rightarrow 2Fe^{3+} + H_2O \quad (10.1)$$

$$S^0 + \tfrac{1}{2}O_2 + H_2O \rightarrow H_2SO_4 \quad (10.2)$$

The ferric iron reacts with mineral sulphides to produce ferrous iron and sulphur, according to the following general scheme of reaction:

$$MS + 2Fe^{3+} \rightarrow M^{2+} + 2Fe^{2+} + S^0 \quad (10.3)$$

A lot of work has been done in this field by many workers. Several authors have proposed different hypotheses on mechanism of bioleaching. The traditional hypothesis that bacteria oxidise sulphides by either a direct mechanism or an indirect mechanism (Silverman and Ehrlich 1964) has evolved into a complex chemical/electrochemical/biochemical description of the interactions of bacteria with sulphide minerals. Tributsch (2001) proposed that the term 'contact' leaching can be used in place of 'direct' leaching because it described the association of the bacteria with the mineral surface rather than the by means of attack. This approach has been refined further and summarised by Crundwell (2003), who described three mechanisms by which microorganisms (specifically *At. ferrooxidans*) might interact with a sulphide mineral, as follows a) contact and b) non-contact mechanism.

According to Rohwerder, microbial processes facilitating mineral bio-oxidation and bioleaching are defined in terms of the contact mechanism, the noncontact mechanism and the cooperative mechanism. Metal leaching is now recognised as being mainly a chemical process in which ferric iron and protons are responsible for carrying out the leaching reactions. The role of the microorganisms is to generate the leaching chemicals and to create the space in which the leaching reactions take place. Microorganisms typically form an exo-polysaccharide layer when they adhere to the surface of minerals but not when growing as planktonic cells. It is within this EPS layer rather than in the bulk solution that the reactions take place most rapidly and efficiently and therefore the EPS serve as the reaction space.

In the contact mechanism, the bacterial cells attach with the aid of extracellular polymeric substance (EPS) layer to the mineral surfaces, resulting in dissolution of the sulphide minerals at the interface by an electrochemical process. In the noncontact mechanism, the ferric iron, produced through bio-oxidation of ferrous iron, comes in contact with the mineral surfaces, oxidises the sulphide mineral and releases ferrous iron back into the cycle, while in the cooperative mechanism, planktonic iron and sulphur oxidisers oxidise colloidal sulphur, other sulphur intermediates and ferrous iron in the leaching solution, releasing protons and ferric iron which is further used in noncontact leaching (Rohwerder et al. 2003).

In both indirect leaching and indirect contact leaching, microorganisms catalyse the oxidation of ferrous ions to ferric ions. The oxidation of a sulphide mineral by ferric ions may proceed via a thiosulphate intermediate or a polysulphide intermediate depending upon its solubility in acid.

(i) Acid-insoluble sulphides (e.g. pyrite, molybdenite, tungstenite) are oxidised to metal ions and sulphate via a thiosulphate intermediate. Bacteria catalyse the thiosulphate oxidation. The main role of the microorganisms in this mechanism is to catalyse the regeneration of the consumed ferric irons by means of aeration.

(ii) Acid-soluble sulphides (e.g. pyrrhotite, sphalerite, chalcopyrite) dissolve to form metal ions and H_2S. In this way, the metal–sulphide bond in the mineral lattice is broken prior to sulphur oxidation. Bacteria catalyzes the soluble polysulphides through the oxidation to sulphate and generate acid.

The microorganism's role in this mechanism is twofold:

- To catalyse the regeneration of the ferric ions consumed for the chemical oxidation of the intermediary hydrogen sulphide into elemental sulphur via formation of polysulphides
- To catalyse the generation of sulphuric acid in order to maintain the supply of protons required in the first reaction step for the dissolution of the mineral

10.3 Energy Production

Irrespective of temperature, the microorganisms obtain their energy from oxidation of either iron- or sulphur-containing minerals. In both the ways, they help in oxidation of complex minerals which otherwise become hard for metal extraction. It is believed that high-temperature, anaerobic acidic environment favours leaching of metals. These organisms use the reduced forms of iron and sulphur compounds as sole source of energy by a variety of mechanisms.

10.3.1 Iron Oxidation Model

Ferrous iron is oxidised to ferric iron by releasing an electron which in turn serves as a donor for bacterial growth. It is believed that the oxidation of ferrous iron by *Acidithiobacillus ferrooxidans* follows uphill and downhill electron pathway. Rather it would not be wrong to say that both are interrelated. However, it also depends upon the Fe(II)/Fe(III) ratio suggesting higher this ratio; higher is the redox potential value. Because of this, a relative amount of Fe (II) is oxidised, and each electron is allowed to transfer to the terminal electron acceptor, i.e. oxygen. There have been a number of models suggested by quite a number of workers explaining the individual roles of each protein molecule in the pathway system until now. These models may differ slightly in respect to the proteins involved during the process but also give an initial view about how they might have been regulated and processed. Unlike many others quires, several unsolved questions still remains mysterious.

Furthermore, the proteins involved in the pathway also depend upon the energy source on which they are grown. For example, it has been seen that *rus* gene becomes predominate in Fe (II)-grown cells, whereas *iro* genes in the case of sulphur-grown cells of *Acidithiobacillus ferrooxidans*. Some groups have proposed models based on different types of protein involved and their order in respiratory chain (Ingledew et al. 1977; Yamanaka et al. 1991; Yamanaka and Fukumori 1995; Blake and Shute 1994; Bruschi et al. 1996; Giudici-Orticoni et al. 2001). Another different approach was followed by some other group based on genetic organisation, gene regulation and subcellular localisation (Appia-Ayme 1998; Appia-Ayme et al. 1999; Yarzabal et al. 2002, 2004; Kusano et al. 1992; Fukumori et al. 1988; Cavazza et al. 1995; Yamanaka and Fukumori 1995; Bruschi et al. 1996; Yamanaka et al. 1991), have suggested the presence of a high potential redox iron–sulphur protein called *iro* gene (HiPIP). Kusano et al. showed the presence of this gene between *pur* A and a transfer RNA-encoding gene which transcribed independently. Bruscella et al. investigate the possible role of *iro* in three different type strains by probing the *iro* gene in various pure strains of *Acidithiobacillus ferrooxidans*. A HiPIP-encoding gene has been detected in five pure strains of *Acidithiobacillus ferrooxidans*. The nucleotide sequence of this gene presents 56.2 % identity with the *iro* gene encoding a HiPIP previously characterised in *Acidithiobacillus ferrooxidans* strain Fe-1(Kusano et al. 1992).

10.3.1.1 Downhill Electron Pathway

In this pathway, the *rus* operon regulates the flow of electrons and is highly expressed in Fe (II)-grown cells. In addition to rusticyanin, it consists of three electron transfer proteins, two cytochromes *c*, namely, Cyc1 and Cyc2, and a aa_3-type cytochrome oxidase (Cox ABCD) (Appia-Ayme et al. 1999). It was demonstrated that all seven genes encoding the corresponding structural protein of the above redox components are located in a single operon and are co-transcribed (Appia-Ayme et al. 1999; Yarzabal et al. 2004). The cytochrome *c* Cyc2 (46 kDa) located in the outer membrane serves as the primary electron acceptor and passes the electron to 21 kDa cytochrome *c* Cyc1 bound to the inner membrane via a protein carrier molecule called rusticyanin present in the periplasmic space (Giudici-Orticoni et al. 1999). Then cytochrome *c* Cyc1 handed over the electron to aa_3-type cytochrome oxidase. It was believed that the catalytic site for oxygen reduction is to be located in the cytoplasmic site. However, the optimum pH of oxidase is around 3.5 which prove that oxygen is reduced at the periplasmic site (Kai et al. 1989; Yamanaka et al. 1991). Starting from receiving of electrons from the outer membrane till reducing oxygen into water, majority of protons entraining the cells via ATP synthase complex embedded in the inner membrane are consumed. *Acidithiobacillus ferrooxidans* generates ATP using proton motive force due to proton gradient across the inner membrane, and electrons from oxidation of iron are used to neutralise the incoming protons.

10.3.1.2 Uphill Electron Pathway

In this pathway, bc_1 complex plays a major role to regenerate NAD(P)H. The redox midpoint (E_m) of the couple Fe^{3+}/Fe^{2+} (650 mV at pH 2) is much more positive than that of couple $NAD(P)^+/NAD(P)H$ (−305 mV at pH 6.5) which require energy from Fe^{2+} to push up the electrons against the potential gradient. Nevertheless, it remain unclear regardless Fe^{2+} plays an active role. Ingledew (1982) prospoed that the energy required for uphill electron transport, influx of protons from proton motive force is required across the membrane (At. ferrooxidans). He also added while doing so there a huge difference in the pH environment was found i.e. the exterior pH was 2 and inside the membrane it was 6.5. Further he suggested that this pathway (uphill) is from Fe(II) via bc_1 complex and a membrane soluble quinine to NAD^+ reductase complex.

Moreover, the reverse flow of electrons from cytochrome *c* through cytochrome bc_1 complex to quinine to NAD(P)H dehydrogenase is clearly demonstrated in *Acidithiobacillus ferrooxidans* (Elbehti et al. 2000; Brasseur et al. 2002). The operon cytochrome bc_1 complex encodes three subunits of bc_1 complex, namely, Pet A1B1C1, a cytochrome *c* (CycA1) and a dehydrogenase (SdrA1). Giudici-Orticoni et al. (2000) proposed that the electron is received by *cyc*A1 and transferred to rusticyanin, whereas Bruscella et al. (2005) and Quatrini et al. (2005) suggest it to belong to bc_1 complex that takes part in reverse flow of electrons between Fe(II) and NAD(P). However, there exist two pathways: one exergonic that involves aa_3-type oxidase as a terminal electron acceptor and reduces oxygen and another endergonic that involves bc_1 complex and reduces NAD(P).

Compiling all the reports and datas, it was believed that rusticyanin transfers electrons to in two different cytochromes C_4: CycA1 encoded by *petI* operon by endergonic pathway and Cyc1 encoded by *rus* operon by exergonic pathway.

10.3.2 Sulphur Oxidation Model

A very complicated chemistry resides during oxidation of reduced inorganic sulphur compounds. Sulphur exists in different oxidation states which makes the pathway more complex and complicated.

The energy produced during oxidation of one sulphur atom is much higher than oxidation of iron atom (Pronk et al. 1990). If we see the sulphur chemistry, several intermediates are formed during the oxidation. These intermediates are more or less produced by combination of spontaneous and enzymatic reactions (Rohwerder and Sand 2003; Schippers and Sand 1999).

Based on studies done by Brasseur et al. (2004), Rohwerder and Sand (2003) and Schippers and Sand (1999), Rawlings proposed a composite model for oxidation sulphur by *Acidithiobacillus ferrooxidans*. As per the model, a thiol-bearing outer membrane protein mobilses elemental sulphur and transport it into the periplasmic space as persulphidic sulphur, a sulphur dioxygenase oxidises persulphide sulphur to sulphite (Silver and Lundgren 1968a; Rohwerder and Sand 2003; Rohwerder et al. 2003), a sulphite oxidoreductase which oxidises sulphite to sulphate (Vestal and Lundgren 1971), a sulphide–quinine oxidoreductase oxidises sulphide to sulphur (Wakai et al. 2004), a thiosulphate oxidase catalyses the oxidation of thiosulphate to tetrathionate (Silver and Lundgren 1968b), a rhodanase splits the thiosulphate to sulphur and sulphite (Tabits et al. 1969), and a tetrathionate hydrolase that hydrolyses tetrathionates to thiosulphate, sulphur and sulphate (De Jong et al. 1997). After a long debate, it was proposed that S^o is transported into the periplasmic space as persulphide sulphur and got oxidised to sulphite by sulphur dioxygenase and later on got oxidised to sulphate by an enzyme sulphite oxidoreductase.

Now the question arises: how are these electrons transferred into the respiratory chain? Pronk et al. (1991), Harahuc et al. (2000) suggest that a diversion occurs at quinol pool where one leads to *bd*-type oxidase while another at ba_3-type oxidase via bc_1 complex (Brasseur et al. 2004). However, the work done by Brasseur et al. (2002, 2004) suggests that there exist two bc_1 complexes depending on the growth substrates. One which grew on Fe (II) ions has downhill mode of electron (operated by *Pet II* operon), whereas the cells grown on S^o function as uphill electron transfer mode (operated by *Pet I* operon).

In *Pet II* operon, bc_1 complexe receives electrons from quinol pool and transfer them to a membrane-bound cytochrome *c* and/or to a soluble redox protein or to a HiPIP protein. This later on transfers the electrons to terminal oxidase where oxygen reduced into water (Trumpower 1990; Bonora et al. 1999; Pereira et al. 1999). On the other side, in case of *Pet I* operon system, a reverse flow of electrons was observed by cytochrome *c* to bc_1 complexes to quinone and NAD (P) is reduced to NADP (H).

10.3.3 Anaerobic Sulphur Oxidation

Anaerobic respiration is seen in many prokaryotes which uses ferric iron as an electron acceptor (Lovely 1995; Lonergan et al. 1996; Johnson and Roberto 1998). It is believed that at an extreme acidic condition, solubility of ferric iron is more as compared to pH > 4. Fe^{2+}/Fe^{3+} ratio is related to pH and most positive in extremely acidic environments (E*h* = +770 mV, pH 2). This value is close to O_2/H_2O couple (+840 mV), suggesting ferric iron can also act as an electron acceptor at low pH conditions (Johnson 2001).

Ferric ion also acts as an electron acceptor in the oxidation of elemental sulphur by *Acidithiobacillus ferrooxidans* (Brock and Gustafson 1976; Sugio et al. 1985), and the enzymes involved are sulphur (sulphide)–ferric-ion oxidoreductase ($S^o + 4Fe^{3+} + 3H_2O = H_2SO^{3+} + 4Fe^{2+} + 4H^+$) (Sugio et al. 1987, 1989) and sulphite–ferric-ion oxidoreductase ($H_2SO_3 + 2Fe^{3+} + H_2O = H_2SO_4 + 2Fe^{3+} + 2H^+$) (Sugio et al. 1992). Ferric ion reduction by sulphur- and iron-oxidising acidophilic bacteria *Acidithiobacillus ferrooxidans*, *Acidithiobacillus thiooxidans* and *Sulfolobus acidocaldarius* was first reported by Brock and Gustafson (1976). They proposed the following equations for the oxidation of elemental S^o under aerobic and anaerobic condition as given below:

$$S^0 + 3O_2 + 2H_2 \leftrightarrow 2HSO_4{}^{2-} + 2H^+ \quad \text{aerobic} \tag{10.4}$$

$$S^0 + 6Fe^{3+} + 4H_2O \leftrightarrow HSO_4{}^{2-} + 6Fe^{2+} + 7H^+ \quad \text{anaerobic} \tag{10.5}$$

Pronk et al. (1991) studied the uptake of glycine under anaerobic conditions as an indicator of respiratory processes, and it was found that glycine was taken up when S^o was the electron donor and Fe^{3+} the acceptor; therefore, it was concluded that the reduction of Fe^{3+} was a respiratory process. In later work, Pronk et al. reported that a linear relationship exists between cell density and the accumulation of Fe^{2+}, although the biomass yield per electron from S^o substrate was two orders of magnitude less under anaerobic than aerobic conditions. Das et al. (1992) compared the specific activity of *At. ferrooxidans*, μmol Fe^{2+}/h/mg cell protein, under aerobic and anaerobic conditions. When grown aerobically on Fe^{2+} and S^o substrates, the specific activities were the same; under anaerobic conditions, the activity was doubled. *At. ferrooxidans* can utilise a variety of electron donors and acceptors, including H_2/S^0, H_2S/Fe^{3+}, H_2/Fe^3 and tetrathionate/Fe^{3+} (Ohmura et al. 2002). Ohmura et al. (2002) reported that anaerobic Fe^{3+} respiration induces *At. ferrooxidans* to synthesise a c-type cytochrome which was responsible for the F^{3+} reduction. Later work by Sugio et al. suggests the Fe^{2+}-reduction enzyme pathways are more complicated than initially suggested, with some dependence on the nature of the electron donor substrate.

Anaerobic bacterial oxidation of elemental sulphur to sulphuric acid may occur in the environment of sulphide minerals (Johnson and Hallberg 2009). So, in perspective of biotechnological and environmental, the detail process has to be understood properly. Unless and until the complete pathway is clearly understood, the role of this bacterium will not be clear. However, in the past, lots of work have been initiated in this direction including *Acidithiobacillus ferrooxidans*, *Leptospirillum* spp. (Tyson et al. 2004, 2005; Ram et al. 2005), *Ferroplasma* spp. (Tyson et al. 2004; Golyshina and Timmis 2005) and *Metallosphaera sedula* (Kappler et al. 2005).

10.4 Application of Anaerobic Bioleaching Process

Bioleaching is an alternative to simple acid leaching, as the microorganisms used can specifically target the minerals that are to be dissolved. Bioleaching of nonferrous ores such as copper sulphides has long been carried out very successfully using aerobic microorganisms that catalyse mineral oxidation. This approach is particularly successful in sulphide mineral leaching, as the products of oxidising sulphide minerals are significantly more soluble than the original sulphides (Brombacher et al. 1997). There are several examples where reductive bioleaching process can be effectively and efficiently considered. Minerals especially oxidic in nature are best target materials for the process, for example, Ni laterites, Fe-oxides, Mn-oxide, etc.

However, such oxidative leaching does not assist in dissolving iron oxides due to the fact that it is the reduced (Fe^{2+}) form of iron that has higher solubility, and oxidation would only produce the low-solubility Fe^{3+} compounds. A solution to this problem is to use a reductive leaching process, converting iron to the Fe^{2+} (ferrous) state. This has the advantage that ferrous iron salts are considerably more soluble than ferric iron salts, and so will dissolve under only mildly acidic conditions (Eisele and Gabby 2014).

10.4.1 On Nickel Laterites (Limonite)

Reductive acid leaching of limonite (goethite in particular), resulting in nickel release, has been demonstrated in mildly acidic conditions (pH range 1.5–2) at atmospheric pressure and ambient temperature (25–30 °C) using *Acidithiobacillus ferrooxidans* (Hallberg et al. 2011). The leaching process is based on the reduction of ferric iron contained in the goethite host mineral, where the electrons are derived from elemental sulphur. However, any effective hydrometallurgical processing requires acidic conditions, typically below pH 2, to facilitate metal dissolution and solubility and a low-cost electron donor. Therefore, for the use of an inorganic electron donor, these chemolithotrophic bacteria would be the best candidate rather than heterotrophic species.

In the case of limonite, the iron-enriched and magnesium-depleted version of laterites, the host minerals are mainly ferric iron oxides, typically goethite, FeO (OH) (Landers et al. 2009). In order to solubilise nickel from this host mineral, the strong bond between oxygen and ferric iron has to be broken. Therefore, bacterially catalysed reductive dissolution would be worthwhile to be considered. Dissolution of goethite was achieved under mild acidic condition (pH 1.8) and at ambient temperature (30 °C), atmospheric pressure, via catalysing the transfer of electrons from elemental sulphur to ferric iron in goethite. This biocatalytic process may enable new potential hydrometallurgical opportunities because of the fact that goethite is one of the major host minerals of nickel and cobalt and because its reduction has been demonstrated to be feasible utilising a low-cost electron donor. The fact that the resulting leachate contains iron in the ferrous state is of further significance as it facilitates direct recovery of valuable metals (Du Plessis et al. 2011; Hallberg et al. 2011). Plessis et al. have proposed a conceptual flow sheet for targeting tropical limonitic laterites and called it as Ferredox process. These process components are designed to (a) facilitate simplified and low-intensity processing, (b) be amenable to modularization, (c) reduce technical process implementation risk and (d) reduce capital costs for tropical limonite projects.

Therefore, acidophiles such as *Acidithiobacillus ferrooxidans* as a potential candidate use ferric iron as an electron acceptor (an alternative to oxygen when this is absent or present at low concentrations) and require an electron donor (sulphur, hydrogen or an organic compound, depending on the species) to drive iron reduction. This bacterium has the ability to oxidise sulphur under anaerobic condition and can be used for the process to recover valuable metals.

10.4.2 On Petroleum Refinery Spent Catalyst

Catalysts used in petroleum refining require periodic replacement due to fouling of the support pore structure by coke build-up and accumulation of metals, notably Ni, V, Mo and Co, on active surfaces. The source of the contaminant metals is the petroleum feed stock. The quantity of spent catalyst discarded by refiners has recently increased for a number of reasons: firstly, to meet market demand, refiner's have increased their use of heavier crude and 'dirtier' feedstocks which have a greater propensity to foul catalysts; secondly, the legislative requirement to reduce the sulphur content of fuels; and thirdly, disposal to landfill is becoming increasingly costly due to the tightening of regulatory requirements for site engineering to mitigate the potential for leaching of metals and entrained oil into the environment (USEPA 2002; Marafi and Stanislaus 2003; Ahmed and Menoufy 2012). In the past, several works have been reported, but the problems remain somewhere neglected and unsolved. However, one approach to this problem could be by enabling the catalyst by decoking and then subjected to a selective leach under mild conditions to remove contaminant metals and restore the material's activity sufficient for reuse.

Of the three metals V, Ni and Mo commonly found in spent hydrotreating catalyst (Marafi and Stanislaus 2003), V and Ni are relatively easily recovered by acid leaching in consideration of metal solubility and the leach medium, E and pH as per the Pourbaix diagram (Pourbaix 1974). Reported recoveries of Ni and V are in the range of 80–89 % and 32–90 % respectively with rates indicative of diffusion control; the recovery of Mo is much lower 30–50 % (Mishra et al. 2007; Beolchini et al. 2010). Low Mo recovery rates have been attributed to a number of factors: impervious S layer, the inherent insolubility of MoO_2 and the re-precipitation of Mo as MoO_3 (Mishra et al. 2007; Pradhan et al. 2010).

It would be advantageous to remove the entrained oil and elemental S prior to the application of bioleach methods. If the catalyst is roasted under oxidising conditions, the remaining metals are likely to report as oxides. The efficacy of a bioleach processes to roasted catalyst may be enhanced by use of anaerobic conditions to reduce the metal oxides back to a valence state in which they are soluble in an aqueous solution.

10.4.3 On Mn Nodules

Ocean manganese nodules offer a rich if not readily accessible source of valuable metals, notably Mn, Ni, Co and Cu. Nodules can form at any depth, but the highest concentration is found between 4,000 and 6,000 m and varies in size from microscopic to more than 20 cm in diameter (Poppe et al. 1984; International Seabed Authority 2004). Occurrence is dependent on sedimentation rates and ocean currents; generally the highest concentrations of nodules are found in the deep ocean basins where sedimentation rates are low (Frazer and Fisk 1981). Nodule composition can be variable and is thought to be dependent on the proximity to metal sources such as volcanic vents, ridges and metal rich crusts (Frazer and Fisk 1981). The average composition is Mn 18.6 %, Fe 12.4 %, Cu 0.66 % and Co 0.27 % with smaller amounts of other metals (Ghosh and Mukhopadhyay 1999). A number of estimates have been made of the resource potential (Mero 1977; Cronan 1983; Morgan 2000). Morgan (2000) estimated the size of the resources in Pacific Clarion–Clipperton Fracture Zone alone to be 78 billion tonnes of Mn, 340 million tonnes of Ni and 78 million tonnes of Co. Despite the apparent large size of the resources, the economically viable resources are likely to be much smaller (Glasby 2002). Techniques for recovery of the nodules from the ocean floor and subsequent mineral processing strategies have been the intensely studied and have been the subject of a numerous articles (Senanayake 2011; Kumari and Natarajan 2001a, b; Mehta et al. 2003; Vu et al. 2005; Mehta et al. 2010; Kanungo and Jena 1988) and a conference series facilitated by the International Seabed Authority.

Application of bioleach processes to Mn ocean nodules for recovery of metals has been investigated by a number of research groups. The parameters examined include the effect on leach efficacy of acid generation by biological oxidation of sulphur and the effect of by-products that aid dissolution by chelate activity (Mehta et al. 2010). It has been recognised that reduction of the Fe and Mn oxides and their subsequent dissolution are necessary to release Cu, Ni and Co entrained within the oxide matrix. This can be achieved by either addition of an organic reductant such as glucose or starch in conjunction with bacterial activity (Kumari and Natarajan 2001a, 2002a, b; Kanungo and Jena 1988; Mehta et al. 2003, 2010) or by addition of an inorganic reductant, typically Fe^{2+} ions (Vu et al. 2005; Das et al. 1992). To our knowledge, the use of acidophilic lithotrophic bacteria to couple the oxidation of elemental sulphur to the reduction of $Fe^{3+} \rightarrow Fe^{2+}$ and $Mn^{4+} \rightarrow Mn^{2+}$ has not been reported.

Given that lithotrophic acidophilic bacteria have the ability to reduce oxidised metal species, it would be logical to consider the application of these processes to oxide minerals.

10.5 Summary

There are different electron donors for acidophilic microorganisms which play a crucial role in leaching mechanism. This chapter discussed elaborately about the possible mechanisms, but all of them are hypothetically true. Nonetheless, debates and arguments could not be made on the exact mechanism going on. Until now Fe (II) oxidation pathway and the role of several enzymes/proteins regulating the transport mechanism are well established, but several pros and cons regarding sulphur metabolism still exist. As discussed above, bioleaching is an efficient process compared to conventional leaching operations; therefore, the urgency on detail mechanism becomes essential. There are wide ranges of acidophilic microbes sharing some common physiological properties. Recent genomic and biochemical data could suggest the substantial role of these biomolecules playing active role in energy transport mechanism of these microbes. But to answer the wide range of questions generated during the studies, a wide approach genetic technique 'micro-array' can be used.

References

Ahmed HS, Menoufy MF (2012) New trends in hydroprocessing spent catalysts utilization. In: Vivek Patel (ed) Petrochemicals. InTech, Rijeka, Available from: http://www.intechopen.com/books/petrochemicals/reuse-and-treatment-of-hydrotreating-spent-catalyst. ISBN 978-953-51-0411-7

Appia-Ayme C (1998) Caracterisation d'un operon codant pour sept proteins transporteurs d'electrons chez *Thiobacillus ferrooxidans*. Universite de la Mediterranee, Aix-Marseille II

Appia-Ayme C, Guiliani N et al (1999) Characterization of an operon encoding two *c*-type cytochromes, an *aa*3-type cytochrome oxidase, and rusticyanin in *Thiobacillus ferrooxidans* ATCC33020. Appl Environ Microbiol 65:4781–4787

Beolchini F, Fonti V, Ferella F, Veglio F (2010) Metal recovery from spent catalysts by means of biotechnological strategies. J Hazard Mater 178:529–534

Blake RC 2nd, Shute EA (1994) Respiratory enzymes of *Thiobacillus ferrooxidans*. Kinetic properties of an acid-stable iron:rusticyanin oxidoreductase. Biochemistry 33:9220–9228

Bonora P, Principi II et al (1999) On the role of high-potential iron-sulfur proteins and cytochromes in the respiratory chain of two facultative phototrophs. Biochim Biophys Acta 1410:51–60

Brasseur G, Bruscella P et al (2002) The *bc*1 complex of the iron-grown acidophilic chemolithotrophic bacterium *Acidithiobacillus ferrooxidans* functions in the reverse but not in the forward direction. Is there a second *bc*1 complex? Biochim Biophys Acta 1555:37–43

Brasseur G, Levican G et al (2004) Apparent redundancy of electron transfer pathways via *bc*1 complexes and terminal oxidases in the extremophilic chemolithoautotrophic *Acidithiobacillus ferrooxidans*. Biochim Biophys Acta 1656:114–126

Brock TD, Gustafson J (1976) Ferric iron reduction by sulfur- and iron-oxidizing bacteria. Appl Environ Microbiol 32:567–571

Brombacher C, Krebs W, Bosshard PP, Bachofen R, Brandl H (1997) FEMS Microbiol Rev 20:605

Bruscella P, Cassagnaud L, Ratouchniak J, Brasseur G, Lojou E, Amils R, Bonnefoy V (2005) The HiPIP from the acidophilic Acidithiobacillus ferrooxidans is correctly processed and translocated in Escherichia coli, in spite of the periplasm pH difference between these two micro-organisms. Microbiology 15:1421–1431

Bruschi M, Cavazza C et al (1996) Biooxidation des mineraux sulfures et dissolution de metaux par la bacterie acidophile: *Thiobacillus ferrooxidans*. Dechets 4:27–30

Cavazza C, Guigliarelli B et al (1995) Biochemical and EPR characterization of a high potential iron-sulfur protein in *Thiobacillus ferrooxidans*. FEMS Microbiol Lett 130:193–200

Cronan DS (1983) Criteria for recognition of areas of potentially economic manganese nodules and encrustations in the CCOP/SOPAC region of the Central and Southwest Tropical Pacific. Committee for Coordination of Joint Prospecting for Mineral Resources in South Pacific and

Offshore Areas (CCOP/SOPAC), United Nations Development Programme Economic and Social Commission for Asia and the Pacific, Technical Report No. 30

Crundwell FK (2003) How do bacteria interact with minerals? Hydrometallurgy 71:75–81

Das A, Mishra AK Pradosh R (1992) Anaerobic growth on elemental sulfur using dissimilar iron reduction by autotrophic *Thiobacillus ferrooxidans*. FEMS Microbiol Lett 97:167–172

De Jong GA, Hazeu W et al (1997) Polythionate degradation by tetrathionate hydrolase of *Thiobacillus ferrooxidans*. Microbiology 143:499–504

Du Plessis CA, Slabbert W, Hallberg KB, Johnson DB (2011) Ferredox: a biohydrometallurgical processing concept for limonitic nickel laterites. Hydrometallurgy 109:221–229

Eisele TC, Gabby KL (2014) Review of reductive leaching of iron by anaerobic bacteria. Mineral Process Extr Metall Rev Int J 35(2):75–105

Elbehti A, Brasseur G et al (2000) First evidence for existence of an uphill electron transfer through the *bc*1 and NADH-Q oxidoreductase complexes of the acidophilic obligate chemolithotrophic ferrous ion-oxidizing bacterium *Thiobacillus ferrooxidans*. J Bacteriol 182:3602–3606

Frazer JZ, Fisk MB (1981) Geological factors related to characteristics of sea-floor manganese nodule deposits. Deep Sea Res 28A(12):1533–1551

Fukumori Y, Yano T et al (1988) Fe-II oxidizing enzyme purified from *Thiobacillus ferrooxidans*. FEMS Microbiol Lett 20:169–172

Ghosh AK, Mukhopadhyay R (1999) Mineral wealth of the ocean: a treatise on distribution, origin, exploration, mining, and management of sea floor non-living resources. Oxford and IBH, New Delhi

Giudici-Orticoni MT, Appia-Ayme C, Bengrine A, Cavazza C, Bruschi M, Chippaux M, Bonnefoy V (1999) Characterization and expression of the co-transcribed cycl and cyc2 genes encoding the cytochrome c4 (c552) and a highmolecular- mass cytochrome c from Thiobacillus ferrooxidans ATCC 33020. FEMS Microbiol Lett 167:171–177

Giudici-Orticoni MT, Leroy G, Nitsck W, Bruschi M (2000) Biochemistry 39:7205–7210

Giudici-Orticoni MT, Leroy G et al (2001) Two *c*-type cytochromes in *Thiobacillus ferrooxidans*: structure, comparison and functional role. In: Ciminelli VST, Garcia O Jr (eds) Biohydrometallurgy: fundamentals, technology and sustainable development, vol A. Elsevier, Amsterdam, pp 291–298

Giudici-Orticoni MT, Pronk JT, De Bruyn JC, Bos P, Kuenen JG (1992) Anaerobic growth of Thiobacillus ferrooxidans. Appl Environ Microbiol 58:2227–2230

Glasby GP (2002) Deep sea mining: past failures and future prospects. Mar Georesour Geotechnol 20:161–176

Golyshina OV, Timmis KN (2005) *Ferroplasma* and relatives, recently discovered cell wall-lacking archaea making a living in extremely acid, heavy metal-rich environments. Environ Microbiol 7:1277–1288

Hallberg KB, Hedrich S, Johnson DB (2011) Acidiferrobacter thiooxydans, gen. nov. sp. nov.; an acidophilic, thermo-tolerant, facultatively anaerobic iron- and sulfur-oxidizer of the family Ectothiorhodospiraceae. Extremophiles 15:271–279

Harahuc L, Lizama HM et al (2000) Selective inhibition of the oxidation of ferrous iron or sulfur in *Thiobacillus ferrooxidans*. Appl Environ Microbiol 66:1031–1037

Ingledew WJ (1982) Thiobacillus ferrooxidans, the bioenergetics of an acidophilic chemolithotroph. Biochim Biophys Acta 683:89–117

Ingledew WJ, Cox JC et al (1977) A proposed mechanism for energy conservation during Fe^{2+} oxidation by *Thiobacillus ferrooxidans*: chemiosmotic coupling to net H+ influx. FEMS Microbiol Lett 2:193–197

International Seabed Authority (2004) Marine mineral resources: scientific advance and economic perspectives. United Nations Division for Ocean Affairs and the Law of the Sea, Office of Legal Affairs, United Nations

Johnson DB (2001) Importance of microbial ecology in the development of new mineral technologies. Hydrometallurgy 59:147–157

Johnson DB, Hallberg KB (2009) Carbon, iron and sulfur metabolism in acidophilic microorganisms. Adv Microb Physiol 54:202–256

Johnson DB, Roberto FF (1998) Heterotrophic acidophiles and their roles in the bioleaching of sulfide minerals. In: Rawlings DE (ed) Biomining. Theory, microbes and industrial processes. Springer, Berlin, pp 259–279

Kai M, Yano T, Fukumori Y, Yamanaka T (1989) Cytochrome oxidase of an acidophilic iron-oxidizing bacterium, Thiobacillus ferrooxidans, functions at pH 3.5. Biochem Biophys Res Commun 160:839–843

Kanungo SB, Jena PK (1988) Reduction leaching of manganese nodules of Indian Ocean origin in dilute hydrochloric acid. Hydrometallurgy 21:41–58

Kappler U, Sly LI, McEwan AG (2005) Respiratory gene clusters of *Metallosphaera sedula* differential expression and transcriptional organization. Microbiology 151:35–43

Kumari A, Natarajan KA (2001a) Bioleaching of ocean nodules in presence of reducing agent. Eur J Miner Process Environ Prot 1:10–24

Kumari A, Natarajan KA (2001b) Electrobioleaching of polymetallic ocean nodule. Hydrometallurgy 62:125–134

Kumari A, Natarajan KA (2002a) Development of a clean bioelectrochemical process for leaching of ocean manganese nodules. Miner Eng 15:103–106

Kumari A, Natarajan KA (2002b) Cathodic reductive dissolution and surface adsorption behavior of ocean manganese nodules. Hydrometallurgy 64:247–255

Kusano T, Takeshima T et al (1992) Molecular cloning of the gene encoding *Thiobacillus ferrooxidans* FeII oxidase. High homology of the gene product with HiPIP. J Biol Chem 267:11242–11247

Landers M, Gilkes RJ, Wells M (2009) Dissolution kinetics of dehydroxylated nickeliferous goethite from limonitic nickel ore. Appl Clay Sci 42:615–624

Lonergan DJ, Jenter HI, Coates JD, Philips EJP, Schmidt TM, Lovely DR (1996) Phylogentic analysis of dissmilatory Fe(III)-reducing bacteria. J Bacteriol 178:2402–2408

Lovely DR (1995) Microbial reduction of iron, manganese, and other metals. Adv Agron 54:175–231

Marafi M, Stanislaus A (2003) Options and processes for spent catalyst handling and utilization. J Hazard Mater B101:123–132

Mehta KD, Pandey BD, Mankhand TR (2003) Studies on kinetics of biodissolution of metals from Indian Ocean nodules. Miner Eng 16:523–527

Mehta KD, Chitrangada Das, Pandey BD (2010) Leaching of copper, nickel and cobalt from Indian Ocean manganese nodules by Aspergillus niger. Hydrometallurgy 105:89–95

Mero JL (1977) Economic aspects of mineral deposits. In: Glasby GP (ed) Marine manganese deposits. Elsevier, Amsterdam, pp 327–355

Mishra D, Kim DJ, Ralph DE, Ahn JG, Rhee YH (2007) Bioleaching of vanadium rich spent refinery catalysts using sulphur oxidizing lithotrophs. Hydrometallurgy 88:202–209

Morgan CL (2000) Resource estimates for the Clarion-Clipperton manganese deposits. In: Cronan DS (ed) Handbook of marine mineral deposits. CRC, Boca Raton, FL, pp 145–170

Ohmura N, Sasaki K et al (2002) Anaerobic respiration using Fe^{3+}, S^0, and H_2 in the chemolithoautotrophic bacterium *Acidithiobacillus ferrooxidans*. J Bacteriol 184:2081–2087

Pereira MM, Carita JN et al (1999) Membrane-bound electron transfer chain of the thermohalophilic bacterium *Rhodothermus marinus*: characterization of the iron-sulfur centers from the dehydrogenases and investigation of the high-potential iron-sulfur protein function by in vitro reconstitution of the respiratory chain. Biochemistry 38:1276–1283

Poppe LJ, Commeau RF, Commeau JA, ManHeim FT, Aruscavage PJ (1984) Ferromanganese micronodules from the surficial sediments of Georges Bank. J Mar Res 42:463–472

Pourbaix M (1974) Atlas of electrochemical equilibria in aqueous solutions. National Association of Corrosion Engineers, Huston, TX

Pradhan D, Kim DJ, Ahn JG, Chaudhury GR, Lee SW (2010) Bio-dissolution of Ni, V and Mo from spent petroleum catalyst using iron oxidizing bacteria. J Environ Sci Health A Tox Hazard Subst Environ Eng 45:476–482

Pronk JT, Meulenberg R et al (1990) Oxidation of reduced inorganic sulfur compounds by acidophilic thiobacilli. FEMS Microbiol Rev 75:293–306

Pronk JT, Meijer WM et al (1991) Energy transduction by anaerobic ferric iron respiration in *Thiobacillus ferrooxidans*. Appl Environ Microbiol 57:2063–2068

Quatrini R, Appia-Ayme C et al (2005) Global analysis of the ferrous iron and sulfur energetic metabolism of *Acidithiobacillus ferrooxidans* by microarrays transcriptome profiling. In: Harrison STL, Rawlings DE, Petersen J (eds) 16th International biohydrometallurgy symposium proceedings, Cape Town, pp 761–771

Ram RJ, VerBerkmoes NC et al (2005) Community proteomics of a natural microbial biofilm. Science 308:1915–1920

Rohwerder T, Sand W (2003) The sulfane sulfur of persulfides is the actual substrate of the sulfur-oxidizing enzymes from *Acidithiobacillus* and *Acidiphilium* spp. Microbiology 149:1699–1710

Rohwerder T, Gehrke T, Kinzler K, Sand W (2003) Bioleaching review part A: progress in bioleaching: fundamentals and mechanisms of bacterial metal sulfide oxidation. Appl Microbiol Biotechnol 63(3):239–248

Schippers A, Sand W (1999) Bacterial leaching of metal sulfides proceeds by the indirect mechanisms via thiosulfate or via polysulfides and sulfur. Appl Environ Microbiol 65 (1):319–321

Senanayake G (2011) Acid leaching of metals from deep-sea manganese nodules—a critical review of fundamentals and applications. Miner Eng 24:1379–1396

Silver M, Lundgren DG (1968a) Sulfur-oxidizing enzyme of *Ferrobacillus ferrooxidans* (*Thiobacillus ferrooxidans*). Can J Biochem 46:457–461

Silver M, Lundgren DG (1968b) The thiosulfate-oxidizing enzyme of *Ferrobacillus ferrooxidans* (*Thiobacillus ferrooxidans*). Can J Biochem 46:1215–1220

Silverman MP, Ehrlich HL (1964) Microbial formation and degradation of minerals. Adv Appl Microbiol 6:153–206

Sugio T, Domatsu C, Munakato O, Tano T, Imai K (1985) Role of a ferric ion-reducing system in sulfur oxidation of Thiobacillus ferrooxidans. Appl Environ Microbiol 49:1401–1406

Sugio T, Mizunashi W et al (1987) Purification and some properties of sulfur:ferric ion oxidoreductase from *Thiobacillus ferrooxidans*. J Bacteriol 169:4916–4922

Sugio T, Katagiri T et al (1989) Actual substrate for elemental sulfur oxidation by sulfur:ferric ion oxidoreductase purified from *Thiobacillus ferrooxidans*. Biochim Biophys Acta 973:250–256

Sugio T, White KJ et al (1992) Existence of a hydrogen sulfide:ferric ion oxidoreductase in iron-oxidizing bacteria. Appl Environ Microbiol 58:431–433

Tabita R, Silver M et al (1969) The rhodanese enzyme of *Ferrobacillus ferrooxidans* (*Thiobacillus ferrooxidans*). Can J Biochem 47:1141–1145

Tributsch H (2001) Direct versus indirect bioleaching. Hydrometallurgy 59:177–185

Trumpower BL (1990) Cytochrome *bc*1 complexes of microorganisms. Microbiol Rev 54:101–109

Tyson GW, Chapman J et al (2004) Community structure and metabolism through reconstruction of microbial genomes from the environment. Nature 284:37–43

Tyson GW, Lo I et al (2005) Genome-directed isolation of the key nitrogen fixer *Leptospirillum ferrodiazotrophum* sp. nov. from an acidophilic microbial community. Appl Environ Microbiol 71:6319–6324

USEPA (2002) Hazardous waste management system; identification and management of hazardous waste: spent catalysts from dual purpose petroleum hydroprocessing reactors. Fed Regist 67(89):30811

Vestal JR, Lundgren DG (1971) The sulfite oxidase of *Thiobacillus ferrooxidans* (*Ferrobacillus ferrooxidans*). Can J Biochem 49:1125–1130

Vu H, Jandova J, Lisa K, Vranka F (2005) Leaching of manganese deep ocean nodules in FeSO4–H2SO4–H2O solutions. Hydrometallurgy 77:147–153

Wakai S, Kikumoto M, Kanao T, Kamimura K (2004) Involvement of sulfide:quinone oxidoreductase in sulfur oxidation of an acidophilic iron-oxidizing bacterium, *Acidithiobacillus ferrooxidans* NASF-1. Biosci Biotechnol Biochem 68:2519–2528

Yamanaka T, Fukumori Y (1995) Molecular aspects of the electron transfer system which participates in the oxidation of ferrous ion by *Thiobacillus ferrooxidans*. FEMS Microbiol Rev 17:401–413

Yamanaka T, Yano T et al (1991) The electron transfer system in an acidophilic iron-oxidizing bacterium. In: Mukohata Y (ed) New era of bioenergetics. Academic, Tokyo, pp 223–246

Yarzabal A, Brasseur G et al (2002) The high-molecular-weight cytochrome *c* Cyc2 of *Acidithiobacillus ferrooxidans* is an outer membrane protein. J Bacteriol 184:313–317

Yarzabal A, Appia-Ayme C et al (2004) Regulation of the expression of the *Acidithiobacillus ferrooxidans rus* operon encoding two cytochromes *c*, a cytochrome oxidase and rusticyanin. Microbiology 150:2113–2123

Chapter 11
Microbial Processing for Valorization of Horticultural Wastes

Sandeep K. Panda and Ramesh C. Ray

11.1 Introduction

Rigorous agriculture/horticulture and agro-industries have resulted in increased output due to rise in world population and demand of the products. Every year there is a rise in food production. Huge surplus of cereals, vegetables, fruits and other horticultural crops such as potato, sweet potato, sugar beet, etc., are produced and processed for value-added products. Most of the starchy crops such as cassava, sweet potato and potato and fruits such as mango, pineapple, apple and pears are perishable and need huge infrastructure for storage. It is required therefore to dispose them off as quickly as possible or preserve them by canning, freezing, drying and other methods (i.e. lacto-pickling) or process into value-added products. During the phase of processing, a lot of wastes (both solid and liquid) are generated. Horticultural wastes contain mainly lignocelluloses, besides soluble sugars, starch (in case of root crops), fibres, phenols and other hydrolysable materials (Tengerdy and Szakacs 2003) that can be metabolized by a wide range of microorganisms into value-added products. Biovalorization is a popular and attractive technology that facilitates cleaning up of the environment coupled with value addition by means of metabolic potential of novel microorganisms (Watanabe 2001; Thassitou and Arvanitoyannis 2001). Low-valued fruit and vegetable wastes have been successfully processed into many value-added products via fermentation technology (Bhalla and Joshi 1994; Zheng and Shetty 1998; Krishna 2005; Ward et al. 2006). For example, sweet potato (*Ipomea batatas* L.) tubers, bael (*Aegle marmelos* L.), fruits and minor fruits such as

S.K. Panda
Department of Biotechnology and Food Technology, Faculty of Science, University of Johannesburg, South Africa

R.C. Ray (✉)
ICAR- Central Tuber Crops Research Institute (Regional Center), Indian Council of Agricultural Research, Bhubaneswar 751019, Odisha, India
e-mail: rc_rayctcri@rediffmail.com

L.B. Sukla et al. (eds.), *Environmental Microbial Biotechnology*, Soil Biology 45,
DOI 10.1007/978-3-319-19018-1_11

sapota (*Achras sapota* Linn) and tendu (*Diospyros melanoxylon* L.) have been fermented into wine (Panda et al. 2012, 2014a, b; Sahu et al. 2013).

The objective of the current chapter is to review recent developments in the emerging areas of biovalorization of horticultural wastes. The chapter includes comprehensive discussion on high-valued products (except mushrooms) developed using horticultural [fruit, vegetable and plantation crop (coconut, oil palm and coffee)] wastes as substrate and selected microorganisms with possible applications in agrobiological applications, agro-food industries, fermentation industries, bioremediation, biodegradation and bioresource reclamation.

11.2 Biovalorization Process

Fallow and Wheelock (1982) described wastes produced from horticultural processing into three types:

1. Waste produced before storage on processing site
2. Waste produced during storage on processing site
3. Waste produced during processing

All these wastes whether solid or liquid, contain starch, cellulose, sugars or carbohydrates which serve as primary carbon source during biological treatment processes. In case of starchy waste water, success of the treatments depends on the capacity of the microorganism to catabolize starch, which has to be hydrolysed to easily assimilable sugars, i.e. maltose and glucose.

11.3 Definition

Valorization is the transformation of knowledge into a specific new finished product or process. Biovalorization refers to the process of value creation from biological especially microbiological knowledge by making a product or waste or less-valued matter suitable and/or available for economic and/or social use (Chandrasekaran and Bahkali 2013). Various microbial technologies have been adopted to valorize the horticultural wastes, depending upon its type for high-end finished products.

11.4 Biovalorized Products

The important biovalorized products derived from horticultural wastes (Table 11.1) are discussed below with selective examples.

Table 11.1 Demonstrative examples of various value-added products (excluding mushrooms) by microbial processing of horticultural wastes

Product	Substrate	Microorganism	Reference
Enzymes			
Amylase	Mango kernel Cassava waste	*Fusarium solani*, *Aspergillus niger*, *Rhizopus stolonifer*	Kumar et al. (2013), Pothiraj et al. (2006)
Cellulase and hemicellulase	Banana waste Palm oil fibre Waste newspaper	*Cellulomonas carte*, *Pseudomonas fluorescence*, *Pseudomonas putida*, *Bacillus megaterium*	Dabhi et al. (2014), Xin and Geng (2010)
Pectinase	Wastes of cashew apple, pineapple, banana and grapes; sapota peel	*Aspergillus foetidus*, *Aspergillus oryzae*	Venkatesh et al. (2009), Akbar and Prasuna (2012)
Tannase	Grape peel	*Aspergillus niger*, *Penicillium chrysogenum*, *Trichoderma viride*	Paranthaman et al. (2009)
Organic acids			
Lactic acid	Peels of potato, green peas, sweet corn, mango, orange and cassava fibrous residue	*Lactobacillus casei*, *Lactobacillus plantarum*	Mudaliyar et al. (2012), Jawad et al. (2013)
Citric acid	Banana peel, pineapple pulp waste	*A. niger*, *Yarrowia lipolytica*	Kareem and Rahman (2013), Bezalwar et al. 2013
Acetic acid	Pineapple peel	*S. cerevisiae* + *Acetobacter aceti*	Raji et al. (2012)
Colours			
Carotenoid	Apple pomace Mung bean waste flour	*Rhodotorula* sp., *Chromobacter* sp., *Micrococcus* sp.	Joshi and Attri (2006)
Torulene	Apple pomace	*Rhodotorula glutinis* + *Debaryomyces castellii*	Buzzini (2001)
Flavours			
α-Terpineol	Orange peel	*Ceratocystis fimbriata*, *Penicillium digitatum*	Badee et al. (2011)
Vanillin	Coffee husk	*Aspergillus niger*, *Pycnoporus cinnabarinus*	Vaithanomsat and Apiwatanapiwat (2009)
Agriculture-oriented products			
Biopesticide	Dry cassava root pieces	*Beauveria bassiana*	Maza et al. (2000)
Indole-3-acetic acid	Cassava fibrous residue	*Bacillus subtilis*	Swain and Ray (2008)
Single-cell protein	Cassava tubers Cucumber peel Orange peel	*Saccharomyces* sp., *Endomycopsis fibuligera*, *Pichia burtonii*, *Rhizopus* sp.	Oliveira and Reis (2001), Yang (1988), Yang et al. (1993)

(continued)

Table 11.1 (continued)

Product	Substrate	Microorganism	Reference
Biofuels			
Bioethanol	Whey Cassava plant residue Sweet potato	*Saccharomyces cerevisiae*, *Zymomonas mobilis*	Zhang et al. (2013)
Biomethane	Tomato processing wastes Vegetable wastes	*Sporotrichum* sp. *Methanothrix soehngenii*, *Methanococcus voltae*, *Fusarium* sp.	Thauer and Shima (2006), Roati et al. (2012)
Biohydrogen	Effluents of dairy, winery and brewery	*Clostridium butyricum*, *Enterobacter aerogenes*	Yokoi et al. (2001), Kapadan and Kargi (2006)

11.4.1 Single-Cell Protein

Single-cell protein refers to proteins extracted from mixed or pure dead and dry yeast, algae, fungi or bacterial cells. Yeast and yeast-like organisms were chosen for the single-cell protein fermentation of food waste, starch, potato and cassava bagasse as well as waste water (Oliveira and Reis 2001). The protein-rich single cells are especially grown on horticultural or agricultural residues. Several studies have been carried out to establish successful production of single-cell protein from horticultural wastes. A mixed culture of *Candida utilis* and *Endomycopsis fibuligera* efficiently and rapidly utilized both starch and free sugars present in cassava starch factory effluents. After 28 h of submerged fermentation, the protein content of the biomass was 22 % (w/v) (Manilal et al. 1991). Protein enrichment of sweet potato residue with amylolytic yeasts has also been attempted using solid-state fermentation (Yang 1988). Sweet potato residues fermented with the yeast strain, *Saccharomyces* sp. IFO1426 could produce an animal feed product containing 16–21 % protein. Yeasts such as *Candida utilis*, *Endomycopsis fibuligera*, *Pichia burtonii* and *Saccharomyces* spp. are used for single-cell protein production (Yang 1988; Yang et al. 1993). Sweet potato distillery wastes are enriched with yeast protein and the feed is utilized for red carp (*Cyprinus carpio* L.) (Mokolensang et al. 2003). In all these processes, production of SCP or protein-enriched biomass from solid starchy wastes proceeds in two steps, i.e. liquefaction and saccharification of starch (polysaccharide) to sugar (monosaccharide) and then utilization of sugar for production of biomass (Ward et al. 2008). In another study the production of single-cell protein from cucumber peel and orange peel was compared by using *Saccharomyces cerevisiae*. The study revealed that cucumber peel generates higher amount of protein (53.4 % per 100 g of substrate) followed by that of orange (30.5 % per 100 g of substrate). Addition of glucose to the supplemented fruit hydrolysate medium enhanced the protein content (60.31 %) within the yeast cell (Mondal et al. 2012). Thus, the single-cell protein production by yeast depends on the growth substrates or media composition.

11.4.2 Enzymes

Enzymes are biological molecules that catalyse reactions in the microorganisms and higher living systems to sustain life. With the advent of the microbiological and biotechnological knowledge, the enzymes are produced in industrial scale and are being applied in various industries. For example, amylases are used in food processing and detergent industries, cellulases and pectinases are applied to clarify juices and papain function to tenderize meat, respectively. Re-engineering of horticultural wastes by certain microorganisms to produce enzymes is one of the major focuses of modern researchers. The production of enzymes from horticultural wastes makes the produce cheaper, hence commercially viable. Enzymes have been categorized in various groups depending upon its structure and function. The important groups are discussed below.

11.4.2.1 Amylases

Starch consists of two types of glucose polymers, namely, amylose (linear chains of unbranched D-glucose residues connected by α-1-4 linkages) and amylopectin (highly branched D-glucose residues; branching occurs in every 24–30 residues as α-1-6 linkages). Amylolytic enzymes catalyse hydrolysis of starch and similar oligo- and polysaccharides. Depending upon the mode of hydrolysis, amylase may be grouped into endoamylase and exoamylase. Endoamylase hydrolyses α-1-4 bonds and bypasses α-1-6 linkage in amylopectin and related polysaccharides. Alpha-amylase is the best known endoamylase and produces oligosaccharides of varying length from starch as the end product. Exoamylase also cleaves α-1-4 bonds; some exoamylase like glucoamylase cleaves both α-1-4 bonds and α-1-6 linkage to generate low molecular weight products of starch (maltose and glucose). They target the substrate bond from the nonreducing end of starch (Kar and Ray 2008).

Mango kernel, a low valued agro solid waste obtained after decortication of mango stone, is a rich source of starch (up to 60 %). Kumar et al. (2013) demonstrated the production of enzyme (0.889 U/g) that was obtained at substrate (mango kernel) concentration of 5 % (w/v), pH 4, temp 30 °C on the 9th day of incubation with *Fusarium solani*. Solid waste from cassava sago industries have potential to be processed into amylase. A study was carried out to estimate the amylase activity by fermentation of cassava waste using three fungal cultures, *Aspergillus niger*, *A. terreus* and *Rhizopus stolonifer*. *R. stolonifer* was observed to be more efficient than the other two fungi with a saccharifying activity of 70 % starch and could release 44.5 % reducing sugar in 8 days of solid-state fermentation (Pothiraj et al. 2006).

11.4.2.2 Cellulases

These are very important group of enzymes having a role for degradation of cellulosic and lignocellulosic plant biomass (Ray 2011). Cellulase and hemicellulase are the two enzymes used for the lignocellulose degradation purpose. An experiment was conducted to investigate the enzyme activity of 15 selected microorganisms (13 fungi and 2 bacteria) on different lignocellulosic substrates. Most microorganisms produced endoglucanase and endoxylanase from all substrates analysed. Only two fungi exhibited high enzyme activities ranging from 0.05 to 0.08 U/ml for endoglucanase and 0.06 to 2.4 U/ml for endoxylanase (Guisado et al. 2004).

Banana waste was used as a solid substrate for the production of cellulolytic enzymes using bacterial consortium. Four bacteria, i.e. *Cellulomonas cartae*, *Pseudomonas fluorescens*, *Pseudomonas putida* and *Bacillus megaterium*, were used to prepare the consortium. The consortium was proved to be successful as it showed high titres of filter paper assay (FPase) (0.178 U/ml on the 20th day), carboxymethyl cellulose assay (CMCase) (1.716 U/ml on the 20th day) and β-D-glucosidase (0.602 U/ml on the 25th day) (Dabhi et al. 2014). Horticultural waste (wood chips, wood dust, palm oil fibre and waste newspaper) in chips form collected from a landscape company in Singapore was utilized as the substrate for the production of cellulase and hemicellulase under solid-state fermentation by *Trichoderma reesei* RUT-C30. The enzyme mixture obtained at the optimal conditions contained FPase (15.0 U/g substrate dry matter, SDM), CMCase (90.5 U/g SDM), β-glucosidase (61.6 U/g SDM), xylanase (52.1 U/g SDM), and β-xylosidase (10.4 U/g SDM). The soluble protein concentration in the enzyme complex was 26.1 mg/g SDM (Xin and Geng 2010). The performance of the crude enzyme complex was better than the commercial enzyme blend.

11.4.2.3 Pectinases

Pectinase is an important enzyme used in food processing industries. Pectinase is used in the processes involving the degradation of plant materials such as expediting juice extraction from fruits (Swain and Ray 2010). *Aspergillus foetidus* was used to produce pectinase from various tropical fruit wastes. Among the different studied media compositions, the medium containing 5 g tropical fruit waste (cashew apple, pineapple, banana and grapes) +0.05 g urea +0.25 g ammonium sulphate facilitated better growth of the fungus. Grape waste containing media was the ideal medium for the production of pectinase with incubation temperature of 40 °C for 8 days (Venkatesh et al. 2009). In another study, 20 moulds were isolated from decaying fruits and municipal solid waste; among them *Aspergillus oryzae* was screened for pectinase production. Different production media were formulated using various fruit waste (peel) in different compositions. It was observed that sapota peel in composition of groundnut oil cake was the best medium for the production of potential exo-pectinase (42 U/ml). The optimum conditions for enzyme production are temperature, 50 °C; pH 4.5 and substrate concentration, 1.5 % (Akbar and Prasuna 2012).

11.4.2.4 Tannase

Tannin acylhydrolase, commonly known as tannase, is known to hydrolyse tannin (common phenolic compound). Tannase catalyses hydrolysis of tannin into gallic acids and glucose (Mahapatra et al. 2005). It is also applied to tannery waste water to reduce the tannic acid concentration in tannery effluent.

A study was carried out to estimate the production of tannase by using grape peel as substrate and mixed culture of fungi. *Aspergillus niger*, *Penicillium chrysogenum* and *Trichoderma viride* were selected for the coculture experiment. It was observed that the combination of *P. chrysogenum* + *T. viride* produced the highest activity of 84 U/g/min than other microbial combinations in an incubation period of 96 h (Paranthaman et al. 2009).

11.4.3 Organic Acids

Organic acids are organic compounds with weak acidic properties. Lactic acid, citric acid and acetic acid are the common examples of organic acids. These acids have a wide range of applications in food processing and pharmaceutical industries.

Microorganisms, mainly fungus, are natural producers of citric acid. Shaikh and Qureshi (2013) conducted a study in which 47 fungal cultures were isolated on rose bengal agar and potato dextrose agar medium. Out of the 47 isolated cultures, seven showed the potential to produce organic acids as per the results obtained from acid unitage value.

11.4.3.1 Lactic Acid

Lactic acid is most widely available organic acid. Lactic acid has gained importance for its application in food, feed, chemical, pharmaceutical and beverage industries. Apart from it, this organic acid is used for the potential production of biodegradable plastics (Ray and Swain 2011) and products for medical care. Lactic acid is regarded as generally recognized as safe (GRAS) for use in food industries by the US FDA (Datta et al. 1995). One of the major procedures for synthesis of lactic acid is through fermentation. Lactic acid synthesis via fermentation is more efficient as compared to chemical synthesis. Homofermentative lactate fermentation produces pure and high-quality lactate.

A study was conducted by Mudaliyar et al. (2012) to estimate the lactic acid production on different horticultural wastes, namely, peels of potato, green peas, sweet corn, mango and orange as substrates. It was observed that 63.33 g/l of lactic acid was obtained for mango peels by *Lactobacillus casei* followed by 54.54 g/l of lactic acid for mango peels by *Lb. delbrueckii*. In another research work, optimization of lactic acid production by fermentation with mango peel was investigated using factorial design. The operation factors were temperature (15–35 °C), initial medium pH (4–10) and duration of fermentation (3–6 days). The results showed the maximum production of lactic acid (17.484 g/l) at an initial medium pH of 10, incubation time of 6 days and temperature of 35 °C (Jawad et al. 2013).

11.4.3.2 Citric Acid

Citric acid is a weak organic acid popularly used as a natural preservative. *Aspergillus niger* is known to be the best producer of citric acid. Banana peel has been proved to be a potential substrate for the production of citric acid. Research reports revealed that production of 82 g/kg dry wt. citric acid has been facilitated by *Aspergillus niger* by the addition of nutrient and trace element supplements to the main substrate, i.e. dried banana peel. The yield of citric acid was 90 % based on the amount of sugar consumed (Kareem and Rahman 2013). Similarly in another study carried out by Bezalwar et al. (2013) optimization of citric acid production was done using different fruit pulp waste as substrate with *Aspergillus niger*. The production was carried out both in the presence and absence of methanol. It was concluded that maximum production, i.e. 3.25 g/kg of citric acid, was obtained by using pineapple pulp in the medium without methanol and 5.25 g/kg with methanol. Pineapple pulp waste was a better substrate for the production of citric acid as compared to pulp wastes obtained from Indian jujube, beetroot apple, guava, papaya and wood apple.

11.4.3.3 Acetic Acid

Microorganisms produce acetic acid via aerobic and anaerobic fermentation. *Clostridium* and *Acetobacterium* can convert glucose to acetic acid directly via anaerobic fermentation. The most trusted method for production of food grade acetic acid is through the application of *Acetobacter* (acetic acid bacteria, genus *Acetobacter*) for oxidative fermentation. A number of horticultural wastes have been used for the production of acetic acid. Pineapple peel was first fermented by *S. cerevisiae* for 48 h to produce ethanol by conversion of the sugars present in it. Subsequently another fermentation step was carried out by *Acetobacter aceti* for 9 days which transformed the ethanol of the mid fermented product to acetic acid. The maximum yield of acetic acid was 4.77 % at optimum conditions (Raji et al. 2012). Another study was conducted to design a batch-type acetifier for wine-vinegar production from pineapple peel waste. Pineapple peel blended with starter solution was fed to the inner stainless-steel perforated peel, solid separator tank (130 mm diameter having perforations of 50 mm size). The acetifier was designed for 5 kg pineapple peel per batch. The acetifier is equipped with a stirrer, an aerator and a monobloc pump. The microorganisms used in this study were *Saccharomyces cerevisiae and Acetobacter rancens*. The acetifier was able to produce an end product of 3.5 l wine vinegar of 2 % acidity per day (Singh and Singh 2007).

11.4.4 Microbial Colour

Natural colourants are applied to foods and beverages to improve the appearance and sensory property of the product (Malik et al. 2012). Microorganisms are potential natural source to produce scalable biocolours. Furthermore, microbial pigments also possess antioxidative and anticancerous properties. Bacteria, algae, fungi, yeasts and actinomycetes contain coloured pigments of different colours and shades; those can be extracted and used as food-colouring agents. Successful production of microbial pigments from cheaper substrates like horticultural wastes can make the extraction of microbial colourant in industrial scale.

Rhodotorula sp. known for successful production of carotenoid pigments was grown on apple pomace. The yeast was grown at 30 °C for 5 days in batch fermentation in a rotary shaker before extraction of carotenoid pigments (Joshi et al. 2003). It has been studied that coculturing of *Rhodotorula glutinis* along with *Debaryomyces castellii* produces more biomass when fermented in a small bioreactor (1 l). Fed-batch cocultures gave a volumetric production of 8.2 mg total carotenoid/l, about 150 % of that observed in batch cocultures. The productions of different carotenoid pigments (β-carotene, torulene, torularhodin) were observed (Buzzini 2001).

Apple pomace has also been used to produce colourants from microorganisms using solid-state fermentation. Apple pomace-based basic medium was used to produce *Rhodotorula* sp. (pink colour), *Chromobacter* (dark red) and *Micrococcus* sp. (light yellow) (Joshi and Attri 2006).

11.4.5 Microbial Flavour

Flavour and aroma is the outcome of various volatile and non-volatile chemical compounds. Non-volatile compounds present in food/beverage confer the taste, whereas volatile compounds contribute to the aroma (Berger 2007). Biosynthesis of flavouring compounds by microorganisms has proved to be a promising technology. Horticultural wastes are used as a fermentation medium for the production of flavour and fragrances with selected microorganisms.

Seven media were prepared by taking different concentrations of cassava bagasse, apple pomace, soya bean and amaranth. Solid-state fermentation was carried out in these media using *Ceratocystis fimbriata* in 250 ml Erlenmeyer flasks at 30 °C. The study revealed that except for pure amaranth medium which produced pineapple aroma, all other media produced fruity aroma (Bramorski et al. 1998).

Orange peel is a by-product produced in food industries. The main component of orange peel oil is D-limonene which constitutes 96.1 % of the total content. Fungal strain of *Penicillium digitatum* was used for the biotransformation of D-limonene to α-terpineol, which has a floral lilac odour and a coniferous odour characteristic. The highest conversion of D-limonene to α-terpineol was achieved by using malt-yeast broth medium at pH of 6.1. The bioconversion of D-limonene to α-terpineol in malt-yeast broth medium increased with incubation period, i.e. 79 % at 3 h and 95.5 % at 7 h after the second substrate addition on the first day (Badee et al. 2011).

Vaithanomsat and Apiwatanapiwat (2009) successfully produced vanillin in a single fermentation step by applying *Aspergillus niger* and *Pycnoporus cinnabarinus* to a substrate obtained from steam explosion of *Jatropha curcas* stem hydrolysate containing 1.55 g/l ferulic acid. The maximum 0.65 g/l of vanillin was obtained with the conversion rate of 45.2 % based on the initial ferulic acid.

11.4.6 Agriculture-Oriented Products

Agro-processing residues are usually rich in sugar, starch and other nutrients providing a suitable medium for solid substrate fermentation for growth of bacteria and fungi that synthesize bioinsecticides, bioherbicides, biofungicides, growth regulators and other useful chemicals applicable to agricultural operations.

11.4.6.1 Biopesticides

Biopesticides include bioinsecticides, bioweedicides and biofungicides and employ a large number of organisms, i.e. parasitoids, predators and microorganisms (bacteria, fungi, nematodes and viruses). The use of entomopathogenic and mycoparasitic fungi for biological control of insects and pests has received special attention (Prior 1998). Entomopathogenic fungi such as *Beauveria bassiana*, *Metarhizium anisopliae* and *Paecilomyces fumosoroseus* are used to control a wide array of insect pests covering a large number of agricultural and horticultural crops (Castellon et al. 2001; Deshpande 1999).

Attempts have been made to produce entomopathogenic fungi in solid-state fermentation. Desgranges et al. (1993) used solid-state fermentation process for production of *B. bassiana* employing agro-industrial wastes. In Cuba, the naturally occurring local strains of *B. bassiana* are isolated and mass produced using dry cassava root pieces as solid substrate. The technology is simple, has a low cost and has been adapted by many farmers and their families. The fungus has also been shown to be safe to humans and wildlife (Maza et al. 2000). Soccol et al. (1997) developed an SSF-based process employing various agro-industrial substrates (potato, cassava and coffee pulp waste) to produce spores from *B. bassiana* for use in the biological control of pests of banana, sugarcane, soya bean and coffee.

11.4.6.2 Plant Growth Regulators

Gibberellins, kinetin, indole-3-acetic acid, indole-2-butyric acid and other compounds are well-known plant growth regulators which promote seed germination, sprouting, stem elongation and flowering (Devlin 2002). These growth regulators can be potentially produced by solid-state fermentation using microorganisms such as *Bacillus subtilis* (producing indole-3-acetic acid) (Swain et al. 2007; Swain and Ray 2008) and *Gibberella fujikuroi* (producing gibberellic acid) (Pastrana et al. 1995).

11.4.7 Biofuels

Fossil fuel is a limited resource hence there is always a search for alternative fuel. Among most of the nonconventional renewable sources of energy such as pyrolysis, solar, thermal, hydro and wind, biogas and bioethanol production is considered the most favoured, simple and economical method of production of energy from wastes, residues and by-products of agriculture, agro-industrial and food processing wastes (Krishna Nand 1994).

11.4.7.1 Bioethanol

Although different processes for ethanol production from sugar, starch or cellulose are feasible, production costs and energy consumption strongly depend on raw materials. The current research on ethanol production from agricultural waste biomass is focused on reducing the production costs, using suitable low cost feedstocks and increasing energy efficiency by means of energy integration of the plant processes.

Ethanol from Starchy Biomass

Ethanol production from plant biomass is being studied extensively by various research laboratories throughout the world. Brazil has developed processes that utilize cassava roots, palm trees, sugarcane, and babassu coconut for ethanol production. The babassu coconut crop in Brazil alone could enable the production of about eight billion litres of ethanol annually. This coconut (23 % starch) is produced at the rate of some 210 million tonnes a year, and theoretically it could be converted to nearly twice the 1980 ethanol production (4.3 billion litres) in Brazil.

Carioca and Scares (1978) obtained a relative yield at 76 % conversion rate of babassu flour (60 % starch) to ethanol in their fermentation experiments. Their process involved gelatinizing the babassu starch at 80–85 °C and then adding a heat-stable α-amylase. Complete saccharification was accomplished by glucoamylase treatment for 40 h at room temperature. Yeast extract was added to the hydrolysed coconut starch to support microbial growth. The fermentation was conducted at 28–30 °C for 42 h. After distillation, the yield was 90 ml of 92 % purity ethanol from 250 g babassu flour in 1 l H_2O.

Potato bagasse produced during chips manufacturing is an economical source for biomass and bioethanol production. Hashem and Darwish (2010) studied the starch saccharification and ethanol production from potato bagasse. Their results demonstrated that 1 % H_2SO_4 at 100 °C for 1 h was enough to hydrolyse the starch contained in the residues. Two strains of *Saccharomyces cerevisiae* (y-1646 and a commercial one) were able to utilize and ferment the acid-treated residue stream under both aerobic and semi-anaerobic conditions. The maximum yield of ethanol was 5.52 g/l at 35 °C by *S. cerevisiae* y-1646 after 36 h. Banana and cooking banana (*Musa* spp.) production systems accumulate a considerable quantity of discard due to high-quality demands of markets. Ripe fruits have high sugar contents, which can be easily processed to ethanol (Graefe et al. 2011). Velásquez-Arredondo et al. (2010) made a critical energy analysis of ethanol production from banana fruit and its lignocellulosic wastes; the study shows that both acid hydrolysis and enzymatic hydrolysis of amylaceous materials from the banana biomass can be considered energetically feasible for ethanol production.

A study was conducted by Walker et al. (2013) on the application of restaurant waste for ethanol production. Food wastes comprising corn, potatoes and pasta were rich in starch content. The waste was first treated using a two-part enzymatic digestion of starch using α-amylase and glucoamylase for production of sugars; later the sugars were converted into ethanol by fermentation using yeast. Ethanol concentrations of 0.8 % (by mass) were obtained because of the low initial composition of starch in the substrate. Zhang et al. (2013) demonstrated an energy-saving procedure to produce ethanol from uncooked fresh sweet potato. Mutant strain of *Aspergillus niger* isolated from mildewed sweet potato was utilized to produce starch saccharification enzymes. Subsequently fermentation of the enzyme-treated sweet potato *Zymomonas mobilis* yielded 14.4 g of ethanol per 100 g fresh roots. Energy consumption of various operative conditions for bioethanol production from sweet potato was studied. Parameters like dry matter ratio of sweet potato to water, fermentation efficiency and sweet potato sugar content on the energy consumption (steam and electricity) were, respectively, evaluated. The best ratio of dry matter to total water to work with sweet potato was 0.2 kg dry sweet potato/kg water. It shows significant reduction in energy consumption (Ferrari et al. 2013).

11.4.7.2 Biomethane

All types of wastes including those from horticultural processing can be ensiled to produce methane for domestic cooking, generating steam for boilers, etc.

Ensilaging and Methane Generation

Ensilaging is used for storing forage crops, and various other agricultural commodities were used with some modification for the storage of mango peel, orange, lemon and lime peels, pineapple and tomato processing wastes. It helps to hydrolyse polymeric constituents such as cellulose, hemicellulose and pectin, improving the biogas yield and its methane content. Similarly, many of these wastes when used for anaerobic digestion, float on the top of digesters and form a thick scum. Ensilaging of green pea shell eliminates this problem.

Pretreatment of Fruits and Vegetables Processing Wastes

Majority of agro-industrial and food processing wastes especially from fruits and vegetables are recalcitrant to biodegradation and bioconversion due to presence components having complex nature of polymeric constituents such as cellulose, hemicellulose, pectin, fats, proteins and lignin. In fact, these complex constituents delay the microbiological attack and consequently prolong the hydraulic retention time thereby, adversely affecting economics of the process. If these feedstocks of fruit and vegetable origin are treated aerobically by consortia or some selected strains of fungi, improvement in the utilization of feedstock is expected for biogas production, Orange processing wastes treated with *Sporotrichum* sp., *Aspergillus* sp., *Penicillium* sp. and *Fusarium* sp. yield 0.5–0.6 m^3/kg VS added with a methane content of 55 % (Krishna Nand 1994).

High-Rate Methane Generation Processes

High-rate biomethanation processes have been developed in UP, the USA, France and the Netherlands for the treatment of effluents and waste waters. These processes are upflow anaerobic sludge blanket (UASB), expanded granular sludge bed (EGSB), fixed film (FF), fluidized bed (FB) and plug flow. Since there are two distinct stages of anaerobic digestion, viz. acidogenic and methanogenic, fixed-bed reactors having polyurethane foam and burnt coke were used for acidogenesis and methanogenesis separately. Methanogens were avoided in the first set of reactors by using chloroform. As high as 13 g/l of volatile fatty acid yield was achieved with 2 days of HRT. The most striking observation recorded was 90 % methane content of biogas.

Fixed-bed reactors have not been used for solid wastes, and therefore, anaerobic digestion of mango peel for methane generation was attempted by homogenizing a 2-year-old ensiled mango peel in equal volume of water and passing the slurry to the PVC plugs, clay granules and iron ore reactors. Stable digestion was achieved with an HRT of 10 days, and it can be considered a simple and economical method of biogas production from mango peel.

Citrus wastes are difficult to digest anaerobically for biogas production due to limonene which inhibits the growth of microorganisms. Attempts were made to utilize these wastes first by ensiling them and then passing the soluble and suspended solids through the fixed-bed reactors having PVC plugs, snail shells (ground), burnt coconut shells and pumice stones. The pumice stone-packed reactor exhibited the maximum rate of methane production, whereas PVC resulted in the higher conversion of carbon into acetate. The maximum biogas yield recorded was 0.72 m^3/kg VS added with a methane yield of 6.5 % of a 7-day HRT (Krishna Nand 1994).

11.4.7.3 Biohydrogen

Effluents generated from food industries such as dairy industry, olive mill, baker's yeast, brewery, winery and distillery are rich in carbohydrates. Such effluents possess tremendous potential to be bioprocessed for production of biohydrogen, a clean fuel with no emission of carbon dioxide (Kapadan and Kargi 2006). The cellulose-/starch-rich effluents are pretreated for removal of undesirable compounds and hydrolysis of complex sugars. Delignification is an important pretreatment procedure in which lignin is removed from the substrate; as a result it facilitates better microbial growth. Two-stage fermentation processes (anaerobic dark fermentation and photo-fermentation) are used for generation of biohydrogen. Dark fermentation occurs under anaerobic conditions, and compounds other than O_2 serve as electron acceptor. Carbohydrates mainly glucose and fructose are the carbon source for the process that produce H_2 with acetic acid and butyric acid (Nath and Das 2006). Photo-fermentation takes place in the presence of the light energy, nitrogenase enzyme and organic acids by photo-heterotrophic bacteria for production of H_2 using organic waste in batch or continuous cultures (Das et al. 2008). Several microorganisms belonging to genus *Clostridium* are known to produce hydrogen during growth phase. Hydrogen production was carried out by using dried sweet potato starch residue. Mixed culture of *Clostridium butyricum* and *Enterobacter aerogenes* was used in long-term repeated-batch operation, and the yield was 2.4 mol H_2/mol glucose from 2 % starch residue containing waste water (Yokoi et al. 2001).

11.5 Bioremediation of Effluents

EPA (Environmental Protection Agency, USA) defined bioremediation as "treatment that use naturally occurring organisms to breakdown hazardous substances into less toxic or non-toxic substances". Several popular bioremediation technologies include bioaugmentation, rhizofilteration, phytoremediation, bioventing and mycoremediation.

A study was conducted to evaluate the bioremediation potential of three allochthonous microorganisms, *Lactobacillus delbrueckii* subsp. *bulgaricus*, *Saccharomyces cerevisiae* and *Dekkera bruxellensis*, over whey effluent, orange effluent and chocolate effluent. The highest biodegradable efficiency of the chemical oxygen demand (COD), biological oxygen demand (BOD) and total organic nitrogen (TON) of the effluents was observed when treatment was done using both allochthonous and autochthonous microorganisms. *S. cerevisiae* reduced the BOD, COD and TON of whey effluents by 12.35 %, 20 % and 68.42 %, respectively. *Dekkera bruxellensis* was proved as the best utilizer of orange effluent and reduced the BOD, COD and TON by 18 %, 20 % and 53.38 %, respectively. *Lactobacillus* is the best utilizer of chocolate effluent that reduced the BOD, COD and TON by 24.27, 18.3 and 63.63 %, respectively (Mcheik et al. 2013). Waste water to beer production ratio varies from 1.2 to 2.0 m^3/m^3, and part of the water is disposed off as effluent containing sugars, soluble starch, ethanol, volatile fatty acids, etc., leading to a BOD/COD ratio of 0.6:0.7 (Brito et al. 2007). Before discharging the brewery effluents into surface waterbodies, it is essential for the removal of nutrients and organic matter. Rodriguez et al. (2011) suggested anaerobic pretreatment combined with subsequent aerobic post treatment as the best solution for the removal of organic load in the brewery effluent.

11.6 Conclusion and Perspectives

Horticultural wastes possess a major stake in the generation of global solid waste. Unlike other wastes such as mining spillage, scraps, plastics, etc., horticulture waste beholds enormous potential to act as a substrate for growth of benevolent microorganisms to produce desired value-added products. In the current chapter, critical discussions have been conducted to focus on various technologies developed to produce high-ended products such as enzymes, biofuel, biocolours and bioflavours by microbial processing of horticultural wastes. Technologies of microbial remediation for detoxification of horticulture-industrial effluents have also been discussed. Because the horticultural residues and effluent are of higher quality in terms of nutrients based on their biological constituents (i.e. starch, sugar, phenol, vitamins and minerals) as compared to agricultural and other solid wastes, it has become a point of attraction for researchers, particularly in the field of bioprocess engineering. In this chapter, we have discussed several updated laboratory technologies for the production of value-added products (except mushroom cultivation) from horticultural waste, but the real challenge is to transform the laboratory-scale research to pilot scale and further to commercial scale.

References

Akbar S, Prasuna RG (2012) Exploitation of fruit wastes for pectinase production using *Aspergillus oryzae*. Int J Pharm Bio Sci 3(3):756–765

Badee AZM, Helmy SA, Morsy NFS (2011) Utilisation of orange peel in the production of α-terpineol by *Penicillium digitatum* (NRRL 1202). Food Chem 126:849–854

Berger RG (ed) (2007) Flavours and fragrances, chemistry, bioprocessing and sustainability. Springer, Heidelberg

Bezalwar P, Gomashe AV, Sanap HM, Gulhane PA (2013) Production and optimization of citric acid by *Aspergillus niger* using fruit pulp waste. Int J Curr Microbiol Appl Sci 2(10):347–352

Bhalla TC, Joshi M (1994) Protein enrichment of apple pomace by coculture of cellulolytic molds and yeasts. World J Microbiol Biotechnol 10:116–117

Bramorski A, Christen P, Ramirez M, Soccol CR, Revah S (1998) Production of volatile compounds by the edible fungus *Rhizopus oryzae* during solid-state cultivation on tropical agro-industrial substrates. Biotechnol Lett 20:359–362

Brito AG, Peixoto J, Oliveira JM, Oliveira JO, Costa C, Nogueira R, Rodrigues A (2007) Brewery and winery wastewater treatment: some focal points of design and operation. In: Oreopoulou V, Winfred R (eds) Utilization of by-products and treatment of waste in the food industry, vol 3, ISEKI series. Springer, New York

Buzzini P (2001) Batch and fed-batch carotenoid production by *Rhodotorula glutinis-Debaryomyces castellii* co-cultures in corn syrup. J Appl Microbiol 90:843–847

Carioca JOB, Scares JB (1978) Production of ethyl alcohol from babassu. Biotechnol Bioeng 20:443–445

Castellon M, Alcazar J, Morales A, Lagnaoui A (2001) The potential use of *Metarhizium anisopliae* for sweet potato weevils *Cylas formicarius* control in Cuba. In: Proceedings of international symposium on sweet potato: food and health for the future. International Potato Center (CIP), Lima, Peru, 23–29 Nov 2001

Chandrasekaran M, Bahkali AH (2013) Valorization of date palm (*Phoenix dactylifera*) fruit processing by-products and wastes using bioprocess technology—review. Saudi J Biol Sci 20(2):105–120

Dabhi BK, Vyas RV, Shelat HN (2014) Use of banana waste for the production of cellulolytic enzymes under solid substrate fermentation using bacterial consortium. Int J Curr Microbiol Appl Sci 3(1):337–346

Das D, Khanna N, Veziroglu TN (2008) Recent developments in biological hydrogen production processes. Chem Ind Chem Eng Q 14(2):57–67

Datta R, Tsai SP, Bonsignore P, Moon SH, Frank JR (1995) Technological and economic potential of poly(lactic acid) and lactic acid derivatives. FEMS Microbiol Rev 16:221–231

Desgranges C, Vergoignan C, Lereec A, Riba G, Durand A (1993) Use of solid state fermentation to produce *Beauveria bassiana* for the biological control of European corn borer. Biotechnol Adv 11:577–587

Deshpande MV (1999) Mycopesticide production by fermentation: potential and challenges. Crit Rev Microbiol 25:229–243

Devlin RM (2002) Plant physiology, 5th edn. East-West Press, New Delhi

Fallow SJ, Wheelock JV (1982) Byproducts from the U.K. food system. Conserv Recycl 5:163–172

Ferrari MD, Guigou M, Lareo C (2013) Energy consumption evaluation of fuel bio-ethanol production from sweet potato. Bioresour Technol 136:377–384

Graefe S, Dufour D, Giraldo A, Luis Armando Muñoz LA, Mora P, Solís H, Garcés H, Gonzalez A (2011) Energy and carbon footprints of ethanol production using banana and cooking banana discard: a case study from Costa Rica and Ecuador. Biomass Bioenergy 35:2640–2649

Guisado G, Lopez MJ, Vargas-Garcia MC, Suarez F, Moreno J (2004) Cellulase and hemicellulose production by microorganisms isolated from plant wastes composting piles. In: Martinez J, Bernal MP (eds) Municipal and industrial residues in agriculture. FAO European Cooperative Research, Rennes, pp 261–264

Hashem M, Darwish SMI (2010) Production of bio-ethanol and associated by-products from potato starch residue stream by *Saccharomyces cerevisiae*. Biomass Bioenergy 34:953–959

Jawad AH, Alkarkhi AFM, Jason OC, Easa AM, Norulaini NAN (2013) Production of the lactic acid from mango peel waste—factorial experiment. J King Saud Univ Sci 25:39–45

Joshi VK, Attri D (2006) Solid state fermentation of apple pomace for the production of value added products. Nat Prod Rad 5(4):289–296

Joshi VK, Attri D, Bala A, Bhushan S (2003) Microbial pigments. Indian J Biotechnol 3:362–369

Kapadan IK, Kargi F (2006) Biohydrogen production from waste material. Enzym Microb Technol 38:569–582

Kar S, Ray RC (2008) Statistical optimization of α-amylase production by *Streptomyces erumpens* MTCC 7317 cells in calcium alginate beads using response surface methodology. Pol J Microbiol 57(1):49–57

Kareem SO, Rahman RA (2013) Utilization of banana peels for citric acid production by *Aspergillus niger*. Agric Biol J N Am 4(4):384–387

Krishna C (2005) Solid-state fermentation systems—an overview. Crit Rev Biotechnol 25:1–30

Krishna Nand (1994) Biogas from fruit waste. Indian Food Ind 13(3):22

Kumar D, Yadav KK, Muthukumar M, Garg N (2013) Production and characterization of α-amylase from mango kernel by *Fusarium solani* NAIMCC-F-02956 using submerged fermentation. J Environ Biol 34:1053–1058

Mahapatra K, Nanda RK, Bag SS, Banerjee R, Pandey A, Szakacs G (2005) Purification, characterization and some studies on secondary structure of tannase from *Aspergillus awamori nakazawa*. Process Biochem 40(10):3251–3254

Malik K, Tokkas J, Goyal S (2012) Microbial pigments: a review. Int J Microb Resour Technol 1 (4):361–365

Manilal VB, Narayanan CS, Balagopalan C (1991) Cassava starch effluent treatment with concomitant SCP production. World J Microbiol Biotechnol 7:185–190

Maza N, Morales A, Ortiz, O, Winters P, Alcazar J, Scott G (2000) Economic impact of IPM on the sweet potato weevil (Cylas formicarius Fab.) in Cuba. International Potato Center, Lima, Peru, pp 52

Mcheik A, Fakih M, Olama Z, Holail H (2013) Bioremediation of four food industrial effluents. Am J Agric Forest 1(1):12–21

Mokolensang JF, Yamasaki S, Onone Y (2003) Utilization of sweet potato distillery by products as feedstuff for red carp *Cyprinus carpio* L. J World Aquacult Soc 34(4):512–517

Mondal AK, Sengupta S, Bhowal J, Bhattacharya DK (2012) Utilization of fruit wastes in producing single cell protein. Int J Sci Environ Technol 1(5):430–438

Mudaliyar P, Sharma L, Kulkarni C (2012) Food waste management-lactic acid production by *Lactobacillus* species. Int J Adv Biol Res 2(1):34–38

Nath K, Das D (2006) Amelioration of biohydrogen production by two stage fermentation process. Ind Biotechnol 2:44–47

Oliveira MA, Reis EM (2001) Biological treatment of waste water from cassava meal industry. Environ Res 85(2):177

Panda SK, Swain MR, Singh S, Ray RC (2012) Proximate compositions of a herbal purple sweet potato (*Ipomoea batatas* L.) wine. J Food Process Preserv 37(5):596–604

Panda SK, Sahu UC, Behera SK, Ray RC (2014a) Bio-processing of bael [*Aegle marmelos* L.] fruits into wine with antioxidants. Food Biosci 5:34–41

Panda SK, Sahu UC, Behera SK, Ray RC (2014b) Fermentation of sapota (*Achras sapota* Linn.) fruits to functional wine. Nutrafoods. doi:10.1007/s13749-014-0034-1

Paranthaman R, Vidyalakshmi R, Murugesh S, Singaravadivel K (2009) Effects of fungal co-culture for the biosynthesis of tannase and gallic acid from grape wastes under solid state fermentation. Glob J Biotechnol Biochem 4(1):29–36

Pastrana LM, Gonzalez MP, Pintado J, Murado MA (1995) Interactions affecting gibberellic acid production in solid-state culture: a factorial study. Enzyme Microb Technol 17:784–790

Pothiraj C, Balaji P, Eyini M (2006) Enhanced production of cellulases by various fungal cultures in solid state fermentation of cassava waste. Afr J Biotechnol 5(20):1882–1885

Prior C (1998) Biological pesticides for low input agriculture. Biocontrol News Inf 10:17–22

Raji YO, Jibril M, Misau IM, Danjuma BY (2012) Production of vinegar from pineapple peel. Int J Adv Sci Res Technol 3(2):656–666

Ray RC (2011) Solid state fermentation for production of microbial cellulase: an overview. In: Golan AE (ed) Cellulase: types and action, mechanisms and uses. Nova, Hauppauge, NY, pp 135–158

Ray RC, Swain MR (2011) Bio-ethanol, bioplastics and other fermented industrial products from cassava starch and flour. In: Pace CM (ed) Cassava: farming, uses and economic impacts. Nova, Hauppauge, NY, pp 1–32

Roati C, Fiore S, Ruffino B, Marchese F, Novarino D, Zanetti MC (2012) Preliminary evaluation of the potential biogas production of food-processing industrial wastes. Am J Environ Sci 8 (3):291–296

Rodriguez CA, Ramond JB, Welz PJ, Cowan DA, Odlare M, Burton SG (2011) Treatment of high ethanol concentration wastewater by constructed wetlands: enhanced COD removal and bacterial community dynamics. In: Proceedings of the IWA microbes in wastewater and waste treatment conference, Goa, 24–26 Jan 2011

Sahu UC, Panda SK, Mohapatra UB, Ray RC (2013) Preparation and evaluation of wine from tendu (*Diospyros melanoxylon* L.). Int J Food Ferment Technol 2(2):167–178

Shaikh Z, Qureshi P (2013) Screening and isolation of organic acid producers from samples of diverse habitats. Int J Curr Microbiol Appl Sci 2(9):39–44

Singh R, Singh S (2007) Design and development of batch type acetifier for wine-vinegar production. Indian J Microbiol 47:153–159

Soccol CR, Ayala LA, Soccol VT, Krieger N, Santos HR (1997) Spore production by entomopathogenic fungus, *Beauveria bassiana* from de-classified potatoes by solid state fermentation. Rev Microbiol 28:34–42

Swain MR, Ray RC (2008) Optimization of cultural conditions and their statistical interpretations for production of indol-3-acetic acid by *Bacillus subtilis* CM5 using cassava fibrous residue. J Sci Ind Res India 67:622–628

Swain MR, Ray RC (2010) Production, characterization and application of a thermostable exo-polygalacturonase by *Bacillus subtilis* CM5. Food Biotechnol 24:37–50

Swain MR, Naskar SK, Ray RC (2007) Indole-3-acetic acid production and effect on sprouting of yam (*Dioscorea rotundata* L.) minisetts by *Bacillus subtilis* isolated from culturable cowdung microflora. Pol J Microbiol 56(2):103–110

Tengerdy RP, Szakacs G (2003) Bioconversion of lignocellulose in solid substrate fermentation. Biochem Eng J 13:169–179

Thassitou PK, Arvanitoyannis IS (2001) Bioremediation: a novel approach to food waste management. Trends Food Sci Technol 12:185–196

Thauer RK, Shima S (2006) Biogeochemistry: methane and microbes. Nature 440:878–879

Vaithanomsat P, Apiwatanapiwat W (2009) Feasibility study on vanillin production from *Jatropha curcas* stem using steam explosion as a pretreatment. World Acad Sci Eng Technol 3:05–20

Velásquez-Arredondo HI, Ruiz-Colorado AA, De Oliveira S Jr (2010) Ethanol production process from banana fruit and its lignocellulosic residues: energy analysis. Energy 35:3081–3087

Venkatesh M, Pushpalatha PB, Sheela KB, Girija D (2009) Microbial pectinase from tropical fruit wastes. J Trop Agric 47(1):67–69

Walker K, Vadlani P, Madl R, Ugorowski P, Hohn KL (2013) Ethanol fermentation from food processing waste. Environ Prog Sustain Energ 32(4):1280–1283

Ward OP, Singh A, Ray RC (2006) Production of renewable energy from agricultural and horticultural substrates and wastes. In: Ray RC, Ward OP (eds) Microbial biotechnology in horticulture, vol 1. Science, Enfield, NH, pp 517–558

Ward OP, Singh A, Ray RC (2008) Single- cell protein from horticultural and food processing wastes. In: Ray RC, Ward OP (eds) Microbial biotechnology in horticulture, vol 3. Science, Enfield, NH, pp 273–298

Watanabe K (2001) Microorganisms relevant to bioremediation. Curr Opin Biotechnol 12:237–241

Xin F, Geng A (2010) Horticultural waste as the substrate for cellulase and hemicellulase production by *Trichoderma reesei* under solid-state fermentation. Appl Biochem Biotechnol 162:295–306

Yang SS (1988) Protein enrichment of sweet potato residue with amylolytic yeasts by solid state fermentation. Biotechnol Bioeng 32:886–890

Yang SS, Jang HD, Liew CM, Du Preez JC (1993) Protein enrichment of sweet potato residues by solid-state cultivation with mono-and co-cultures of amylolytic fungi. World J Microbiol Biotechnol 9:258–264

Yokoi H, Saitsu AS, Uchida H, Hiroe J, Hayashi D, Takasaki Y (2001) Microbial hydrogen production from sweet potato starch residue. J Biosci Bioeng 91:58–63

Zhang P, Chen C, Shen Y, Ding T, Ma D, Hua Z, Sun D (2013) Starch saccharification and fermentation of uncooked sweet potato roots for fuel ethanol production. Bioresour Technol 128:835–838

Zheng Z, Shetty K (1998) Cranberry processing waste for solid state fungal inoculant production. Process Biochem 33:323–329

Chapter 12
Microbial Interaction in Mining Soil

Smita H. Panda, Santanu Kumar Jena, Sushrirekha Das, Nakulananda Mohanty, and Umaballav Mohapatra

12.1 Introduction

Mining is the extraction (removal) of minerals and metals from the Earth. Mine wastes have been generated from mining activities for several centuries. Mine tailings usually contain high concentrations of metals (Cu, Zn, Fe, Mn, Ni, Pb, Cd, etc.) ranging from 1 to 50 g/kg (Monica et al. 2008). Manganese, tantalum, cassiterite, copper, tin, nickel, bauxite (aluminum ore), iron ore, gold, silver, and diamonds are just some examples. Mining is generally very destructive to the environment. It is one of the main causes of deforestation. In order to mine, trees and vegetation are cleared and burned. With the ground completely bare, large-scale mining operations use huge bulldozers and excavators to extract the metals and minerals from the soil. Metals are a significant toxic factor to biota in the environment. In order to amalgamate (cluster) the extractions, they use chemicals such as cyanide, mercury, or methylmercury. These chemicals go through tailings (pipes) and are often discharged into rivers, streams, bays, and oceans (Peterson and Heemskerk 2001). This pollution contaminates all living organisms within the body of water. Mining occurs in many places around the world, including the USA, South America, Indonesia, South Africa, and other SE Asian countries. In South America, mining is particularly active in the Amazonia region, Guyana, Suriname, and other South diamonds. In Central Africa, mining devastated a national park called Kahuzi–Biega in the eastern Democratic Republic of Congo (DRC).

Microorganisms, also called microbes, are living things that can be observed by microscope. They are the dominant organism in the Earth. Microorganisms live

S.H. Panda (✉) • S. Das • N. Mohanty
Department of Zoology, North Orissa University, Takatpur, Baripada 757003, Odisha, India
e-mail: panda.smita@gmail.com

S.K. Jena • U. Mohapatra
Department of Botany, North Orissa University, Takatpur, Baripada 757003, Odisha, India

L.B. Sukla et al. (eds.), *Environmental Microbial Biotechnology*, Soil Biology 45,
DOI 10.1007/978-3-319-19018-1_12

nearly everywhere: in soil, water, air, animals, humans, plants, and foods. They live under natural conditions, i.e., hot, cold, salty, arid, acidic, alkaline, high pressure, oxygen-free or toxic-hot springs, geysers, volcanoes, and ocean vents. It was once thought that the region far beneath the soil and upper crust of the Earth surface was sterile. However, bacterial populations can be found in petroleum, coal, and mineral deposits containing copper, zinc, gold, and uranium. Moreover, they affect all groups of organisms (bacteria, fungi, yeast, and algae) and ecosystem processes, including microbial-mediated processes and activities. Several remediation approaches have been developed such as excavation, land fill, thermal treatment, electro reclamation, and soil capping, but all are expensive and environmentally destructive. Microbes, however, play the key role in mineralization of biological compounds, especially biopolymers, e.g., lignocellulose and chitin by decomposing. Thus, they are essential for the global biogeochemical cycling of elements. Keeping the above factor in mind, this chapter will discuss the interactions of microbes in soil affected by mining activities.

12.2 Interaction of Microorganism in Mining Soil

There are billions of microorganisms in a mere handful of typical garden soil. Bacterial cells that can accumulate high quantities of precious metals are an efficient and green alternative to traditional recycling methods. Microbes could soon be used to convert metallic wastes into high-value catalysts for generating clean energy. *E. coli* cells are surrounded by nanoparticles of palladium and gold (black deposits), and *Desulfovibrio desulfuricans* was able to reduce palladium in industrial wastes into metallic nanoparticles with biocatalytic activity. Now, the precise molecules involved in the reduction process have been identified. Hydrogenase enzymes located on the surface membrane of the bacterium carry out the reduction of palladium, which results in the accumulation of catalytic nanoparticles. The bacterial cells coated with palladium nanoparticles are known as "BioPd." Some micron-sized bacteria and nanometer (nm)-sized particles of magnetic iron showed that the anaerobic, rod-shaped extremophiles can convert iron hydroxide to magnetic iron or magnetite (Neubauer et al.2002). Due to the phytotoxicity, the soil contains high levels of available metals. The adaptation of microorganisms to heavy metal-rich environments shows activities for biosorption, bioprecipitation, extracellular sequestration, transport mechanisms, and/or chelation (Nanda and Abraham 2013). Such resistance mechanisms are the basis for the use of microorganisms in bioremediation approaches.

12.3 Metals in Soil–Microbe System

Microbes interact with metals and minerals in natural and synthetic environments, altering their physical and chemical state, with metals and minerals also able to affect microbial growth, activity, and survival. In addition, many minerals are biogenic in origin, and the formation of such biominerals is of global geological and industrial significance, and provide important structural components for many organisms, including important microbial groups such as diatoms, foraminifera, and radiolaria (Gadd and Raven 2010). Geomicrobiology can simply be defined as the roles of microbes in geological processes (Konhauser 2007; Ehrlich and Newman 2009).

Metals play an integral role in the life process of microorganisms. Some metals, such as calcium, cobalt, chromium, copper, iron, potassium, magnesium, manganese, sodium, nickel, and zinc, are essential and serve as micronutrients and are used for redox processes (Bruins et al. 2000). Many other metals have no biological role (e.g., silver, aluminum, cadmium, gold, lead, and mercury) and are nonessential and potentially toxic to microorganisms. These metal-resistant potentials were made by the metal-containing environment (waste dumps of mining industry), and they could posses effective bioremediation or metal sorption potentiality. Two bacterial strains such as *P. aeruginosa* absorb heavy metals very effectively (in the order of Cd, Zn, Cr, and Ca) than *T. ferrooxidans*. The *T. ferrooxidans* also reduce the heavy metal contents from the effluents (soil) of mining (magnetite and bauxite) industry in a very short period, even growing at pH 2–3 and temperature 50 °C (Mathiyazhagan and Natarajan 2011). Due to heavy metal tolerance and metal absorption potentiality of these bacterial species, they have a suitable choice for bioleaching and bioremediation processes (Mathiyazhagan and Natarajan 2011). Metal contamination of soil ecosystems could be used as measures of reclamation progress and success. Historically, soil chemical and physical parameters have been used as indicators of soil quality.

12.4 Acid Mine Drainage and Acid Rock Drainage

Mining areas has been found in the formation of acid mine drainage (AMD). The runoff from mining heaps of active and abandoned mines can reach to very low pH values. The microbes are mainly responsible for the formation of AMD are metabolically active even below pH 2 (Rawlings 2002). If the chemical and microbial processes in the exposed overburden are set into motion once, AMD formation is hard to control again and can last for incalculably long times. Chemical and biological oxidation of the abundant mineral pyrite (FeS_2) takes place after the unearthing of pyrite containing rock formations and results in an acidification of the dump material. AMD has a typical orange appearance which is due to the iron hydroxide that is formed during the oxidation. The iron hydroxide precipitates as sludge, coating the bottoms of streams and canals. Under acidic conditions, most heavy metals are leached from the dump waste and are subsequently transported as AMD in stream waters; if they are not collected, conditions required for the generation of AMD are (1) contact with the atmospheric oxygen, (2) an aqueous environment, (3) and the occurrence of iron-oxidizing, acidophilic bacteria. Iron-oxidizing bacteria like members of the genera *Thiobacillus*, *Leptospirillum*, and *Ferroplasma* use Fe^{2+} as electron donor to satisfy their energetic demands. But due to the high energy demand for autotrophic life, supply of reducing power for CO_2 fixation—the energetic yield of the Fe^{+2} to Fe^{+3} oxidation—is relatively scarce for the overall energy requirement of the cell. To satisfy the energy demand and to maintain the vital functions of the cell, the substrate turnover has to be high (uranium mining). The formation of one-gram biomass (dry weight) requires the oxidation of an amount of about 55 g Fe^{+2}. In turn, Fe^{+3} oxidizes pyrite in a fast autocatalytic mechanism in the presence of water under generation of protons which leads to a pH decrease. In the overall reaction, the part of the abiotic oxidation of iron is comparatively slow under acidic conditions.

12.5 Microbial Processes That Can Change Metal Mobility

From a metal point of view, several different situations concerning the metal distribution between mobile and immobile phases in mine waste environment can occur. The presence of bacteria can influence the distribution of metals between these phases in different ways. Free-living bacteria constitute mobile suspended particles which may have a capacity higher than that of the surrounding environment. If the majority of bacteria are growing in biofilms on surfaces, the transport of metals may be reduced. Several bacteria may produce compounds that precipitate metals, e.g., sulfides. The interaction between metals and microorganisms can be disturbed by the presence of other compounds, like clay minerals, inorganic anions, competing cations, complexing organic matter, etc. The metals can become hydrated, chelated, or adsorbed to these compounds which can make the metal less available for microbial interaction. Microorganisms participate in the cycling of carbon and thereby influence the amount and character of the organic matter. This can be of substantial importance for metal mobility, because organic compounds of various sizes can bind metals. Microbial degradation may change immobile metal–organic compounds to mobile, water-soluble metal forms. *Klebsiella oxytoca* was observed when metals such as Al, Cd, Co, Cu, and Zn were bound to the organic anion. Sulfide, which is produced from sulfate by sulfate-reducing bacteria in anoxic environments with an abundance of sulfate and readily degradable organic matter, can interact strongly with many metals to generate barely soluble compounds. Nitrifying bacteria produce nitric acid, and sulfide-oxidizing bacteria produce sulfuric acid. Specific chelating are exudates produced by microorganisms that require iron or other essential metals for growth.

12.6 Survival Strategies of Heavy Metal-Resistant Bacteria

Heavy metals affect the microbial cell in various ways. From mining sites, the resistance toward the different metals seems to be higher than pure cultures. A great number of heavy metal-resistant bacteria such as *Cupriavidus metallidurans* is known to possess efflux transporters that excrete toxic or over-concentrated metals. Well-adapted microbes isolated from these types of habitats can probably support remediation and are of great interest for strategies on environmental conservation. The physiology of metal-adapted microorganisms determines the methodology that has to be applied for isolation and cultivation. With microbes playing a role in the mobilization/immobilization of metals in soils, microbial activity and the effects of the soil matrix itself on mobility and sorption of metals have to be differentiated. From genomic data mining, Actinobacteria presumably can have cation efflux transporters, but they were not functionally identified yet. The ABC transporters of Actinobacteria are, in contrast, well investigated and are responsible for antibiotic resistance. Some of the ABC transporters function as metal efflux pumps as well. Numerous soil microbes, for example, the widespread fungus *Aspergillus niger*, solubilize metals by the release of organic acids, while others—or even the same microorganisms—immobilize metals through the excretion of compounds as, e.g., oxalates (White et al. 1998). If toxic metals have entered the cell and cannot be excreted by efflux transporters, several organisms have developed a cytosolic sequestration mechanism for protection. It has been shown for many metal-resistant organisms that internal inclusion bodies, e.g., polyphosphate granules (volutin), bind large amounts of metal cations. The cell envelope equips the cell with an additional metal resistance feature. The cell wall, in combination with the cell membrane, supports the sorption of metals and facilitates bioreduction as well. It has been shown that, for example, *Penicillium chrysogenum* has the capacity to reduce silver. After reduction, the metallic silver precipitates at the cell wall. Metal resistance of microbes is accomplished by intra- and extracellular mechanisms; metals can be excreted via efflux transport systems; sequestering compounds of the cytosol can bind and detoxify metals inside the cell; the release of chelators into the extracellular milieu leads to bound and fixed metals; the structure of the cell envelope is prone to bind large amounts of metals by sorption thus preventing influx.

12.7 Role of Microorganism in Mining Soil

The role of the microorganisms in total ecosystem function and the sensitivity of soil microbial communities to disturbance, biological indicators may also be useful as measures of mine soil reclamation (Sousa et al. 2007). The microbial community is an integral component of soil quality and plays critical role in the cycling of nutrients and formation of soil structure. Soil microorganisms are also highly sensitive to disturbance in the soil ecosystem, and changes in soil microbe activities may be effective at early signals of degradation or improvement of soil (Harris 2003). Microbes can potentially accumulate metals either by a metabolism-independent, passive, or a metabolism-dependent mechanism in the mining areas. It is a clear mutual influence: microbes in soil are not only affected by their environment directly or indirectly, but they also control particular soil parameters. Growth and metabolism can lead to changes in pH, redox potential, and ionic strength of the soil. For example, oxidation of pyrite by members of the genus *Acidithiobacillus* resulted a strong pH decline and higher mobility of heavy metals (Gadd 2010). This process of metal mobilization, in turn, determines the species composition within the habitat to a great extent. The microflora, again, strongly participates in processes like decomposing soil constituents as well as particle aggregation and influences soil texture and availability of nutrients for plants. This means that the food web in the soil is constituted to a high degree by microbes, which (1) produce substances that change the microenvironment by solubilization of minerals and subsequent rock breakdown (Cole 1979); (2) modify the soil structure, by production of extracellular polysaccharides; and (3) influence the biogeochemical cycling of elements, e.g., sulfur. The impairment of the biological activity of soils due to metal loading leads basically to a reduction in decomposition and turnover rates of organic matter (Haferburg and Kothe 2007). The bioavailability of metals in the habitat is influenced by the constitution of the soil matrix, climatic conditions, microbial activity, and especially the water flow. The metals contained in soil minerals are released into the soil solution as a result of weathering processes. Among the many parameters that govern the behavior of a metal in the soil, the hard–soft character of a metal is not to underestimate as it determines the ligand preferences of the metal (Haferburg and Kothe 2007). The ligand preference, in turn, affects the distribution and speciation of the metal, thereby influencing the organisms of the habitat. Biologically essential metals, like nickel, prefer oxygen ligands and usually form ionic bonds with the ligands. On the other hand, many toxic metals, e.g., cadmium, are often associated with environmental pollution, have a higher affinity for nitrogen and sulfur-containing ligands, and form bonds of covalent character. Microbes probably lack highly specific uptake systems for most metals. The surface characteristics of the bacteria determine their metal-adsorption properties. The differences in cell wall construction of Gram-positive and Gram-negative bacteria have minor influence on the sorption behavior of different metals. Many microorganisms are capable of enzymatic oxidation or detoxification of cyanide, which is used in processing gold ores. *Pseudomonas paucimobilis is* used for oxidation of both thiocyanate and cyanide to ammonia.

12.7.1 Interaction of Microorganism in Manganese-Bearing Mining Soil

Many different organisms have the ability to catalyze Mn oxidation, including a diverse array of bacteria, fungi, algae, and even eukaryotes (Ghiorse 1984). Among the prokaryotes, the ability to oxidize Mn is also quite widespread, included are members of many phylogenetic and physiological groups, e.g., cyanobacteria, a diversity of heterotrophic rods and cocci, the sheathed-like (*Leptothrix*) and budding-like (*Hyphomicrobium*) bacteria, some purported autotrophic strains related to *Pseudomonas* species, and the still-controversial *Metallogenium* group. The anaerobic lactobacilli, which utilize the Mn oxidation reaction as a protection against oxygen toxicity, are not included, as they do not precipitate extracellular Mn oxides, but rather accumulate millimolar levels of protein-associated Mn in the cytoplasm. So many bacteria have been identified as Mn oxidizers.

12.7.2 Interaction of Microorganism in Copper-Bearing Mining Soil

Copper has been one of mankind's most important metal resources, since the beginning of civilization, and to this day it holds an important role in the functioning of modern society. Its main uses include piping, coinage, electronics, and even antibiotics. Copper is rarely found in its native form; most often it exists as ores made up of various copper sulfides and oxides that have little practical value of their own resources. As such, methods for mining copper involve long and complex chemical processes to produce metal from the ores. This involves using harsh chemical reagents to pull the ore from the rock and subsequent treatment to further isolate the metal and extract pure copper. These processes are energetically expensive, and they create a significant amount of pollutant chemicals such as sulfur dioxide and cyanide. A viable alternative to these harmful procedures is biomining or the use of bacterial metabolic processes to assist in the solubilization of metal ores. *Acidithiobacillus ferrooxidans* is a lithoautotroph famous for using ferrous iron (Fe^{2+}) as an electron donor along with elemental sulfur. They grow in subterranean mineral deposits and even sewer pipes, and their activity has been linked to the processing of mineral ores. Copper is not a metabolite of *A. ferrooxidans*, so the bacteria are not directly responsible for the solubilization of copper. Rather, it occurs through an abiotic reaction between the copper ores and the ferric ions and protons produced by the metabolic activities of the bacteria. These reactions regenerate the reduced iron and sulfur compounds that the bacteria reuptake to start a new cycle. The copper-solubilizing activity of *A. ferrooxidans* strains isolated from a low-grade copper mine in Chile revealed that when grown in a chalcopyrite-containing medium, a particular strain called D3-2 had a greatly enhanced solubilizing activity in relation to other strains. The reason for this increased activity is connected to the D3-2 strain's resistance to sulfite (SO_3^{2-}), a toxic intermediate in sulfur oxidation. Sulfite hinders the activity of iron oxidase, an important enzyme in the bacterium's iron oxidation pathway which provides ferric ions for the solubilization of sulfides (Sugio et al. 1994). Most organisms including *A. ferrooxidans* possess a sulfite oxidase enzyme that converts sulfite to sulfate under low-sulfite conditions (Feng et al. 2007), but this activity tends to be inhibited at higher concentrations. However, the D3-2 strain has a sulfite reductase that is more resistant to sulfite and thus can oxidize it faster, resulting in a lowered sulfite concentration that doesn't inhibit the iron oxidation pathway (Sugio et al. 2008).

12.7.3 Interaction of Microorganism in Arsenic-Bearing Mining Soil

The toxic element arsenic (As) occurs widely in solid and liquid mine wastes. Microbes play a significant role in the precipitation and dissolution of these As-bearing miners. The weathering of solid mine wastes contaminates waters with potentially toxic components, such as arsenic (As), in concentrations that may pose serious hazards to human and ecosystem health (Hudson-Edwards and Santini 2013). The stability and potential mobility of As in mine wastes is controlled by their uptake in minerals. Many of the reactions that result in the dissolution of primary As-bearing minerals, and formation of secondary products, are mediated by prokaryotes, i.e., bacteria. The two soluble inorganic forms of arsenic, arsenate [As (V)] and arsenite [As (III)], occur widely in mining-affected environments. Even though these forms of As are highly toxic, some microbes can gain energy from their reduction and oxidation, respectively (Rhine et al. 2005). Microorganisms they use As (V) as the terminal electron acceptor in anaerobic respiration are known as arsenate respirers, and they can couple As (V) reduction to the oxidation of both inorganic (e.g., sulfide) and organic (e.g., acetate) electron donors (Stolz 2012). Other organisms can use As (III) as the electron donor coupled to the reduction of oxygen, nitrate, or chlorate (Osborne and Santini 2012). Most microbes they oxidize As (III) or respire As (V) are able to survive and grow in As-contaminated mining soil, and many have acquired additional As-resistance mechanisms (Stolz et al. 2006). Several anaerobic arsenate-respiring bacteria have also been isolated from arsenic-contaminated mining soil, for example, *Chrysiogenes arsenatis*. Several organisms found in mining-impacted soil which are involved in the metabolism of other constituents of the minerals (e.g., oxidation of Fe, S, etc.) are often resistant to As either via the Ars system or another unknown mechanism, but do not metabolize it. These include iron-oxidizing (e.g., *Acidithiobacillus ferrooxidans*), iron-reducing (e.g., *Shewanella putrefaciens*), and sulfur-oxidizing organisms (e.g., *Acidithiobacillus caldus*). Some species of *Shewanella* have, however, been shown to respire As (V). Bacteria are closely involved in the oxidative and reductive dissolution of As-bearing mine waste minerals.

For example, As-bearing sulfide minerals such as arsenopyrite, enargite, orpiment, and realgar can be oxidized by aerobic bacteria such as *Acidithiobacillus ferrooxidans* and *Leptospirillum ferrooxidans* (Hudson-Edwards and Santini 2013). As-bearing minerals found in mine wastes have been shown to undergo microbially assisted reductive dissolution. The dissimilatory arsenate reducers *Shewanella* sp. O23S and *Aeromonas* sp. O23A play a significant role in releasing As (V) from As-bearing minerals in the Zloty Stok gold mine.

12.7.4 Role of Microorganism in Arsenic-Bearing Soil

Most prokaryotes that oxidize As (III) or respire As (V) are able to survive and grow in As-contaminated mining soils. Some organisms also detoxify inorganic As by methylation. This produces monomethyl arsonate [MMA (V)], methylarsonite [MMA (III)], dimethylarsinate [DMA (V)], dimethylarsenite [DMA (III)], and trimethylarsine oxide (TMAO), as well as volatile arsines (As-III) (Stolz et al. 2006). Microbial oxidation of arsenite to arsenate is much more rapid than chemical oxidation of arsenite to arsenate (Tamaki and Frankenberger 1992). Therefore, rapid arsenite to arsenate oxidation in the environment can be attributed to arsenite-oxidizing bacteria. Microbes solubilize minerals directly (through attachment) to secure essential nutrients or indirectly (without attachment) by changing aqueous solution chemistry so that reactive solutes are produced. Mineral dissolution rates can then be increased or decreased due to the production of pH gradients at the mineral surface or to the improvement of the kinetics of secondary mineral formation (Welch et al. 1999). Experimental studies have demonstrated that bacteria are closely involved in the oxidative and reductive dissolution of As-bearing mine waste minerals. For example, As-bearing sulfide minerals such as arsenopyrite, enargite, orpiment, and realgar can be oxidized by aerobic bacteria such as *Acidithiobacillus ferrooxidans* and *Leptospirillum ferrooxidans*. Such bacteria, which typically occur in acid mine drainage environments, are described in detailed reviews. Islam et al. (2013) have also demonstrated that native arsenic can be dissolved at alkaline pH (~8) with the assistance of microbes. The involvement of bacteria in the oxidation of arsenopyrite has been comprehensively reviewed by Corkhill and Vaughan (2009). The studies cited in their work demonstrated that oxidation of arsenopyrite by acidophilic Fe- and S-oxidizing bacteria is considerably more effective than abiotic oxidation. For example, the oxidative dissolution of arsenopyrite was shown to be much more rapid and extensive when it occurred in the presence of *Leptospirillum ferrooxidans* than abiotically. This bacterium produced dissolution in the mineral surface, and As was oxidized more rapidly than S or Fe. The oxidation of arsenopyrite and As-bearing pyrite by *Acidithiobacillus ferrooxidans* can also result in pitting on the mineral surfaces and of thin coatings of jarosite. It should be noted, however, that in these cases, the bacteria are oxidizing the Fe and/or S, making the As more accessible to abiotic oxidation.

12.7.5 Interaction of Microorganism in Zinc-Bearing Mining Soil

Zinc is an essential metal, since it is a constituent of many enzymes and proteins. However, excessive concentrations of this metal are well known to be toxic to most living organisms. Elevated concentrations of Zn exist in many agricultural soils from management practices including application of sewage sludge or animal manure and from mining activities, and this may represent a risk to environmental quality and sustainable food production (Li and Christie 2001). Many evidences suggest that microorganisms are far more sensitive to heavy metal stress than animals or on the same soils. In recent years, several studies have shown the harmful effects of metals in high concentrations on microbial diversity and their activity in the soil (McGrath et al. 1995). Zinc only occurs as the divalent cation Zn^{2+}, which does not undergo redox changes under biological conditions. Zinc is a component in a number of enzymes and DNA-binding proteins, for example, zinc-finger proteins, which exist in bacteria. However, zinc is apparently less toxic than copper. In *Escherichia coli*, the toxicity of zinc is similar to that of copper, nickel, and cobalt. Most commonly, the mechanism of resistance in bacteria is an efflux of the toxic metals by the action of P-type ATPases or secondary efflux systems. An important mechanism on this respect is the synthesis of extracellular polymeric substances, a mixture of polysaccharides, mucopolysaccharides, and proteins which can bind significant amounts of potentially toxic metals and entrap precipitated metal sulfides and oxides. In bacteria, peptidoglycan carboxyl groups are main cationic-binding sites in Gram-positive species. Zn toxicity initially inhibited the growth rate of two bacterial strains isolated from Zn-polluted soil and represented the two most abundant cultivable bacterial groups in such soil. However, after few hours, *Brevibacillus* sp. (strain B-I) recovered its ability to grow in a Zn-polluted medium. The development of resistance against heavy metal ions is a generally observed phenomenon.

12.7.6 Interaction of Microorganism in Lead-Bearing Mining Soil

Heavy metal pollution is a serious environmental problem. Heavy metals have an obvious effect on or are potentially harmful to biota. Among heavy metals, lead is commonly associated with soil pollution, is considered to be particularly toxic, and is responsible for significant decreases in biological activities in soils (Smith et al. 1997). Discharge of heavy metal has detrimental effects on human health and the environment. Lead mines are major sources that release significant amounts of Cd, lead (Pb), and chromium (Cr) into the natural environment. Recently, the effect of metal toxicity on microorganisms has received special attention because microorganisms are key components for recycling of nutrients. The effects of heavy metals on the soils around lead mine are complex and very difficult to study because of the diversity of the microorganisms that are present in the soils. Some microorganisms can tolerate toxicity of heavy metals, whereas others cannot. However, microorganisms that have the ability to resist the toxicity are not yet well known. Some microorganisms have the ability to remove heavy metals. Many studies have shown that even small amounts of heavy metals in the soil have detrimental effects on all living organisms and decrease litter decomposition and subsequently nutrient cycling in the whole ecosystem. Heavy metal pollution inhibits microbial processes such as mineralization in soil and litter decomposition. Therefore, the normal functioning of a microbial community in soils around the lead mine tailing might have been weakened (Qing et al. 2007).

12.7.7 Interaction of Microorganism in Iron-Bearing Mining Soil

Iron like aluminum is a commonly occurring metallic element. The typical range of iron concentrations in soils is form 0.2 to 55 %. Iron occurs in either the divalent (Fe^{+2}) or trivalent (Fe^{+3}) states. Iron occurs predominately as Fe^{+3} oxides in soils. Significant amount of iron are released into environment as a result of coal mining. Iron is an important element for growth of almost all living microorganisms since it acts as a catalyst in various enzymatic processes, oxygen metabolism, electron transfer, and DNA and RNA synthesis (Touati 2000; Verkhovtseva et al. 2001). Microorganisms play an essential role in the mining soil. Microorganisms produce chelating agents like siderophores of low molecular mass (200–2,000 Da), especially in iron-limiting conditions (Schwyn and Neilands 1987). Siderophores in soil increase Fe bioavailability by promoting the dissolution of iron-bearing minerals. The role of siderophores is primarily to scavenge iron and also form complexes with other elements (Mo, Mn, Co, and Ni) from the surrounding environment and make them available for microbial cells. Siderophores have three main functional groups—hydroxamate, catecholate, and carboxylate—forming very strong complexes with iron. Siderophores play an important role in the extracellular solubilization of iron from minerals and make it available to microorganisms (Lamont et al. 2002; Dale et al. 2004). Most of the bacterial siderophores are catecholates, and few are hydroxamates and carboxylates, whereas most fungal siderophores are hydroxamates.

12.7.8 Interaction of Microorganism in Chromium-Bearing Mining Soil

Chromium is highly toxic nonessential metal for microorganisms, and its occurrence is rare in nature. Lower to higher chromium-containing effluents and solid wastes are released by activities such as mining, metal plating, wood preservation, ink manufacture, and dye, pigment, glass, and ceramic, tanning, and textile industries (Dhal et al. 2013). Besides natural process like weathering, biochemical processes also contribute to the mobility of chromium which enters into the soil affecting the plant growth and metabolic function of living species. The heavy metal contamination is a serious problem to the environment, because the anthropogenic activities from mining, processing, and applications of these metals have increased enormously during the fast decades and have become a challenge for life on Earth. Hence, the removal or remediation has become all the more necessary. Chromium is one of the most used contaminants. Chromium-bearing ores are found in many forms, but the economically extractable form is the mineral chromite.

Chromium can exist in several oxidation states, ranging from Cr^{2+} to Cr^{6+}, but in soils the most stable forms are trivalent Cr (III) and hexavalent Cr (VI) species, which display quite different chemical properties and affect organisms in different ways. Trivalent chromium tends to be absorbed on soil surface or precipitates as chromium hydroxide in a slightly acidic and alkaline environment. Chromium is an essential micronutrient in the diet of animals and humans, as it is indispensable for the normal sugar, lipid, and protein metabolism of mammals. Its deficiency in the diet causes alteration to lipid and glucose metabolism in animals and humans. Chromium is included in the complex named glucose tolerance factor.

12.7.9 Role of Microorganism in Chromium-Bearing Mining Soil

Chromium in mining soil has a low microbial population due to chromium toxicity. The bacterial species are able to grow in the toxic conditions and are generally assumed to be tolerant/resistance to chromium. The terms resistance and tolerance are often used interchangeable ably, but their significance is different. Gadd (2005) defined "resistance" as "the ability of a microorganism to survive toxic effects of metal exposure by means of a detoxification mechanism produced in direct response to the metal species concerned" and defined tolerance as "the ability of a microorganism to survive metal toxicity by means of intrinsic properties and environmental modification of toxicity." The first chromium-reducing bacteria are *Pseudomonas* spp.

The positive effects of chromium are known in plants and microorganisms. However, elevated levels of chromium are always toxic, although the toxicity level is related to the oxidation state of chromium. Cr (VI) is not highly toxic to all forms of living organisms, mutagenic in bacteria, and mutagenic in humans and animals. It has been difficult to assess the toxicity of chromium to soil microorganisms. In a soil chronically polluted with chromium by leather tannery activity, the oxygenic phototropic microorganisms and heterotrophic bacterial communities are both affected by chromium. The size of the cultivable heterotrophic bacterial community is generally not affected by chromium pollution, but there is a relationship between the percentage of chromate-tolerant bacteria and the level of chromium in the soil. The Gram-positive bacteria are more chromate tolerant than Gram-negative bacteria. (Shi et al. 2002) established that chromium negatively affected the soil microbial activity and accumulation of soil organic carbon.

12.7.10 Role of Microorganism in Aluminum-Bearing Mining Soil

Aluminum is the third most abundant element in the Earth's crust after silicon and oxygen. Various microbes are involved in the formation of some aluminum-containing minerals through bioweathering. The formation of bauxite (bauxitization) involves two stages where microbes are involved. The major constituents of bauxite are Al_2O_3, Fe_2O_3, and SiO_2 or aluminosilicate in various forms, and the source material for bauxitization may be volcanic and other aluminosilicate rocks, limestone, and alluvium. Weathering of source rock (formation of protobauxite) is promoted by those activities of bacteria and fungi that mobilize aluminum, iron, and silicon, which are then subsequently precipitated as oxides, silica, and silicate minerals. Maturation of protobauxite to bauxite is promoted by iron-reducing and fermentative bacteria under anaerobic conditions, which selectively mobilize iron oxides and silica or silicate and enrich the bauxite in aluminum (Ehrlich and Newman 2009).

12.8 Remediation Technologies

Various physiochemical and biological remedial technologies have been developed over the last decades, and selection of each technology is site specific (Mulligan et al. 2001). Biochemical process such as bioleaching involving *Thiobacillus* spp. and *Aspergillus niger*, biosorption of low concentrations of metals in water by algal or bacterial cells, bio-oxidation or bioreduction of metal contaminants by *Bacillus subtilis* and sulfate-reducing bacteria, and biomethylation of metals such as arsenic, cadmium, mercury, or lead have shown some promises and could be used for soil sediment treatments (Mulligan et al. 2001). Several technologies exit for in situ chemically enhanced soil flushing by extracting solutions such as organic and inorganic acids, and complexation agents have also been proposed for remediation.

12.9 Bioremediation

Bioremediation, a strategy that uses living microorganisms, is essentially proposed to clean up the environment from organic pollutants. Microbial resistance mechanisms toward heavy metals are essential for the potential applications of microorganisms in bioremediation. The understanding of the resistance phenomena at the molecular level is the prerequisite necessary for the biological treatment of solid mining waste and resulting effluents. The combination of genomic approaches with geochemical and hydrological models is the ultimate goal to accelerate bioremediation (Lovley 2003). Bioremediation is the application of biological systems to the cleanup of organic and inorganic pollution, with bacteria and fungi being the most important organisms for reclamation, immobilization, or detoxification of metallic and radionuclide pollutants. Some biominerals or metallic elements deposited by microbes have catalytic and other properties in nanoparticle, crystalline, or colloidal forms, and these are relevant to the development of novel biomaterials for technological and antimicrobial purposes. On the negative side, metal and mineral transformations by microbes may result in spoilage and destruction of natural and synthetic materials; rock and mineral-based building materials (e.g., concrete); acid mine drainage and associated metal pollution; biocorrosion of metals, alloys, and related substances; and adverse effects on radionuclide speciation, mobility, and containment, all with immense social and economic consequences. The ubiquity and importance of microbes in biosphere processes make geomicrobiology one of the most important concepts within microbiology and one requiring an interdisciplinary approach to define environmental and applied significance and underpin exploitation in biotechnology. Many microbial metal and mineral transformations have potential for the treatment of environmental pollution, and some processes are in commercial operation (Gadd 2005; Lloyd and Renshaw 2005). Detailed knowledge of microbe–metal-related reactions may allow further optimization of the desired process by altering the physicochemical conditions of the contaminated area. A combination of genetic engineering with appropriate eco-engineering of polluted sites may be relevant to some future bioremediation strategies (Valls and de Lorenzo 2002) although subject to significant legal and sociopolitical barriers. In addition to bioremediation, microbe–metal–mineral transformations have applications in other areas of biotechnology and bioprocessing, including biosensors, biocatalysts, electricity generation, and nanotechnology.

12.10 Biomining

Biomining entails the use of acidophilic microbes to facilitate the recovery process of metals from sulfide minerals in the processes of bioleaching and bio-oxidation. The microbes found in these environments are (extreme) acidophiles growing at a pH of 3 or lower and span a wide range of different phyla. The majority belong to the bacterial and archaea domains; microbial consortia, however, and unicellular eukaryotes have also been reported (Baker and Banfield 2003; Bonnefoy and Holmes 2011). Most organisms including *A. ferrooxidans* possess a sulfite oxidase enzyme that converts sulfite to sulfate under low-sulfite conditions (Feng et al. 2007), but this activity tends to be inhibited at higher concentrations. However, the D3-2 strain has a sulfite reductase that is more resistant to sulfite and thus can oxidize it faster, resulting in a lowered sulfite concentration that doesn't inhibit the iron oxidation pathway (Sugio et al. 2008).

12.11 Bioleaching

Bioleaching is the solubilization of metals of interest such as cobalt, copper, and nickel from sulfide minerals. The two processes are industrially well established and are commercially applied worldwide (Rawlings et al. 2003). Microorganisms oxidize both sulfur and iron of sulfide minerals, such as pyrite. It is generally accepted that leaching takes place via an "indirect" mechanism, which can be divided into the "contact" and "noncontact" mode (Baker and Banfield 2003). The "indirect" mechanism assumes that chemoautotrophic iron-oxidizing microorganisms like *Acidithiobacillus ferrooxidans* and *Leptospirillum ferrooxidans* generate ferric ions by oxidation of ferrous iron (Rawlings 2002). *Leptospirillum ferrooxidans* generate ferric ions by oxidation of ferrous iron (Rawlings 2002). During the "noncontact" mode, planktonic microbes oxidize aqueous ferrous ions to ferric ions, which in turn attack the mineral surface by chemical oxidation. The "contact" mode assumes a small reaction space between the microbial cell wall, and the mineral surface where ferric ions are concentrated in biofilms for a localized attack of the sulfide mineral. Either mode yields different intermediate and final sulfur species depending on the ore leached. Microbiological solubilization of metals from solid minerals (bioleaching) is a well-established process in the mining industry. In addition, other metal and mineral wastes, including contaminated soil, may be subjected to microbial bioleaching for metal recovery, recycling, and bioremediation purposes. Metals such as Cd, Cu, Ni, and Zn can be solubilized from fly ash (originating from incineration of municipal solid waste) by bacterial and fungal activity (Brandl 2001; Brandl and Faramarzi 2006). HCN-forming bacteria, e.g., *Chromobacterium violaceum* and *Pseudomonas fluorescens*, can mobilize Ni, Au, Pt, and Cu as various cyanide complexes and compounds from solid materials such as copper-containing ores, electronic scrap, and spent automobile catalytic converters (Brandl and Faramarzi 2006. Industrial-scale bioleaching of metals is used to extract metals such as copper, gold, uranium, and others from their ores (Suzuki 2001; Rawlings 2002; Rawlings et al. 2003; Jerez 2009). This process, also termed biomining, employs chemolithoautotrophic microbes, with the most common leaching bacteria belonging to the genus *Acidithiobacillus*, e.g., *A. ferrooxidans* and *A. thiooxidans*. Thermophilic archaea capable of oxidizing sulfur and Fe (II), including *Sulfolobus*, *Acidianus*, *Metallosphaera*, and

Sulfurisphaera spp., as well as mesophilic archaea, have also been described, e.g., *Ferroplasma* spp. Most bioleaching operations for gold or copper use some kind of irrigation process involving percolation of leaching solutions through the crushed ore contained in a column, heap, or dump (Rawlings 2002; Rawlings et al. 2003). The leachate containing, e.g., copper sulfate generated by microbial solubilization of insoluble copper sulfides in the ore is subjected to solvent extraction to produce a highly concentrated copper sulfate solution from which the copper is recovered in an electrowinning plant to generate electrolytic copper of high purity. Bioleaching bacteria can also be used for gold recovery. Gold is usually associated with minerals containing arsenic and pyrites (arsenopyrites) (Rawlings et al. 2003). During gold bioleaching, iron- and sulfur oxidizers attack the arsenopyrite, releasing the gold particles. Following release, the gold is complexed with cyanide prior to recovery using standard gold-mining procedures. Most leaching bacteria grow as a biofilm attached to the surface of solid substrates such as elemental sulfur and metal sulfides and can "pit" the surface of mineral ores such as pyrite (FeS_2) (Suzuki 2001), although other studies have shown that attachment is not necessary for pitting to occur with pyrite (Edwards et al. 2001). Cells are embedded in a matrix of extracellular polymeric substances in which an indirect mechanism generates Fe (II) and thiosulfate, which is finally oxidized to sulfuric acid, with elemental sulfur being a side product. The oxidation of some metal sulfides, such as chalcopyrite ($CuFeS_2$), generates elemental sulfur as a side product instead of thiosulfate. Insoluble metal sulfides are oxidized to soluble metal sulfates by the chemical action of Fe (III) (Edwards et al. 2001), the main role of the microbes being the reoxidation of generated Fe (II) to additional Fe (III). In an environmental context, acidophilic microbes can mobilize metals from waste ores, abandoned mines, and dumps and generate acid mine drainage, itself a serious environmental problem (Jerez 2009). Bioremediation, a strategy that uses living microorganisms, is essentially proposed to clean up the environment from organic pollutants. However, since there is an evidence that several microorganisms possess the capability to reduce Cr (VI) to relatively toxic Cr (III), bioremediation gives immense opportunities for the development of technologies for the detoxification of soil contaminated with Cr (VI) as an alternative to existing physical–chemical remediation technologies (Cervantes et al. 2001).

12.12 Future Commercial Applications

Prediction of future long-term commercial applications of biological processes in the mining area industry is risky at best and foolhardy at worst. Nevertheless, it is almost certain that some of today's research will lead to innovative processes for commercial application. The predominant microbial system for research, process development, and commercial application has been the members of the genus *Thiobacillus* and *T. ferrooxidans* in particular. Recently, the *Leptospirillum* have been included in the stable of useful microorganisms. Future process developments will and must include thermophilic bacteria that will have an increasingly important role in bio-oxidation of minerals. Thermophilic *Archaea*, *Sulfolobus* species, *Acidianus brierleyi*, and *Metallosphaera sedula*, which grow at 68–75 °C, are particularly adapted in bioleaching of copper from the highly refractory chalcopyrite.

References

Baker BJ, Banfield JF (2003) Microbial communities in acid mine drainage. FEMS Microbiol Ecol 44:139–152

Bonnefoy V, Holmes DS (2011) Genomic insights into microbial iron oxidation and iron uptake strategies in extremely acidic environments. Environ Microbiol. doi:10.1111/j.1462_2011_2920

Brandl H (2001) Heterotrophic leaching. In: Gadd GM (ed) Fungi in bioremediation. Cambridge University Press, Cambridge, pp 383–423

Brandl H, Faramarzi MA (2006) Microbe-metal-interactions for the biotechnological treatment of metal-containing solid waste. China Particuol 4:93–97

Bruins MR, Kapil S, Oehme FW (2000) Microbial resistance to metals in the environment. Ecotoxicol Environ Saf 45:198–207

Cervantes C, Campos-García J, Devars S et al (2001) Interactions of chromium with microorganisms and plants. FEMS Microbiol Rev 25(3):335–347

Cole MA (1979) Solubilization of heavy metal sulfides by heterotrophic soil bacteria. Soil Sci 127:313–317

Corkhill CL, Vaughan DJ (2009) Arsenopyrite oxidation—a review. Appl Geochem 24:2342–2361

Dale SE, Doherty-Kirby A, Lajoie G et al (2004) Role of siderophore biosynthesis in virulence of *Staphylococcus aureus*: identification and characterization of genes involved in production of a siderophore. Infect Immun 72:29–37

Dhal B, Thatoi HN, Das NN et al (2013) Chemical and microbial remediation of hexavalent chromium from contaminated soil and mining/metallurgical solid waste. J Hazard Mater 250:272–291

Edwards KJ, Hu B, Hamers RJ, Banfield JF (2001) A new look at microbial leaching patterns on sulfide minerals. FEMS Microbiol Ecol 34:197–206

Ehrlich HL, Newman DK (2009) Geomicrobiology, 5th edn. CRC/Taylor & Francis, Boca Raton, FL

Feng C, Tollin G, Enemark JH (2007) Sulphite oxidizing enzymes. Biochim Biophys Acta 1774(5):527–539

Gadd GM (2005) Microorganisms in toxic metal polluted soils. In: Buscot F, Varma A (eds) Microorganisms in soils. Roles in genesis and functions. Springer, Berlin, pp 325–356

Gadd GM (2010) Metals, minerals and microbes: geomicrobiology and bioremediation. Microbiology 156:609–643

Gadd GM, Raven JA (2010) Geomicrobiology of eukaryotic microorganisms. Geomicrobiol J 27:491–519

Ghiorse WC (1984) Biology of iron-and manganese-depositing bacteria. Annu Rev Microbiol 38:515–550

Haferburg G, Kothe E (2007) Microbes and metals interaction in the soil. J Basic Microbiol 47(6):453–467

Harris JA (2003) Measurements of the soil microbial community for estimating the success of restoration. Eur J Soil Sci 54:801–808

Hudson-Edwards KA, Santini JM (2013) Arsenic-microbe-mineral interactions in mining-affected environments. J Miner 3:337–351
Islam ABMR, Maity JP et al (2013) Arsenic mineral dissolution and possible mobilization in mineral-microbe-groundwater environment. J Hazard Mater 262:989–996. doi:10.1016/j.jhazmat.2012.07.022
Jerez CA (2009) Metal extraction and biomining. In: Schaechter M (ed) Encyclopedia of microbiology, 3rd edn. Elsevier, Amsterdam, pp 407–420
Konhauser K (2007) Introduction to geomicrobiology. Blackwell, Oxford, p 425
Lamont IL, Beare PA, Ochsner U et al (2002) Siderophore-mediated signaling regulates virulence factor production in *Pseudomonas aeruginosa*. Proc Natl Acad Sci U S A 99:7072–7077
Li XL, Christie P (2001) Changes in soil solution Zn and pH and uptake of Zn by arbuscular mycorrhizal red clover in Zn-contaminated soil. Chemosphere 42:201–207
Lloyd JR, Renshaw JC (2005) Bioremediation of radioactive waste: radionuclide-microbe interactions in laboratory and field-scale studies. Curr Opin Biotechnol 16:254–260
Lovley DR (2003) Cleaning up with genomics: applying molecular biology to bioremediation. Nat Rev Microbiol 1:35–44
Mathiyazhagan N, Natarajan D (2011) Bioremediation on effluents from magnesite and bauxite mines using *Thiobacillus Spp* and *Pseudomonas Spp*. J Bioremed Biodegrad 2:115
McGrath SP, Chaudri AM, Giller KE (1995) Long-term effects of metals in sewage sludge on soils, microorganisms and plants. J Ind Microbiol 14:94–104
Monica OM, Julia WN, Raina MM (2008) Characterization of a bacterial community in an abandoned semiarid lead-zinc mine tailing site. Appl Environ Microbiol 74:3899–3907
Mulligan CN, Young RN, Gibbs BF (2001) Remediation technologies for metal-contaminated soils and ground water: an evaluation. Eng Geol 60:193–207
Nanda S, Abraham J (2013) Remediation of heavy metal contaminated soil. Afr J Biotechnol 12 (21):3099–3109
Neubauer SC, Emerson D, Megoniga JP (2002) Life at the morphology of an iron plaque on the roots of Typha latifolia grown energetic edge: kinetics of circumneutral Fe oxidation by litho-in solution culture. Am J Bot 71:666–675
Osborne TH, Santini JM (2012) Prokaryotic aerobic oxidation of arsenite. In: Santini JM, Ward SA (eds) The metabolism of arsenite; arsenic in the environment, vol 5. CRC, London, pp 61–72
Peterson GD, Heemskerk M (2001) Deforestation and forest regeneration following small-scale gold mining in the Amazon. The case of suriname. Environ Conserv 28(2):117–126
Qing HU, Hong-yan QI, Zeng J et al (2007) Bacterial diversity in soils around a lead and zinc mine. J Environ Sci 19:74–79
Rawlings DE (2002) Heavy metal mining using microbes. Annu Rev Microbiol 56:65–91
Rawlings DE, Dew D, du Plessis C (2003) Biomineralization of metal-containing ores and concentrates. Trends Biotechnol 21:38–44
Rhine ED, Garcia-Dominguez E et al (2005) Environmental microbes can speciate and cycle arsenic. Environ Sci Technol 39:9569–9573
Schwyn B, Neilands JB (1987) Universal chemical assay for the detection and determination of siderophores. Anal Biochem 160:47–56
Shi W, Bischoff M, Turco R et al (2002) Long-term effects of chromium and lead upon the activity of soil microbial communities. Appl Soil Ecol 21:169–177
Smith E, Leeflang P, Wernars K (1997) Detection of shifts in microbial community structure and diversity in soil caused by copper contamination using amplified ribosomal DNA restriction analysis. J FEMS Microbiol Ecol 23:249–261
Sousa DMG, Miranda LN, Oliveira SA (2007) Acidez do solo e sua correção (Soil acidity and its correction). In: Novais RF, Alvarez VH, Barros NF et al (eds) Fertilidade do solo. Sociedade Brasileira de Ciência do Solo, Viçosa, MG, Brazil, pp 205–275 (in Portuguese)

Stolz JF (2012) Overview of microbial arsenic metabolism and resistance. In: Santini JM, Ward SA (eds) The metabolism of arsenite; arsenic in the environment vol. 5. CRC, London, pp 55–60
Stolz JF, Basu P, Santini JM et al (2006) Selenium and arsenic in microbial metabolism. Annu Rev Microbiol 60:107–130
Sugio T, Iwahori K, Uemura S et al (1994) Inhibition of iron-oxidizing activity by bisulfite ion in *Thiobacillus ferrooxidans*. Biosci Biotechnol Biochem 58:2109–2110
Sugio T, Wakabayashi M, Kanao T et al (2008) Isolation and characterization of *Acidithiobacillus ferrooxidans* strain D3-2 active in copper bioleaching from a copper mine in Chile. Biosci Biotechnol Biochem 72(4):998–1004
Suzuki I (2001) Microbial leaching of metals from sulfide minerals. Biotechnol Adv 19:119–132
Tamaki S, Frankenberger WT Jr (1992) Environmental biochemistry of arsenic. Rev Environ Contam Toxicol 124:79–110
Touati D (2000) Iron and oxidative stress in bacteria. Arch Biochem Biophys 373:1–6
Valls M, de Lorenzo V (2002) Exploiting the genetic and biochemical capacities of bacteria for the remediation of heavy metal pollution. FEMS Microbiol Rev 26:327–338
Verkhovtseva YV, Filina YY, Pukhov DE (2001) Evolutionary role of iron in metabolism of prokaryotes and in biogeochemical processes. J Evol Biochem Physiol 37:444–450
Welch SA, Barker WW et al (1999) Microbial extracellular polysaccharides and plagioclase dissolution. Geochim Cosmochim Acta 63:1405–1419
White C, Sharman AK, Gadd GM (1998) An integrated microbial process for the bioremediation of soil contaminated with toxic metals. Nat Biotechnol 16:572–575

Chapter 13
A Strategic Scheme for Resource Recovery from Sulfurous Industrial Wastes Through Plant–Microbe Interaction

Sanchita Kukde and Bijaya Ketan Sarangi

13.1 Introduction

Pollution of the natural resources is one of the alarming global issues which need large-scale and coordinated efforts to cope up and sustain. Countless human activities are continuously adding contaminants to water. Designing cleaner technologies and processes to generate fewer pollutants and treating the existing levels of pollutants to rejuvenate the contaminated resources are priorities. Numerous research and development investigations are being carried out by environmentalists and scientists to develop sustainable processes. There is a constant demand of newer cleanup technologies which are contaminant specific and result in long-term effectiveness. Approach to deal this global issue needs to be targeted, cost-effective, sustainable, and coordinated to achieve the objective. This chapter proposes a strategy to treat sulfurous industrial wastes through reduction by concerted plant–microbe activity and reuse the regenerated sulfur as plant nutrient for economic and ecological benefits.

Sulfur exists as sulfates, sulfites, and sulfides in liquid and solid wastes as sulfurous compounds. Various industries use sulfur-containing compounds like thiosulfates, sulfonates, etc. Sulfurous pollutants are thus found in the wastes generated from these processes and known to increase the acidity and oxygen demand of wastewater, solid wastes, and leachates (Lens et al. 2008). The presence of these pollutants in the form of sulfides is also known to generate hydrogen sulfide gas (Haneklaus et al. 2003; Haq and Ali 2003) which is toxic and poses a threat to aquatic life. Moreover, increased concentration of sulfide and sulfite compounds in wastewater poses problems in biological treatment of the wastewater. On the other hand, these sulfurous pollutants when converted to sulfates are essential nutrient for

S. Kukde • B.K. Sarangi (✉)
Environmental Biotechnology Division, National Environmental Engineering Research Institute, Nehru Marg, Nagpur 440020, Maharashtra, India
e-mail: bk_sarangi@neeri.res.in

L.B. Sukla et al. (eds.), *Environmental Microbial Biotechnology*, Soil Biology 45,
DOI 10.1007/978-3-319-19018-1_13

plants and microbes for their growth and metabolism (Chen et al. 2008, 2013). This chapter portrays a strategy to oxidize the sulfites and sulfides present in the wastewater to sulfates employing specific microbes, subsequently feeding the sulfate-containing liquid or solid wastes to plants as nutrient for their growth (Chen et al. 2008) and generating resources (Calderwood and Kopriva 2014; Takahashi et al. 2011).

13.2 Industrial Pollution Due to High Sulfite- and Sulfate-Containing Wastes

Due to the reactive potentiality, sulfites and sulfiting agents like sulfur dioxide, bisulfites, and metabisulfites are a major class of industrial chemicals. A number of industrial processes such as leather tanning, paper milling, photography, and food industry use them (Bajpai 2013; Pundir and Rawal 2013; Dhami et al. 2013; Zhang 2014). These processes use various sulfurous compounds like sulfuric acid, sulfate-rich feedstock, and less oxidized sulfur compounds (sulfide, sulfite, thiosulfate, dithionite, etc.). Scrubbing of sulfur dioxide-containing air emissions in water also leads to the formation of hydrogen sulfite (HSO_3^-) and sulfite (SO_3^{2-}) (Chen et al. 2008; Chandra et al. 2013). Decomposition of reduced sulfur compounds such as thiosulfate, polythionates, etc., results in production of sulfites which persist in the environment in the absence of oxygen (Hayes et al. 2006; Kappler 2008, 2011). The resulting sulfite anion from the abovementioned different processes, in excess, is released in wastewaters that act as a reducing agent (Hille 2003). Similarly, dry scrubbing of flue gases or lime neutralization of acid waste effluents and manufacturing of superphosphate fertilizers results in formation of calcium sulfite and sulfate wastes. These solid wastes, when untreated, lack bearing strength and are prone to leach objectionable amounts of the sulfate ion into the groundwater (Ince et al. 2011; Gómez-Ramirez et al. 2014). Sulfite salts also pose a problem in wastewater pretreatment systems, surface discharges, and other disposal avenues.

Huge volumes of sulfate-rich wastewaters are also generated from industries involved in flue gas desulfurization where sulfur dioxide is liberated as one of the major pollutants (Chen et al. 2008; Yan et al. 2014). Secondary lead smelters stand as a typical example which produce several tons of sulfur dioxide routinely. The gas is removed from the flue gas by packed bed caustic scrubbers. The contact liquid is usually a high-pH solution of sodium carbonate or sodium hydroxide. Scrubbing with sodium bases produces wastewater with high concentrations of sodium sulfite. Sulfite salts are oxygen scavengers which are stable in basic solutions. However, changes in pH during wastewater treatment render instability to sulfite anions which decompose into sulfur dioxide gas under acidic conditions. Various processes in the dye and detergent manufacturing, bleaching, and photographic industries also generate sulfite- and thiosulfate-rich wastewaters. Sulfur compounds are also found in wastewaters of food industry where sulfur dioxide and sodium

Table 13.1 Concentrations of sulfur-containing compounds in wastewaters released during various processes in pulp and paper industry

Source	SO_3^{2-} (g/l)	SO_4^{2-} (g/l)	pH	S^{2-} (g/l)	COD (g/l)	Reference
Chemical pulping wastewater	0.04	-	-	0.05–0.01	-	Lin et al. (2014)
Neutral sulfite semichemical spent liquor	2.7	0.1	-	-	-	Guiot et al. (1991), Pöykiö et al. (2014)
Thermomechanical pulping wastewaters	-	0.2–0.7	4.5–5.0	-	2–5	Habets and de Vegt (1991), Rintala et al. (1991), Zheng and Liao (2014)
Chemi-thermomechanical pulping	-	1.2–1.5	7–9	0.05–0.20	7.5–10.4	Habets and de Vegt (1991), Meyer and Edwards (2014)
Sulfite pulp mill evaporator wastewater	0.45–0.80	-	7–9	-	-	Salkinoja-Salonen et al. (1985), Wang et al. (2014)

bisulfite are used in the food processing (Wolicka 2010; Yu et al. 2013; Zhang et al. 2014).

The pulp and paper industry also generates wastewaters with high sulfur load (Ince et al. 2011). Various sulfur compounds like sulfate, sulfur dioxide, sulfite, and dithionate are used in pulp and paper industries for various pulping processes (Ince et al. 2011; Tripathi et al. 2014). Sulfur compounds are also present in wood and in raw water due to use of fillers and whiteners in the chemical pulp treatment. Examples of usual concentrations of various oxides of sulfur used in different processes of the pulp and paper industry are presented in Table 13.1.

Sulfurous wastes in water are known to create nuisance during anaerobic biological treatments of these waters. Basic sulfite salt solutions should be oxidized to sulfate salts which are less troublesome to dispose or can serve as a sulfate source. In anaerobic conditions sulfur-reducing bacteria are not able to use sulfate as energy source, and thus, hydrogen sulfide is liberated which is toxic and corrosive (Postgate 1984; Hamilton 2003). Wastewaters of high sulfur content, when used for land application or discharged in water bodies, lead to sulfur buildup and give rise to S-polluted lands (Tripathi et al. 2014). Remediation of sulfurous water and land resources is an important issue to be addressed.

This chapter discusses the scope and approach for conversion of sulfurous wastes in waters or lands to plant-available nutrients through plant–microbe interaction and further utilization of the sulfates as a resource for enrichment of S-deficient soil and nutrient for enhanced plant growth and crop productivity.

13.3 Sulfur Metabolism in Microbes

Sulfur is designated as the sixth most abundant element in microbial biomass. In microbial biochemistry sulfur plays versatile role as a structural element, redox center, or carbon carrier, as an energy source and as a reactant in biochemical processes. The redox reactions carried out for interconversion of reduced and oxidized sulfur compounds serve as a source of biochemical energy in microbial metabolism. In some genera of *Archaea* and lithotrophic bacteria, H_2S serves as an energy source in the oxidizing sulfur pathway (Österberg 1997; Nisbet and Sleep 2001). These biochemical reactions carried out by microbial biomass as a part of their metabolic process have profound effect on biogeochemical sulfur cycle.

Microbes take up the sulfate form from extracellular pool. Transport of sulfate into the cell is catalyzed either by ATP-binding cassette (ABC)-type transporters (SulT family) or by permease-type transporters (SulP family) (Kopriva et al. 2008; Takahashi et al. 2011). The role of microbes in sulfur cycle can be described under three categories: (1) assimilatory sulfate reduction to synthesize organic sulfur compounds, (2) dissimilatory sulfate reduction to liberate hydrogen sulfide, and (3) oxidation of less oxidized forms of sulfur to sulfate in more stable compound as depicted in Fig. 13.1. Assimilatory sulfate reduction pathway in microbes is similar to that in plants. It results in synthesis of cysteine which further gets converted to methionine and other organic derivatives as per metabolic requirements of microbe (Dahl and Friedrich 2008). This mechanism is commonly found in microorganisms, but sulfur utilization through this pathway is very less; hence, this pathway is less significant in biogeochemical sulfur cycle. The more significant pathway of sulfate utilization is the dissimilatory pathway of S reduction confined to specific microbes which are mostly anaerobes like sulfate-reducing bacteria (SRB), e.g., *Desulfovibrio* and *Archaea*. In this pathway microbe utilizes sulfate as terminal electron acceptor in electron transport chain to fulfill its energy demand (Shen and Buick 2004; Klotz et al. 2011). The reduced hydrogen sulfide in this pathway cannot be trapped in the organic form and thus is liberated out of the cell. SRBs have been studied widely in relation to wastewater treatment (Okabe et al. 1999, 2003; Wolicka 2010). In typical domestic wastewaters due to low dissolved oxygen, sulfate reduction can be the dominant terminal electron-accepting process and accounts for up to 50 % of mineralization of organic matter in wastewater biofilms (Kuhl and Jørgensen 1992; Okabe et al. 2003; Wolicka 2010). A major drawback of sulfate reduction in wastewater treatments by SRBs is the liberation of H_2S, which produces nuisance of odor. The liberated H_2S also causes serious corrosion of metals in industrial water systems and concrete sewer pipes in wastewater treatment facilities, which is an universal problem (Postgate 1984; Hamilton 1985, 2003; Nielsen et al. 1993; Lee et al. 1994; Hamilton and Lee 1995).

Some genera of microbes possess the ability of oxidizing different compounds to much stable forms (Dahl and Friedrich 2008; Hell et al. 2008). Oxidation of sulfides to elemental S is carried out by *Thiothrix* and *Beggiatoa* aerobically and by purple sulfur bacteria anaerobically. Plants cannot uptake elemental S as a nutrient

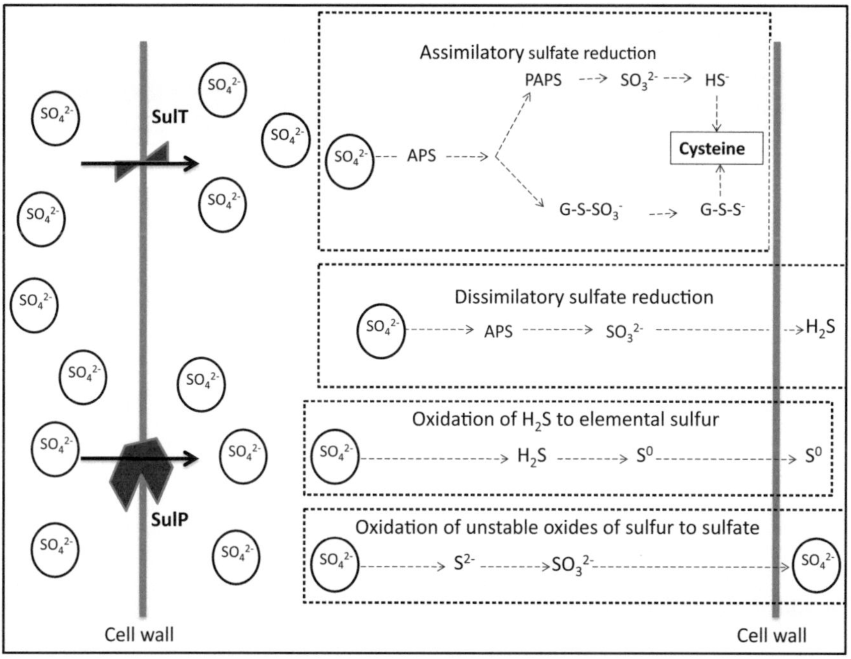

Fig. 13.1 S metabolism in microbes (SulT, ABC-type sulfate transporter family; SulP, permease transporter family)

resource unless converted into available form and tend to persist in the environment. On the contrary, some members of genera *Thiobacillus*, *Thiobacterium*, *Thiospira*, *Macromonas*, and *Achromatium* have the ability to oxidize elemental S to more oxidized sulfate form. Plants have the mechanism to uptake sulfate from the soil and water matrix and further use as a nutrient source.

13.4 Sulfur Metabolism in Plants as Sink for Sulfurous Compounds

Sulfur is considered as one of the most versatile elements in living organisms. The versatility of S is attributed to multiple stable oxidation states (Klotz et al. 2011). The oxidation states are −2, 0, +2, +4, and +6 which can get the forms like hydrogen sulfide, elemental S, sulfur monoxide, sulfur dioxide, and sulfur trioxide, respectively. In living systems S plays key role by engaging in plenty of metabolic processes and their intermediates like disulfide bridges in proteins, iron sulfur clusters involved in electron transport, secondary metabolites, and catalytic sites of several enzymes like urease and coenzyme A (Klotz et al. 2011).

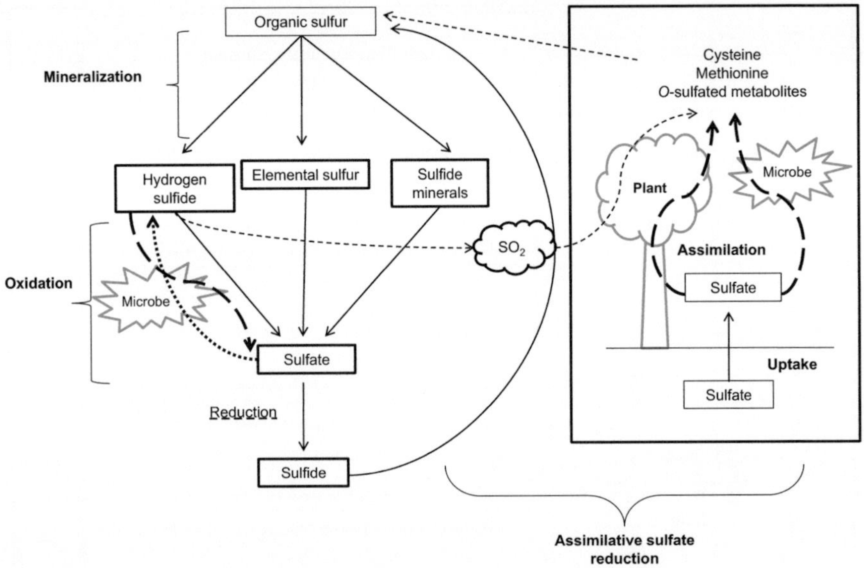

Fig. 13.2 Role of plant and microbes in sulfur cycle

Total S flux in ecosystem involves terrestrial and atmospheric fluxes together. Living world or biota forms the bridge between these two fluxes by converting inorganic S to organic one and *vice versa*. In the sulfur cycle, microbes and plants play an essential and unavoidable role by converting the inorganic S to organic form (Fig. 13.2). Taking advantage of multiple stable oxidation states, sulfur is cycled in the ecosystem through a characteristic biogeochemical cycle involving assimilative sulfate reduction, desulfurization, and dissimilative S reduction to hydrogen sulfide followed by oxidation of hydrogen sulfide to elemental S and oxidation of elemental S to sulfates which is assimilated as a plant nutrient and fixed into organic form.

Sulfur is an essential mineral in plant nutrition classified as secondary macronutrient. Sulfur uptake takes place through roots mostly in the form of sulfates from soil. Uptake and assimilation of sulfates resembles that of phosphates and nitrates (Kopriva and Rennenberg 2004; Hesse et al. 2004; Salon et al. 2013). The average concentration of sulfate ion in external medium is 0.25 mmol/l, whereas internal concentration of sulfate ions in root cells is found to be 19 mmol/l (in pea roots) (Higinbotham et al. 1967). This concentration variation suggests S uptake is active against the concentration gradient. Most of the S is derived from sulfate which is absorbed via H^{+}–SO_4^{-} symporters from the soil solution. These are high-affinity transporters, and the maximal sulfate uptake rate is achieved when sulfate level in the rhizosphere is about 0.1 mM and lower (Hawkesford 2000; Hawkesford and Wray 2000; Hawkesford et al. 2003a, b). Concentration of available S in the rhizosphere of a plant may vary considerably depending on numerous factors.

Sulfate taken up by root is further mobilized to different plant organs via xylem. Loading of sulfate to the sieve element is accordingly an important physiological process for the effective recycling of S nutrients (Yoshimoto et al. 2003). Plants respond to fluctuating S availabilities by regulating acquisition and utilization. Acquisition is regulated by means of sulfate transporters, and utilization is regulated by various metabolic processes. Sulfate is very stable and undergoes through various processes and subsequent reactions (Leustek et al. 2000) with sequential steps and cysteine is synthesized. The steps are summarized in Fig. 13.3. The reduction of sulfate to cysteine changes the oxidation number of S from +6 to −2. Cysteine is used in plastids for synthesis of methionine, and both of these amino acids get incorporated in proteins. Other S-containing compounds are thus synthesized with the help of these proteins (Klotz et al. 2011).

Plants experience the scenario of receiving excess S through natural sources in areas like coastal regions with higher sea sprays or regions having volcanic activity. Various anthropogenic activities also introduce high levels of S compounds in the vicinity of plant rhizosphere as well as phyllosphere. High-S-containing industrial wastes are the major sources for S eutrophication in land and water matrix (Chen et al. 2008, 2013). The generalized way of plants to deal with this scenario is to uptake the excess S within the threshold limit; in exceptional cases avoidance at uptake level is one of the strategies to prevent excess entry into the plant body

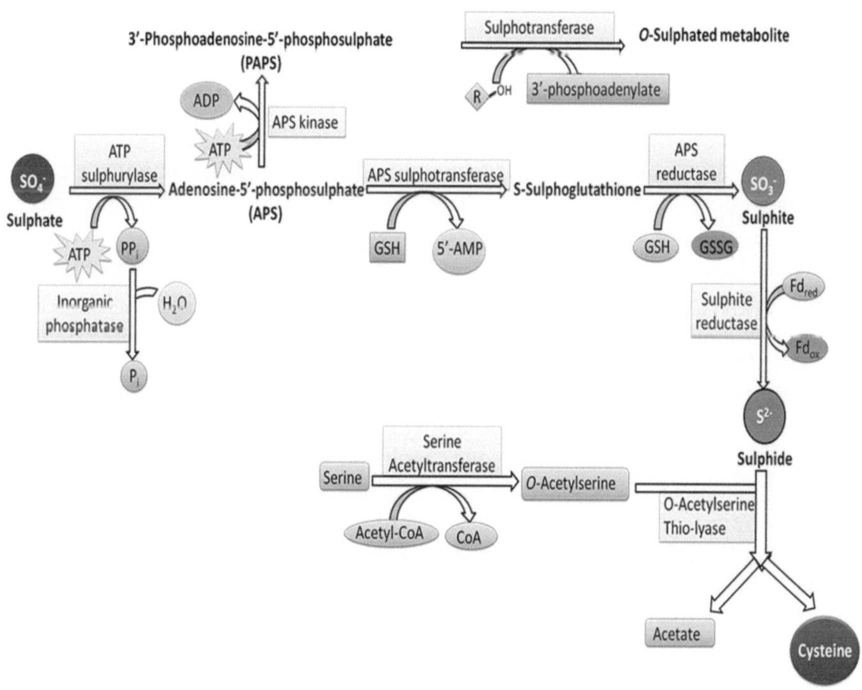

Fig. 13.3 Sulfur assimilation in plants

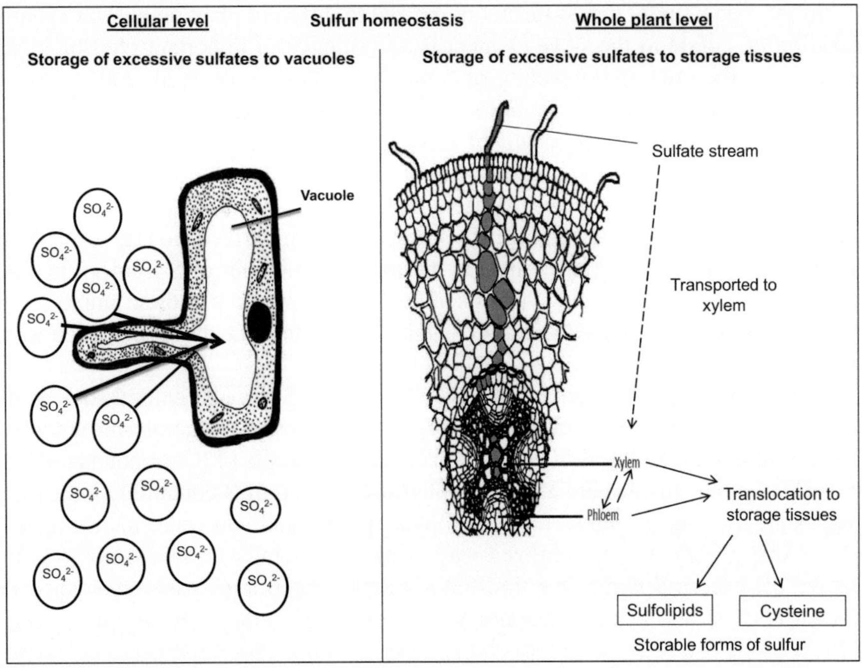

Fig. 13.4 Channeling of excess sulfur in plants

(Rennenberg 1984; Randewig et al. 2013). The excess S taken up is converted to suitable form for translocation to aerial and other parts for storage in appropriate form. In *Arabidopsis*, SULTR4 group transporters are identified to be responsible for vacuolar efflux of sulfates (Kataoka et al. 2004b) which are important candidates for remobilization of stored S. The channeling of excess sulfur in plants is demonstrated in Fig. 13.4.

The mechanism of storage of excess mineral nutrients helps plants to cope up the nutritional demand in unfavorable environmental conditions where uptake is hampered (Kataoka et al. 2004a; Takahashi 2010). The major organic storage form of excessive S is sulfolipid, i.e., sulfoquinovosyl diacylglycerol (SQDG) (Benning 1998; Harwood and Okanenko 2003; Frentzen 2004). SQDG forms a large fraction of organic S in the biosphere and a major S sink in plant sulfur status like glutathione. Plant sulfolipids are known to play a crucial role in stabilizing photosynthetic complexes. They are localized in the photosynthetic membranes of plastids which renders the negative charge to the thylakoid membrane (Benning et al. 2008). SQDG biosynthesis takes place in chloroplast with the help of UDP-SQ synthase (SQD1) and SQDG synthase (SQD2). Sulfite and UDP-glucose are acted upon by SQD1 to synthesize UDP-SQ which is further converted to SQDG with sequential steps (Kleppinger-Sparace and Mudd 1990; Benning et al. 2008).

Sulfite is synthesized from sulfate in the chloroplast during S assimilation. The enzyme adenosine-5′-phosphosulfate reductase (APR, EC 1.8.4.9) is responsible

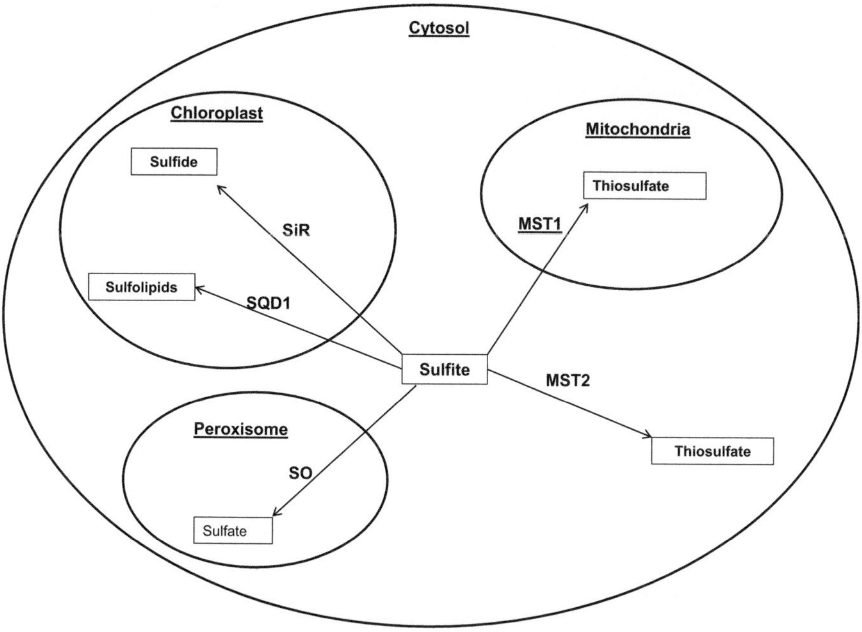

Fig. 13.5 Cellular level sulfite homeostasis [SiR (sulfite reductase; EC 1.8.7.1); SQD1 (UDP-sulfoquinovose synthase; EC 3.13.1.1); MST1 and MST2 (mercaptopyruvate sulfurtransferases; EC 2.8.1.2); SO (sulfite oxidase; EC 1.8.3.1)]

for this conversion. At cellular level, removal of excess sulfite is achieved by assimilation, incorporation into metabolites, and detoxification (Fig. 13.5). Sulfite enters S assimilatory pathway when acted upon by chloroplast-localized sulfite reductase (SiR; EC 1.8.7.1) catalyzing ferredoxin-dependent sulfite reduction to sulfide (Khan et al. 2010). Another pathway to incorporate sulfite is sulfolipid synthesis where chloroplast-localized UDP-sulfoquinovose synthase (SQD1; EC 3.13.1.1; Sanda et al. 2001) utilizes sulfite. The detoxification mechanism for sulfite involves the mitochondrion and cytosol-localized *b*-mercaptopyruvate sulfurtransferases (STs; EC 2.8.1.2), MST1 and MST2, respectively, which catalyze the synthesis of the less toxic compound thiosulfate in the presence of *b*-mercaptopyruvate and sulfite (Papenbrock and Schmidt 2000; Tsakraklides et al. 2002). Also, in the peroxisomes, re-oxidization to sulfate by the molybdenum cofactor-containing enzyme, sulfite oxidase (SO; EC 1.8.3.1), reduces the toxicity and converts it in storable form (Eilers et al. 2001). Plant tries to cope up with excess S levels by cellular as well as whole-plant-level homeostasis. But, under high available S levels, like that in S-rich soil or water matrix, its entry to the plant becomes higher and results in toxicity above the tolerance limit. Most detrimental effects are observed due to accumulation of S compounds in the form of sulfites. Although sulfite is a less oxidized form of sulfate and one of the intermediates in S assimilation, still high concentrations of sulfite are known to be potentially cytotoxic at cellular level (Sanda et al. 2001; Davidian and Kopriva 2010) as well as whole plant (Murray 1997; Brychkova et al. 2007).

13.5 Biological Sulfur Fixation Through Plant–Microbe Association

13.5.1 Strategy for Selection of Microbes Appropriate for Specific Ecological Niche

Sulfur-assimilating microbes are potential candidates which could be employed to facilitate conversion of different sulfurous wastes to sulfates for plant uptake and fixation in the biomass. Oxidation of S-containing compounds is one of the metabolic characteristics of many microorganisms, but the product of oxidation differs among the microbes as described in Sect. 13.4. Purple sulfur bacteria are known to oxidize the S compounds like thiosulfates. Two different pathways exist for oxidation of S compounds in purple bacteria. In phototrophic purple sulfur bacteria and many chemotrophic sulfur oxidizers like magnetotactic bacteria, *Beggiatoa* sp., or *Thiothrix*, the formation of globules of polymeric, water-insoluble S appears to be an important step during thiosulfate oxidation (Nelson and Castenholz 1981; Dahl 1999; Howarth et al. 1999; Dahl and Prange 2006; Hensen et al. 2006; Williams et al. 2006), whereas in numerous facultative chemolithotrophic or photolithotrophic organisms, like *Paracoccus pantotrophus* or *Rhodovulum sulfidophilum*, both S atoms of thiosulfate are oxidized to sulfate without the appearance of S deposits as intermediates (Appia-Ayme et al. 2001; Friedrich et al. 2001, 2005). Another important group possessing the ability to oxidize S compounds is green sulfur bacteria. Inorganic compounds including sulfide (S^{2-}), elemental sulfur (S^0), polysulfides (Sn^{2-}), thiosulfate ($S_2O_3^{2-}$), tetrathionate ($S_4O_6^{2-}$), hydrogen (H_2), and ferrous iron (Fe^{2+}) are known to be oxidized by green sulfur bacteria (Brune 1995; Heising et al. 1999; Garrity and Holt 2001). In microbial cells sulfite can arise from a variety of reactions, such as dissimilatory S compound oxidation, amino acid degradation, S assimilation, or sulfonate breakdown. Microbial cells can also become exposed to sulfite present in the environment, e.g., in the anaerobic regions of the Black Sea (Sorokin 1995). The common strategy for sulfite detoxification in Bacteria and *Archaea* seems to involve an oxidation of sulfite to sulfate, which can proceed either via a direct formation of sulfate from sulfite or via the indirect APS reductase pathway (Kappler and Dahl 2001). Despite the fact that direct oxidation pathway does not allow for a conservation of energy via substrate-level phosphorylation, it appears to be the more common one (Kappler and Dahl 2001).

Oxidation of sulfite to sulfate is the key step for regeneration of assimilative sulfate from sulfurous pollutants which is mediated through the sulfite oxidase enzyme. Two types of enzymes are recognized catalyzing this step: sulfite oxidase

(SO; EC 1.8.3.1) and sulfite dehydrogenases (SDH; EC 1.8.2.1). Both enzymes are metalloproteins that possess a molybdenum-containing redox center. Additional redox-active centers (e.g., heme groups) may also be present (Eilers et al. 2001; Hemann et al. 2005; Kappler 2008). SOs transfer electrons to oxygen, ferricyanide, and sometimes cytochrome c, while SDHs use either ferricyanide or cytochrome or both, but do not transfer electrons to oxygen. Typical SOs are the plant SO (known electron acceptors are oxygen, ferricyanide) (Eilers et al. 2001; Hemann et al. 2005; Kappler 2008) and the well-studied vertebrate SOs that can use all three electron acceptors listed above but appear to use a cytochrome c as their natural electron acceptor (Rajagopalan 1980; Enemark and Cosper 2002). Bacterial enzymes can be divided into two groups on the basis of their preference for either ferricyanide or cytochrome c as an electron acceptor (Kappler and Dahl 2001). SDHs have been reported in many bacteria, including the soil bacterium *Starkeya novella* (Kappler et al. 2000), *Sulfitobacter* species (Pukall et al. 1999; Sorokin 1995), various *thiobacilli*, and alkanesulfonate-degrading and photosynthetic bacteria (Kappler and Dahl 2001; Cook et al. 2006). While for some bacterial SDHs a monoheme cytochrome c has been established as the natural electron acceptor (Yamanaka et al. 1981; Kappler et al. 2000), in many cases this acceptor is not known. Enzymes that use a cytochrome c as their natural electron acceptor clearly have to be located in an extracellular compartment or, if they were membrane proteins, they would have to possess a periplasmic/extracellular domain that transfers electrons to cytochrome c.

It is essential to identify and select potential microbial species which are able to sustain in the polluted niche in sulfurous waste effluent streams or dumps. These microbial species will be more befitting to carry out the bioconversion of sulfurous compounds to sulfates for further assimilation by the suitable plant species.

13.5.2 Strategy for Selection of Appropriate Plant Species for Specific Ecological Niche

All plants require sulfur for their growth and development, but depending on the industrial product and process, the wastewaters may be rich in many other pollutants. Hence, selection of appropriate plant species is important that can sustain the other pollutants and detoxify and utilize the sulfurous compounds as a nutrient resource (Ernst 1998; Takahashi et al. 2011; Calderwood and Kopriva 2014; Xia et al. 2014). Identification of plant species for sulfate uptake, assimilation, and utilization under sulfate-rich conditions is essential for reuse of sulfurous compounds. Biotechnological and molecular tools may help to identify and modify the potential plant species to enhance their ability of sulfur utilization through activation of the mentioned features. It is desirable that suitable plant species should have the following features:

(a) Ability of plants to take up excess sulfate through roots: efficient root-level sulfate transporters
(b) Ability of plants to store excess S in storable form: metabolic pathway for sulfolipids biosynthesis
(c) Ability of plants to detoxify excess sulfite: highly active sulfite-oxidizing enzyme
(d) Ability of plants for efficient distribution of stored S: efficient vacuolar sulfate transporters

Sulfate uptake through roots is an active process mediated by high-affinity H^+/SO_4^{2-} symporters. In most plants, the genes encoding putative sulfate transporters are classified into five subfamilies (SULTR1 to SULTR5) on the basis of their protein sequence similarities (Hawkesford 2003). In *Arabidopsis thaliana*, 14 members had been historically identified. One of the two genes of the AtSULTR5 subfamily (AtSULTR5;2) has been demonstrated to encode a high-affinity molybdate transporter and renamed MOT1 (Tomatsu et al. 2007). The first plant root high-affinity sulfate transporter SHST1, which belongs to the group SULTR1 and cloned from the pulse *Stylosanthes hamata* (Smith et al. 1995), has demonstrated to be involved in molybdate uptake at the nanomolar range when expressed in yeast cells, revealing a potent double transport function (Fitzpatrick et al. 2008). The high-affinity sulfate root membrane transporter, AtSULTR1;2, is considered the main candidate to control root sulfate uptake (El Kassis et al. 2007). AtSULTR1;2 is reported to be expressed in the root epidermal and cortical plasma membranes and in the root apex, playing important role in plant sulfur nutrition (Shibagaki et al. 2002). Significant decrease in root SO_4^{2-} uptake capacity and shoot SO_4^{2-} accumulation also has been reported when AtSULTR1;2 is mutated (Shibagaki et al. 2002; Yoshimoto et al. 2002; El Kassis et al. 2007; Barberon et al. 2008). The AtSULTR1;1 is another high-affinity sulfate transporter reported (Vidmar et al. 2000), but it is much less expressed in root as compared to AtSULTR1;2 (Rouached et al. 2008). The gene expression for both of these transporters is shown to be upregulated by sulfate limitation (Yoshimoto et al. 2002, 2007; El Kassis et al. 2007). AtSULTR1;3 is also one of the high-affinity sulfate transporters specifically expressed in phloem and responsible for source-to-sink transport of sulfate (Yoshimoto et al. 2003). Detailed study of these transporters may help in identifying the plant species capable of S accumulation at high concentrations.

Excess sulfate taken up by plants is generally transported to various plant organs for storage and further reutilized for metabolism when soil sulfate is unavailable. A major storage form of excess sulfates is sulfolipid. The plant sulfolipid SQDG represents an important component of the global sulfur cycle as it is found in most photosynthetic and a few non-photosynthetic organisms (Benning et al. 2008). The relative sulfolipid content in plant and algae tissues is reported to vary between 2 and 50 % of total polar lipids (Harwood 1980; Dembitsky et al. 1990, 1991; Heinz 1993). Sulfolipids are known to be important for photosynthesis and assimilative capacity of the plant (Barber and Gounaris 1986). SQDG content of bacteria and plants can change depending on growth conditions (Benning et al. 2008). Studies on isolated chloroplasts have revealed that chloroplasts are fully capable of synthesizing SQDG from labeled sulfate when energy requirements were met either by photosynthesis or by the supply of nucleotides (Haas et al. 1980; Roy and Harwood 1999; Benning et al. 2008). Therefore, the chloroplasts of plants aimed for excess sulfur accumulation must contain the biosynthetic machinery to provide the S and carbon precursors for sulfolipid biosynthesis. The enzyme responsible for SQDG synthesis, SQDG synthase, is characterized and known to localize inside the inner envelope membrane of chloroplasts (Seifert and Heinz 1992; Tietje and Heinz 1998). Sulfolipid biosynthesis in plants requires at minimum two specific enzymes: (1) the UDP-SQ synthase (SQD1), which is responsible for the biosynthesis of the head group donor, and (2) the SQDG synthase (SQD2) catalyzing the final assembly of sulfolipid. The head group donor of UDP-SQ is sulfoquinovose. The *Arabidopsis* SQD1 protein is localized in the stroma of the chloroplast (Essigmann et al. 1998). The SQD1 protein contains a tightly bound NAD^+ residue participating in the formation of a 4-keto hexosyl group. This 4-keto intermediate is converted to a 4-keto-5,6-glucoseen intermediate to which sulfite is added. Subsequently, the 4-keto group reverts back to the hydroxyl after accepting a hydride from the NADH bound in the active site, which thereby releases UDP-SQ and regenerates NAD^+ (Sanda et al. 2001). A ferredoxin-dependent glutamate synthase (FdGOGAT) is found to be tightly associates with SQD1 (Shimojima et al. 2005). Modeling of the cyanobacterial FdGOGAT and *Arabidopsis* SQD1 into a plausible complex suggests that FdGOGAT, a flavin-containing protein, could intermittently and covalently bind sulfite and deliver it to the active site of SQD1 (Shimojima et al. 2005). Thus, when sulfurous compounds are taken up in excess and toxic intermediate, sulfite is synthesized in excess quantities; the interaction of FdGOGAT and SQD1 could be considered as a potential association which overcomes this impasse by channeling sulfite to SQD1. However, further experiments will be required to support this current working hypothesis. Enzyme responsible for the final assembly of sulfoquinovosyl diacylglycerol (SQDG) was identified to be SQD2, encoding the plant sulfolipid synthase (Yu et al. 2002). Disruption of this gene by T-DNA insertion resulted in complete loss of SQDG in *Arabidopsis*. Moreover, co-expression of the SQD1 and SQD2 cDNAs from *Arabidopsis* in *E. coli* led to sulfolipid biosynthesis in this bacterium normally lacking this lipid (Yu et al. 2002). Thus, these two plant genes are considered sufficient to reconstitute SQDG biosynthesis demonstrating that the ultimate sulfur donor for SQDG biosynthesis must be a common intermediate of the S assimilation pathway, such as sulfite, present in *E. coli*. Consistent with biochemical data showing that the sulfolipid synthase is associated with the inside of the inner envelope membrane (Seifert and Heinz 1992; Tietje and Heinz 1998), the SQD2 protein is also found to localize inside the plastid (Yu et al. 2002). This highlights that plants with high chlorophyll content may serve as potential candidates to accumulate excess S as their ability to store excess sulfurous compounds in the form of sulfolipids may tend to be higher. The genes for SQD1 and SQD2 can serve as potential markers to identify the candidate species.

The major threat imposed by excess S uptake is formation of cytotoxic intermediate, sulfite, during the metabolism (Papenbrock and Schmidt 2000; Tsakraklides et al. 2002; Khan et al. 2010). Channelizing of excess S is shown in the Fig. 13.4. In view of this, utilization of sulfite as a resource, detoxification of sulfite to sulfate, which is the transportable form of S, can be one of the promising way. The enzyme sulfite oxidase, localized in peroxisome, is responsible for this conversion. Sulfite oxidase contains molybdenum as cofactor. The enzyme catalyzes a two-electron transfer reaction in which the electrons from sulfite reduce the molybdenum cofactor redox center. The electrons are subsequently transferred to molecular oxygen with simultaneous formation of hydrogen peroxide in addition to sulfate (Eilers et al. 2001; Hansch et al. 2006). Among the eukaryotes, plant SO is the smallest molybdenum cofactor-containing enzyme known so far. The enzyme lacks contiguous redox-active centers such as FAD, heme, or Fe–S (Eilers et al. 2001). In *Arabidopsis*, AtSO is localized in peroxisomes (Eilers et al. 2001; Nowak et al. 2004) and is thus distinct from the multienzyme S assimilatory pathway localized to the chloroplast. It has been speculated that SO is required for removing excess sulfite that accumulates upon decomposition of sulfur-containing amino acids or sulfated metabolites (Heber and Hüve 1997; Hänsch and Mendel 2005). The cloned full-length cDNA of *Arabidopsis* SO has a single open-reading frame of 1,182 bp, encoding a protein of 393 amino acids (43.3 kDa), with 35 % identity to the Moco domain of plant nitrate reductase. Function of SO in high S stress response has been demonstrated. In maize, peroxisome-localized sulfite oxidase, ZmSO, has been reported to be responsive to sulfite stress at the transcriptional level during germination of maize seeds (Xia et al. 2014). It was demonstrated in previous studies that the SO proteins from *A. thaliana*, *Lycopersicon esculentum*, and *Nicotiana benthamiana* were involved in sulfite detoxification by catalyzing sulfite oxidation to sulfate (Brychkova et al. 2007, 2013; Lang et al. 2007; Xia et al. 2012). In view of this, high ability for conversion of sulfite to the transportable sulfate form is also a desirable trait for efficient S assimilation.

The excess sulfate is also stored in the vacuoles as a part of cellular homeostasis and redistributed to developing organs when environmental fluxes tend to restrict root uptake. The transporters on the tonoplast are thus expected to play important role in the redistribution of stored excess S (Kataoka et al. 2004b; Takahashi et al. 2011). The two members of the *Arabidopsis* SULTR4 family sulfate transporters have been shown to be localized to the tonoplast. AtSULTR4;1 and AtSULTR4;2 are suggested to be involved in vacuolar SO_4^{2-} remobilization to the cytosol, as evidenced by the higher SO_4^{2-} content of isolated vacuoles from a double sultr4;1–sultr4;2 mutant. Accumulation of transcripts for SULTR4;1 and SULTR4;2 under S limitation indicates that the release of the vacuolar sulfate is promoted when demands for S increase (Kataoka et al. 2004b). Despite similar subcellular localization of both proteins, SULTR4;2 expression is more responsive to SO_4^{2-} starvation in both roots and shoots in comparison to SULTR4;1. Expression studies of both SULTR4 transporters in *Brassica napus* leaves revealed BnSULTR4;2 was more highly expressed than BnSULTR4;1 in response to sulfate depletion (Parmar et al. 2007; Dubousset et al. 2009). In the same sulfur-depletion conditions, rapeseed leaf sulfate content decreased significantly in relation to upregulation of BnSULTR4;1, confirming the involvement of both SULTR4 members in vacuolar sulfate remobilization (Dubousset et al. 2009). Plastidic sulfate transporters were also identified in *Chlamydomonas reinhardtii* and shown to belong to the bacterial ABC type of transporter family (Lindberg and Melis 2008). As all the vacuolar transporters studied till date are responsible for vacuolar efflux, investigating vacuolar transporters responsible for influx is an important strategy to enhance S assimilation by plants.

13.5.3 Plant–Microbe Interaction for Enhanced Sulfur Fixation in Plants

Uptake of inorganic nutrients by plants is highly dependent on the various rhizospheric processes. The importance of microbes in plant sulfur nutrition is well established (Hu et al. 2005; Haneklaus et al. 2007) which is also important in the case of S mineral nutrition in plants. Usually, elemental S (S^0) is applied to soil as a fertilizer to meet the S demand of plants. In order to make this reduced S available to plants, it needs to be oxidized to sulfates. Oxidation of sulfur in soils is primarily a microbial process. *Thiobacilli* and some heterotrophic microorganisms are known to be important candidates for oxidation of elemental S (Hu et al. 2005; Haneklaus et al. 2007). A study on the rhizosphere of *Brassica napus* showed that out of 273 bacterial isolates from the rhizosphere, 245 (89.7 %) oxidized elemental S to thiosulfate or tetrathionate and 133 (48.7 %) oxidized S to sulfates. Also all 70 fungal isolates oxidized S to sulfates (Grayston and Germida 1991). Thus, rhizosphere microflora plays a key role in the mechanism of soil S nutrient dynamics. The soil physicochemical properties vary depending on the soil type and environmental factors which affect the microflora dynamics in the soil–root interface that differ considerably from those in the non-rhizosphere soil. It is known that industrial wastes, i.e., soil or liquid wastes, significantly affect the soil physicochemical characteristics since they often do not meet to the disposal limit. Even though they meet the standards for disposal, due to recurrent disposal and their specific nature, soil or water eutrophication is common, and consequently the matrix gets polluted, unsuitable for normal utilization. In high-S-containing wastes the soil physicochemical properties get disturbed. Soil tends to become acidic, alkaline, saline, low in water retention, and physically and physiologically dry (Gómez-Ramirez et al. 2014; Tripathi et al. 2014; Lin et al. 2014) unsuitable for normal growth of microflora Therefore, S mineralization processes in the soil–root interface greatly vary in different ecological niche depending on the soil physicochemical properties. In view of this, the ability of plants for S fixation greatly varies between soils. In order to combat this situation and operate the mineralization process, utilization of the specific microbial consortia that flourish under such situation is an important strategy. Application of these consortia in combination with similar plant species that could also grow under such hostile soil conditions would be the most effective way to manage with such hostile soils. Microbial consortia are potential agents for degradation of natural or synthetic compounds; nitrification by autotrophic bacteria including reactions of nitrate and sulfate takes place in the rhizosphere system, and they have been effectively utilized for management of such polluted matrix (Gómez-Ramirez et al. 2014).

It is known that the biochemical environment of the rhizosphere contains root exudates as well as secretions from rhizobacteria, fungi, and other soil organisms, arising from the dynamic interactions among plant roots, microbial organisms, and other microfauna (Kertesz and Mirleau 2004). These compounds directly or indirectly influence the microbial growth in the rhizosphere. Appropriate plant species that inherit the biochemical mechanisms to sustain S stress (Ernst 1998; Takahashi et al. 2011; Calderwood and Kopriva 2014; Xia et al. 2014) will also facilitate growth of these S-tolerant-specific microfloras through plant–microbe association. In turn, the rhizosphere microbes will serve as plant growth promoters or act as pathogens inhibiting plant growth (Kertesz and Mirleau 2004; Turgay and Bilen 2012; Miransari 2013). The rhizosphere metabolism is thus the result of complex interactions between microflora and microfauna yielding exudations, lysates, chelators, antibiotics, phytostimulators, and extracellular enzymes, making it distinctive from the rest of the plant system. Thus, it is important to select potential plant and microbial candidate in order to achieve a successful rhizosphere interaction. In order to combat this situation and operate the mineralization process, identification of the specific microbial consortia that flourish under such situation and the use of them in combination with similar plant species that could also grow under such hostile soil conditions would be the most effective ways to manage with such S-polluted soil or water (Fig. 13.6).

The sulfite metabolism in microbes and plants is carried out by the abovementioned two sets of biochemical activities. In microbes cytochrome c or ferricyanide or both can act as electron acceptors, whereas in plants, the terminal electron acceptor is oxygen. These attributes in microbes and plants can be complimented for sulfite detoxification under different physicochemical conditions of the soil and water matrix. Under water-inundated anaerobic conditions sulfite detoxification will be efficiently mediated by the microbes, whereas under usual aerobic conditions sulfite assimilation will be carried out efficiently by the plants. Due to this synergistic effect of both organisms, plant–microbe interaction will be a win-win situation for S assimilation more efficiently.

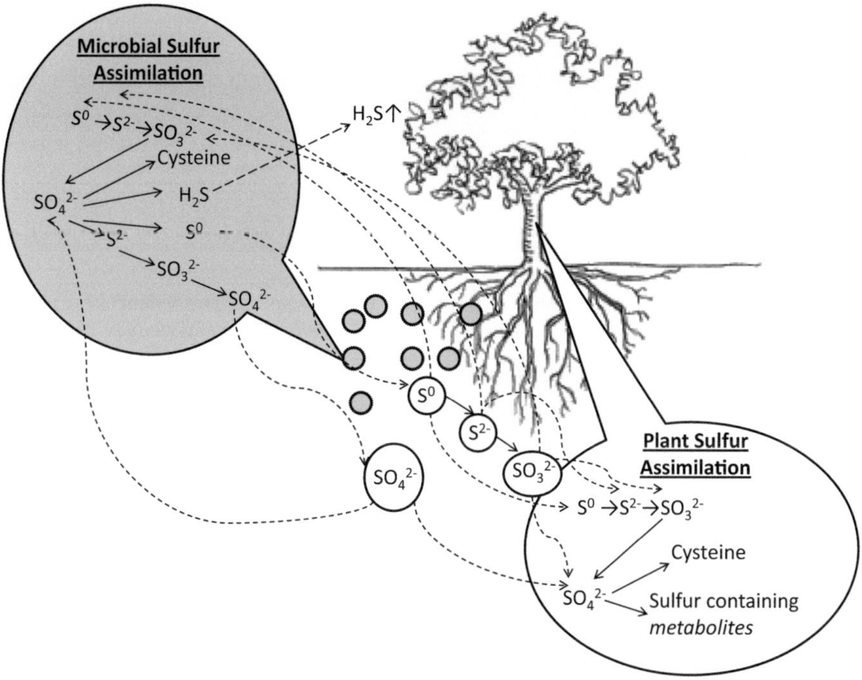

Fig. 13.6 Plant–microbe interaction in the rhizosphere for efficient S assimilation

13.6 Conclusion

The problem of water pollution by sulfurous wastes can be partially answered by application of biotechnology utilizing natural abilities of plants and microbes. Microbes with the ability to oxidize less oxidized sulfur compounds to sulfates are the important candidates which can convert these pollutants to plant-available form, i.e., sulfates. Potential biochemical and molecular markers can be identified which may help in screening the microbes and improving their performance with biotechnological tools. The sulfates formed as a result of microbial activities can be utilized as a nutrient resource for plant growth. As treated wastewater or solid waste is supposed to be sulfate rich, only specific plant species, which have high S requirement and biochemical mechanisms for high sulfate assimilation to accumulate larger quantity of sulfates, can perform well. Biotechnological tools should be employed to identify potential species and improve S accumulation capacity so as to achieve the objective of resource recovery. Microbes play important role in plant nutrient acquisition. Thus, plant–microbe association between selected species is a promising strategy that could result in better management of sulfurous wastes by their synergistic effect. Therefore, R&D strategy needs to be worked out to exploit this potential approach towards management of sulfurous wastes through biological recycling. Further, the generated knowledge can be utilized to augment S fixation in agricultural soils for sustained availability to crop plants as nutrient supplement to agriculture. We are focusing on this strategy so as to achieve objective of S pollution mitigation along with resource generation through fixation in biological sinks.

References

Appia-Ayme C, Little PJ, Matsumoto Y et al (2001) Cytochrome complex essential for photosynthetic oxidation of both thiosulfate and sulfide in *Rhodovulum sulfidophilum*. J Bacteriol 183:6107–6118

Bajpai P (2013) Pulp bleaching and bleaching effluents. In: Bajpai P (ed) Bleach plant effluents from the pulp and paper industry. Springer International Publishing, Cham, pp 13–19

Barber J, Gounaris K (1986) What role does sulfolipid play within the thylakoid membrane? Photosynth Res 9:239–249

Barberon M, Berthomieu P, Clairotte M et al (2008) Unequal functional redundancy between the two *Arabidopsis thaliana* high-affinity sulphate transporters SULTR1;1 and SULTR1;2. New Phytol 180:608–619

Benning C (1998) Biosynthesis and function of the sulfolipid sulfoquinovosyl diacylglycerol. Annu Rev Plant Physiol Plant Mol Biol 49:53–75

Benning C, Garavito RM, Shimojima M (2008) Sulfolipid biosynthesis and function in plants. In: Hell R, Dahl C, Knaff D, Leustek T (eds) Sulfur metabolism in phototrophic organisms. Springer, Dordrecht, pp 185–200

Brune DC (1995) Sulfur compounds as photosynthetic electron donors. In: Blankenship RE, Madigan MT, Bauer CE (eds) Anoxygenic photosynthetic bacteria, vol 2, Advances in photosynthesis. Kluwer, Dordrecht, pp 847–870

Brychkova G, Xia Z, Yang G, Yesbergenova Z, Zhang Z, Davydov O, Fluhr R, Saki M (2007) Sulfite oxidase protects plants against sulfur dioxide toxicity. Plant J 50(4):696–709

Brychkova G, Grishkevich V, Fluhr R, Sagi M (2013) An essential role for tomato sulfite oxidase and enzymes of the sulfite network in maintaining leaf sulfite homeostasis. Plant Physiol 161:148–164

Calderwood A, Kopriva S (2014) Hydrogen sulfide in plants: from dissipation of excess sulfur to signaling molecule. Nitric Oxide 41:72–78

Chandra H, Paliwal S, Tripathi AB (2013) Mitigation of emission in thermal power plant using conventional and non-conventional fuel. Int J Eng Sci Invent 2(4):01–06

Chen L, Kost D, Dick WA (2008) Flue gas desulfurization products as sulfur sources for corn. Soil Sci Soc Am J 72:1464–1470

Chen L, Van Toai T, Dick WA (2013) Effects of flue gas desulfurization products as soil amendments to improve water quality. In: World of Coal Ash (WOCA) conference, Lexington, KY

Cook A, Denger K, Smits T (2006) Dissimilation of C3-sulfonates. Arch Microbiol 185:83–90

Dahl C (1999) Deposition and oxidation of polymeric sulfur in prokaryotes. In: Steinbüchel A (ed) Biochemical principles and mechanisms of biosynthesis and biodegradation of polymers. Wiley-VCH, Weinheim, pp 27–34

Dahl C, Friedrich CG (2008) Microbial sulfur metabolism. Springer, Berlin

Dahl C, Prange A (2006) Bacterial sulfur globules: occurrence, structure and metabolism. In: Shively JM (ed) Inclusions in prokaryotes. Springer, Heidelberg, pp 21–51

Davidian JC, Kopriva S (2010) Regulation of sulfate uptake and assimilation: the same or not the same? Mol Plant 3:314–325

Dembitsky VM, Rozentsvet OA, Pechenkina EE (1990) Glycolipids, phospholipids and fatty acids of brown algae species. Phytochemistry 29:3417–3421

Dembitsky VM, Pechenkina-Shubina EE, Rozentsvet OA (1991) Glycolipids and fatty acids of some seaweeds and marine grasses from the Black Sea. Phytochemistry 30:2279–2283

Dhami JK, Singh H, Gupta M (2013) Industrialization at the cost of environment degradation—a case of leather and iron and steel industry from Punjab economy. Innov J Bus Manag 2 (1):19–21

Dubousset L, Abdallah M, Desfeux A et al (2009) Remobilization of leaf S compounds and senescence in response to restricted sulphate supply during the vegetative stage of oilseed rape are affected by mineral N availability. J Exp Bot 60:3239–3253

Eilers T, Schwarz G, Brinkmann H et al (2001) Identification and biochemical characterization of *Arabidopsis thaliana* sulfite oxidase. A new player in plant sulfur metabolism. J Biol Chem 276 (50):46989–46994

El Kassis E, Cathala N, Rouached H et al (2007) Characterization of a selenate-resistant *Arabidopsis thaliana* mutant. Root growth as a potential target for selenate toxicity. Plant Physiol 143:1231–1241

Enemark JH, Cosper MM (2002) Molybdenum enzymes and sulfur metabolism. Met Ions Biol Syst 39:621–654

Ernst WH (1998) Sulfur metabolism in higher plants: potential for phytoremediation. Biodegradation 9:311–318

Essigmann B, Güler S, Narang RA et al (1998) Phosphate availability affects the thylakoid lipid composition and the expression of SQD1, a gene required for sulfolipid biosynthesis in *Arabidopsis thaliana*. Proc Natl Acad Sci U S A 95:1950–1955

Fitzpatrick KL, Tyerman SD, Kaiser BN (2008) Molybdate transport through the plant sulfate transporter SHST1. FEBS Lett 582:1508–1513

Frentzen M (2004) Phosphatidylglycerol and sulfoquinovosyldiacylglycerol: anionic membrane lipids and phosphate regulation. Curr Opin Plant Biol 7:270–276

Friedrich CG, Rother D, Bardischewsky F et al (2001) Oxidation of reduced inorganic sulfur compounds by bacteria: emergence of a common mechanism? Appl Environ Microbiol 67:2873–2882

Friedrich CG, Bardischewsky F, Rother D et al (2005) Prokaryotic sulfur oxidation. Curr Opin Microbiol 8:253–259

Garrity GM, Holt G (2001) Phylum BXI. Chlorobi *phy. nov.* In: Boone DR, Castenholz RW, Garrity GM (eds) Bergey's manual of systematic bacteriology, vol 1. Springer, New York, pp 601–623

Gómez-Ramirez M, Zarco-Tovar K, Aburto J et al (2014) Microbial treatment of sulfur-contaminated industrial wastes. J Environ Sci Health A 49:228–232

Grayston SJ, Germida JJ (1991) Sulfur-oxidizing bacteria as plant growth promoting rhizobacteria for canola. Can J Microbiol 37:521–529

Guiot SR, Lavoie L, Hawari JA et al (1991) Effect of NSSC spent liquor on granule formation and specific microbial activities in upflow anaerobic reactors. Water Sci Technol 24(3-4):139–148

Haas R, Siebertz HP, Wrage K, Heinz E (1980) Localization of sulfolipid labeling within cells and chloroplasts. Planta 148:238–244

Habets LHA, de Vegt AL (1991) Anaerobic treatment of bleached TMP and CTMP effluent in the Biopaq UASB system. Water Sci Technol 24:331–345

Hamilton WA (1985) Sulfate-reducing bacteria and anaerobic biocorrosion. Annu Rev Microbiol 39:159–217

Hamilton WA (2003) Microbially influenced corrosion as a model system for the study of metal microbe interactions: a unifying electron transfer hypothesis. Biofouling 19:65–76

Hamilton WA, Lee W (1995) Biocorrosion. In: Barton LL (ed) Sulphate-reducing bacteria. Plenum, New York, pp 243–262

Haneklaus S, Bloem E, Schnug E (2003) The global sulphur cycle and its links to plant environment. In: Abrol YP, Ahmad A (eds) Sulphur in plants. Kluwer, Dordrecht, pp 1–28

Haneklaus S, Bloem E, Schnug E (2007) Sulfur interactions in crop ecosystems. In: Hawkesford MJ, DeKok LJ (eds) Sulfur in plants: an ecological perspective. Springer, Dordrecht, pp 17–58

Hänsch R, Mendel RR (2005) Sulfite oxidation in plant peroxisomes. Photosynth Res 86 (3):337–343
Hansch R, Lang C, Riebeseel E et al (2006) Plant sulfite oxidase as novel producer of H_2O_2—combination of enzyme catalysis with a subsequent non-enzymatic reaction step. J Biol Chem 281:6884–6888
Haq K, Ali M (2003) Biologically active sulphur compounds of plant origin. In: Abrol YP, Ahmad A (eds) Sulphur in plants. Kluwer, Dordrecht, pp 375–386
Harwood JL (1980) Sulfolipids. In: Stumpf PK (ed) The biosynthesis of plants, vol 4. Academic, New York, pp 301–320
Harwood JL, Okanenko AA (2003) Sulphoquinovosyl diacylglycerol (SQDG)—the sulpholipid of higher plants. In: Abrol YP, Ahmad A (eds) Sulphur in plants. Kluwer, Dordrecht, pp 189–219
Hawkesford MJ (2000) Plant responses to sulphur deficiency and the genetic manipulation of sulphate transporters to improve S-utilization efficiency. J Exp Bot 51:131–138
Hawkesford MJ (2003) Transporter gene families in plants: the sulfate transporter gene family—redundancy or specialization? Physiol Plant 117:5–163
Hawkesford MJ, Wray JL (2000) Molecular genetics of sulphate assimilation. Adv Bot Res 33:159–223
Hawkesford MJ, Buchner P, Hopkins L, Howarth JR (2003a) The plant sulfate transporter family: specialized functions, integration with whole plant nutrition. In: Davidian JC, Grill D, De Kok LJ, Stulen I, Hawkesford MJ, Schnug E, Rennenberg H (eds) Sulfur transport and assimilation in plants: regulation, interaction and signalling. Backhuys, Leiden, pp 1–10
Hawkesford MJ, Buchner P, Hopkins L, Howarth JR (2003b) Sulphate uptake and transport. In: Abrol YP, Ahmad A (eds) Sulphur in plants. Kluwer, Dordrecht, pp 71–86
Hayes MK, Taylor GT, Astor Y, Scranton MI (2006) Vertical distributions of thiosulfate and sulfite in the Cariaco Basin. Limnol Oceanogr 51:280–287
Heber U, Hüve K (1997) Action of SO_2 on plants and metabolic detoxification of SO_2. Int Rev Cytol 177:255–286
Heinz E (1993) Recent investigations on the biosynthesis of the plant sulfolipid. In: De Kok LJ (ed) Sulfur nutrition and assimilation in higher plants. SPB Academic, The Hague, pp 163–178
Heising S, Richter L, Ludwig W, Schink B (1999) Chlorobium ferrooxidans sp. nov., a phototrophic green sulfur bacterium that oxidizes ferrous iron in coculture with a "Geospirillum" sp. strain. Arch Microbiol 172:116–124
Hell R, Dahl C, Knaff D, Leustek T (2008) Sulfur metabolism in phototrophic organisms. Springer, Berlin
Hemann C, Hood BL, Fulton M et al (2005) Spectroscopic and kinetic studies of *Arabidopsis thaliana* sulfite oxidase: nature of the redox-active orbital and electronic structure contributions to catalysis. J Am Chem Soc 127:16567–16577
Hensen D, Sperling D, Trüper HG et al (2006) Thiosulphate oxidation in the phototrophic sulphur bacterium Allochromatium vinosum. Mol Microbiol 62:794–810
Hesse H, Nikiforova V, Gakière B, Hoefgen R (2004) Molecular analysis and control of cysteine biosynthesis: integration of nitrogen and sulfur metabolism. J Exp Bot 55:1283–1292
Higinbotham N, Etherton B, Foster RJ (1967) Mineral ion contents and cell transmembrane electropotentials of pea and oat seedling tissue. Plant Physiol 42(1):37–46
Hille R (2003) Plants have SOX: the structure of sulfite oxidase from *Arabidopsis thaliana*. Structure 11(10):1189–1190
Howarth R, Unz RF, Seviour EM et al (1999) Phylogenetic relationships of filamentous sulfur bacteria (*Thiothrix spp.* and Eikelboom type 021N bacteria) isolated from wastewater-treatment plants and description of *Thiothrix eikelboomii sp. nov.*, *Thiothrix unzii sp. nov.*, *Thiothrix fructosivorans sp. nov.* and *Thiothrix defluvii sp. nov.* Int J Syst Bacteriol 49:1817–1827
Hu Z, Silvia H, Cao ZH, Ewald S (2005) Chemical behavior of soil sulfur in the rhizosphere and its ecological significance. In: Aspects of sulfur nutrition of plants: evaluation of China's current,

future and available resources to correct plant nutrient sulfur deficiencies—Report of the first Sino-German Sulfur workshop, p 53
Ince BK, Cetecioglu Z, Ince O (2011) Pollution prevention in the pulp and paper industries. In: Broniewicz E (ed) Environmental management in practice. InTech, Rijeka. doi:10.5772/23709. ISBN 978-953-307-358-3
Kappler U (2008) Bacterial sulfite-oxidizing enzymes–enzymes for chemolithotrophs only? In: Dahl C, Friedrich CG (eds) Microbial sulfur metabolism. Springer, Berlin, pp 151–169
Kappler U (2011) Bacterial sulfite-oxidizing enzymes. Biochim Biophys Acta Bioenerg 1807 (1):1–10
Kappler U, Dahl C (2001) Enzymology and molecular biology of prokaryotic sulfite oxidation. FEMS Microbiol Lett 203:1–9
Kappler U, Bennett B, Rethmeier J et al (2000) Sulfite: cytochrome c oxidoreductase from *Thiobacillus novellus*—purification, characterization and molecular biology of a heterodimeric member of the sulfite oxidase family. J Biol Chem 275:13202–13212
Kataoka T, Hayashi N, Yamaya T, Takahashi H (2004a) Root-to-shoot transport of sulfate in *Arabidopsis*: evidence for the role of SULTR3;5 as a component of low-affinity sulfate transport system in the root vasculature. Plant Physiol 136:4198–4204
Kataoka T, Watanabe-Takahashi A, Hayashi N et al (2004b) Vacuolar sulfate transporters are essential determinants controlling internal distribution of sulfate in *Arabidopsis*. Plant Cell 16:2693–2704
Kertesz MA, Mirleau P (2004) The role of soil microbes in plant sulphur nutrition. J Exp Bot 55:1939–1945
Khan MS, Haas FH, Allboje Samami A et al (2010) Sulfite reductase defines a newly discovered bottleneck for assimilatory sulfate reduction and is essential for growth and development in *Arabidopsis* thaliana. Plant Cell 22:1216–1231
Kleppinger-Sparace KF, Mudd JB (1990) Biosynthesis of sulfoquinovosyldiacylglycerol in higher plants: use of adenosine-5′-phosphosulfate and adenosine-3′-phosphate 5′-phosphosulfate as precursors. Plant Physiol 93:256–263
Klotz MG, Bryant DA, Hanson TE (2011) The microbial sulfur cycle. Front Microbiol 2:241
Kopriva S, Rennenberg H (2004) Control of sulfate assimilation and glutathione synthesis: interaction with N and C metabolism. J Exp Bot 55:1831–1842
Kopriva S, Patron NJ, Keeling P, Leustek T (2008) Phylogenetic analysis of sulfate assimilation and cysteine biosynthesis in phototrophic organisms. In: Hell R, Dahl C, Knaff D, Leustek T (eds) Sulfur metabolism in phototrophic organisms. Springer, Dordrecht, pp 31–58
Kuhl M, Jørgensen BB (1992) Microsensor measurement of sulphate reduction and sulfide oxidation in compact microbial communities of aerobic biofilms. Appl Environ Microbiol 58:1164–1174
Lang C, Popko J, Wirtz M et al (2007) Sulphite oxidase as key enzyme for protecting plants against sulphur dioxide. Plant Cell Environ 30:447–455
Lee W, Lewandowski Z, Characklis WG, Nielsen PH (1994) Microbial corrosion of mild steel in a biofilm system. In: Geesey GG, Lewandowski Z, Flemming HC (eds) Biofouling and biocorrosion in industrial water systems. Lewis, Plenum, CRC, Boca Raton, FL, pp 205–212
Lens PN, Meulepas RJ, Sampaio R et al (2008) Bioprocess engineering of sulfate reduction for environmental technology. In: Dahl C, Friedrich CG (eds) Microbial sulfur metabolism. Springer, Berlin, pp 285–295
Leustek T, Martin MN, Bick JA, Davies JP (2000) Pathways and regulation of sulfur metabolism revealed through molecular and genetic studies. Annu Rev Plant Biol 51(1):141–165
Lin CJ, Zhang P, Pongprueksa P et al (2014) Pilot-scale sequential anaerobic-aerobic biological treatment of waste streams from a paper mill. Environ Progr Sustain Energ 33:359–368
Lindberg P, Melis A (2008) The chloroplast sulfate transport system in the green alga *Chlamydomonas reinhardtii*. Planta 228:951–961
Meyer T, Edwards EA (2014) Anaerobic digestion of pulp and paper mill wastewater and sludge. Water Res 65:321–349

Miransari M (2013) Soil microbes and the availability of soil nutrients. Acta Physiol Planta 35 (11):3075–3084
Murray F (1997) Urban air pollution and health effects. In: Brune D, Chapman DV, Gwynne M, Pacyna JM (eds) The global environment—science, technology and management. Wiley-VCH, Weinheim, pp 585–598
Nelson DC, Castenholz RW (1981) Use of reduced sulfur compounds by *Beggiatoa* sp. J Bacteriol 147:140–154
Nielsen PH, Lee WC, Lewandowski Z, Morrison M, Characklis WG (1993) Corrosion of mild steel in an alternating oxic and anoxic biofilm system. Biofouling 7:267–284
Nisbet EG, Sleep NH (2001) The habitat and nature of early life. Nature 409:1083–1091
Nowak K, Luniak N, Witt C et al (2004) Peroxisomal localization of sulfite oxidase separates it from chloroplast-based sulfur assimilation. Plant Cell Physiol 45:1889–1894
Okabe S, Itoh T, Satoh H, Watanabe Y (1999) Analyses of spatial distributions of sulphate-reducing bacteria and their activity in aerobic wastewater biofilms. Appl Environ Microbiol 65:5107–5116
Okabe S, Ito T, Satoh H (2003) Sulphate-reducing bacterial community structure and their contribution to carbon mineralization in a wastewater biofilm growing under microaerophilic conditions. Appl Microbiol Biotechnol 63:322–334
Österberg R (1997) On the prebiotic role of iron and sulphur. Orig Life Evol Biosph 27:481–484
Papenbrock J, Schmidt A (2000) Characterization of a sulfurtransferase from *Arabidopsis* thaliana. Eur J Biochem 267:145–154
Parmar S, Buchner P, Hawkesford MJ (2007) Leaf developmental stage affects sulfate depletion and specific sulfate transporter expression during sulfur deprivation in *Brassica napus* L. Plant Biol 9:647–653
Postgate JR (1984) The sulphate reducing bacteria. Cambridge University Press, Cambridge
Pöykiö R, Manskinen K, Oksanen J et al (2014) Chemical characterisation of biosludge from a wastewater treatment plant in a neutral sulphite semi-chemical pulp mill by single and sequential extraction of heavy metals and micro-/macronutrients—a case study. Chem Pap 68(11):1546–1554
Pukall R, Buntefuß D, Frühling A, Rohde M, Kroppenstedt RM, Burghardt J, Lebaron P, Bernard L, Stackebrandt E (1999) *Sulfitobacter mediterraneus* sp. nov., a new sulfite-oxidizing member of the alphaproteobacteria. Int J Syst Evol Microbiol 49:513–519
Pundir CS, Rawal R (2013) Determination of sulfite with emphasis on biosensing methods: a review. Anal Bioanal Chem 405(10):3049–3062
Rajagopalan KV (1980) Sulfite oxidase (sulfite: ferricytochrome c oxidoreductase). In: Coughlan MP (ed) Molybdenum and molybdenum-containing enzymes. Pergamon, Oxford, pp 243–272
Randewig D, Hamisch D, Eiblmeier M, Boedecker C, Kreuzwieser J, Mendel RR, Hänsch R, Herschbach C, Rennenberg H (2013) Oxidation and reduction of sulfite contribute to susceptibility and detoxification of SO_2 in Populus X canescens leaves. Trees 28:399–411
Rennenberg H (1984) The fate of excess sulfur in higher plants. Annu Rev Plant Physiol Plant Mol Biol 35:121–153
Rintala J, Sanz Martin JL, Lettinga G (1991) Thermophilic anaerobic treatment of sulfate-rich pulp and paper integrate process water. Water Sci Technol 24:149–160
Rouached H, Wirtz M, Alary R et al (2008) Differential regulation of the expression of two high-affinity sulfate transporters, SULTR1.1 and SULTR1.2, in *Arabidopsis*. Plant Physiol 147:897–911
Roy AB, Harwood JL (1999) Re-evaluation of plant sulpholipid labelling from UDP-[14C]glucose in pea chloroplasts. Biochem J 344:185–187
Salkinoja-Salonen MS, Hakulinen R, Silakoski L et al (1985) Fluidized bed technology in the anaerobic treatment of forest industry wastewaters. Water Sci Technol 17:77–88
Salon C, Bataillé M, Gallardo K et al (2014) 34S and 15N labelling to model S and N flux in plants and determine the different components of N and S use efficiency. In: Dieuaide-Noubhani M,

Alonso A (eds) Plant metabolic flux analysis, Method in molecular biology, Vol 1090. Humana, New York, pp 335–346

Sanda S, Leustek T, Theisen MJ, Garavito RM, Benning C (2001) Recombinant *Arabidopsis* SQD1 converts UDP-glucose and sulfite to the sulfolipid head group precursor UDP-sulfoquinovose in vitro. J Biol Chem 276:3941–3946

Seifert U, Heinz E (1992) Enzymatic characteristics of UDP-sulfoquinovose:diacylglycerol sulfoquinovosyl-transferase from chloroplast envelopes. Bot Acta 105:197–205

Shen Y, Buick R (2004) The antiquity of microbial sulfate reduction. Earth Sci Rev 64:243–272

Shibagaki N, Rose A, McDermott JP et al (2002) Selenate-resistant mutants of *Arabidopsis thaliana* identified SULTR1;2 a sulfate transporter required for efficient transport sulfate into roots. Plant J 29:475–486

Shimojima M, Hoffmann-Benning S, Garavito RM, Benning C (2005) Ferredoxin-dependent glutamate synthase moonlights in plant sulfolipid biosynthesis by forming a complex with SQD1. Arch Biochem Biophys 436:206–214

Smith FW, Hawkesford MJ, Prosser IM, Clarkson DT (1995) Isolation of a cDNA from *Saccharomyces cerevisiae* that encodes a high affinity sulphate transporter at the plasma membrane. Mol Gen Genet 247:709–715

Sorokin DY (1995) Sulfitobacter pontiacus gen. nov., sp. nov.—a ne w heterotrophic bacterium from the black sea specialized on sulfite oxidation. Microbiology 64:295–305

Takahashi H (2010) Sulfate transport and assimilation in plants. Int Rev Cell Mol Biol 281:129–159

Takahashi H, Kopriva S, Giordano M et al (2011) Sulfur assimilation in photosynthetic organisms: molecular functions and regulations of transporters and assimilatory enzymes. Annu Rev Plant Biol 62:157–184

Tietje C, Heinz E (1998) Uridine-diphospho-sulfoquinovose: diacylglycerol sulfoquinovosyl transferase activity is concentrated in the inner membrane of chloroplast envelopes. Planta 206:72–78

Tomatsu H, Takano J, Takahashi H et al (2007) An *Arabidopsis thaliana* high affinity molybdate transporter required for efficient uptake of molybdate from soil. Proc Natl Acad Sci U S A 104:18807–18812

Tripathi BM, Kumari P, Weber KP et al (2014) Influence of long term irrigation with pulp and paper mill effluent on the bacterial community structure and catabolic function in soil. Indian J Microbiol 54:65–73

Tsakraklides G, Martin M, Chalam R et al (2002) Sulfate reduction is increased in transgenic *Arabidopsis thaliana* expressing 59-adenylylsulfate reductase from *Pseudomonas aeruginosa*. Plant J 32:879–889

Turgay OC, Bilen S (2012) The role of plant growth-promoting rhizosphere bacteria in toxic metal extraction by Brassica spp. In: Anjum NA, Ahmad I et al (eds) The plant family Brassicaceae. Springer, Dordrecht, pp 213–237

Vidmar JJ, Tagmount A, Cathala N et al (2000) Cloning and characterization of root specific sulfate transporter from *Arabidopsis thaliana*. FEBS Lett 475:65–69

Wang Y, Chen K, Wang Q et al (2014) Comparison between ion-chromatography and titration methods for the determination of sulfite in wastewater containing furfural. Chem Lett 43 (6):885–886

Williams TJ, Zhang CL, Scott JH, Bazylinski DA (2006) Evidence for autotrophy via the reverse tricarboxylic acid cycle in the marine magnetotactic coccus strain MC-1. Appl Environ Microbiol 72:1322–1329

Wolicka D (2010) Application of sulphate-reducing bacteria in biological treatment wastewaters. In: Canton KW (ed) Fluid waste disposal. Nova, New York, pp 49–81

Xia Z, Su X, Wu J et al (2012) Molecular cloning and functional characterization of a putative sulfite oxidase (SO) ortholog from Nicotiana benthamiana. Mol Biol Rep 39:2429–2437

Xia Z, Wu K, Zhang H et al (2014) Sulfite oxidase is essential for timely germination of maize seeds upon sulfite exposure. Plant Mol Biol Rep 1–10

Yamanaka T, Yoshioka T, Kimura K (1981) Purification of sulphite cytochrome c reductase of *Thiobacillus novellus* and reconstitution of its sulphite oxidase system with the purified constituents. Plant Cell Physiol 22:613–622

Yan L, Lu X, Wang Q et al (2014) Research on sulfur recovery from the byproducts of magnesia wet flue gas desulfurization. Appl Therm Eng 65(1):487–494

Yoshimoto N, Takahashi H, Smith FW et al (2002) Two distinct high-affinity sulfate transporters with different inducibilities mediate uptake of sulfate in *Arabidopsis* roots. Plant J 29:465–473

Yoshimoto N, Inoue E, Saito K et al (2003) Phloem-localizing sulfate transporter, Sultr1; 3, mediates re-distribution of sulfur from source to sink organs in *Arabidopsis*. Plant Physiol 131:1511–1517

Yoshimoto N, Inoue E, Watanabe-Takahashi A et al (2007) Posttranscriptional regulation of high-affinity sulfate transporters in *Arabidopsis* by sulfur nutrition. Plant Physiol 145:378–388

Yu B, Xu C, Benning C (2002) *Arabidopsis* disrupted in SQD2 encoding sulfolipid synthase is impaired in phosphate-limited growth. Proc Natl Acad Sci U S A 99:5732–5737

Yu GP, Gong Z, Xu MX, Dou CR (2013) The effect by sodium sulfite on the enzyme activity and stabilization of race bran. Adv Mater Res 634:1219–1224

Zhang S (2014) U.S. Patent 8,715,509. U.S. Patent and Trademark Office, Washington, DC

Zhang JB, Zhang H, Wang HL et al (2014) Risk analysis of sulfites used as food additives in China. Biomed Environ Sci 27(2):147–154

Zheng M, Liao BQ (2014) A comparative study on thermomechanical pulping pressate treatment using thermophilic and mesophilic sequencing batch reactors. Environ Technol 35 (11):1409–1417

Chapter 14
Bioconversion of Cotton Gin Waste to Bioethanol

Shitarashmi Sahu and Krishna Pramanik

14.1 Introduction

Huge quantity of cotton gin waste is generated in cotton mills. The disposal of this waste environmental regulation is one of the biggest problems faced by the cotton ginning industry worldwide (Agblevor et al. 2006). The residues from cotton crop cultivation are of two types: cotton plant trash (CPT) and cotton gin trash (CGT). CPT is the residue that stays in field after harvest of cotton, while CGT is generated from ginning process. Of these two types of wastes, CGT is very important to researchers and cotton producers due to high production and difficulties of disposing of it (Plácido et al. 2013). Raw cotton processing generates cotton gin residue (CGR), which is composed of immature bolls, cotton seed, hulls, burs, sticks, leaves, cotton lint, and dirt (Isci and Demirer 2007). About 218 kg of cotton fiber generates 68–91 kg of CGT (Sharma-Shivappa and Chen 2008). Ginning one bale (227 kg) of spindle-harvested seed cotton lint contributes between 37 and 147 kg of waste (Zabaniotou and Andreou 2010). Worldwide production of this waste is approximately 3.23 million tonnes per year (Jeoh and Agblevor 2001). The USA produces about 1.8 million tonnes of CGT, whereas an average of 800 tonnes of cotton gin trash is produced by Texas in its 30 principal counties. With this large quantity of wastes, the final disposal becomes a major problem to the cotton industry which becomes more critical during winter and rainy seasons when insects use these residues as survival sites (Gavrilescu 2010; White et al. 1996). Availability is one of the most important factors in feasibility of using any product for bioenergy production (Arthe et al. 2008). In this context, though abundance of cotton gin waste throughout the world is a major problem of disposal, it's a

S. Sahu • K. Pramanik (✉)
Department of Biotechnology and Medical Engineering, National Institute of Technology, Rourkela, Orissa, India
e-mail: 509bm604@nitrkl.ac.in; kpr@nitrkl.ac.in

L.B. Sukla et al. (eds.), *Environmental Microbial Biotechnology*, Soil Biology 45,
DOI 10.1007/978-3-319-19018-1_14

simultaneous solution for bioenergy production. These cotton wastes containing minute fibers when suspended in air may cause serious manifestations in human body mainly affecting lungs (Mahalakshmi 2011). The traditional disposal methods including land filling, land application, and incineration of the cotton gin waste have several disadvantages such as environmental pollution and limitation of land supply, etc. (Shen and Agblevor 2008). The current method of choice is incorporation of cotton gin waste into soil. But success of this method critically depends on the climate conditions (Agblevor et al. 2006). There is much concern over the presence of weed seeds, insect infestations, diseases, and excess chemicals in the waste that may degrade the receiving land.

The need for alternative disposal technologies is very pronounced in cotton industry because of the climatic conditions and small ginning plants (Agblevor et al. 2006). The conventional disposal method of cotton gin trash is incineration that produced several health hazards (Aquino et al. 2010) and environmental pollution (Shen and Agblevor 2008). Furthermore, because of the high ash content of the feedstock, there could be potential slagging problems associated with large-scale incineration. Land filling is not a suitable option because tipping fees are very high. On the other hand, emission of green house gases is increasing rapidly with the fast depletion of oil resources. Alternative fuels produced from renewable resources, such as fuel ethanol, provide numerous benefits in terms of environmental protection, economic development, and national energy security (Yang and Lu 2007). In this context, the conversion of this cotton gin waste to ethanol could reduce particulate emission and reduce fire hazards from spontaneous combustion of cotton gin residues (Agblevor et al. 2003). Slight differences in the proportions of components are usually found between varying mechanical harvest methods (Mahalakshmi 2011). Cotton gin waste containing cellulose and hemicellulose, which could be potentially used as a source of monomeric sugars for fermentation to produce high-value products or constituent biopolymers, could be used for the production of microcrystalline cellulose, cellulose derivatives, and plastics (Jeoh and Agblevor 2001). Cotton gin wastes contain about 50 % or more cellulose and hemicellulosic components (Jeoh and Agblevor 2001; Lark et al. 1997) that can be used to produce bioethanol. The conversion of cotton gin waste to biofuel has been investigated by some researchers. Shen and Agblevor (2008) reported a production of 157 l of ethanol produced per ton of cotton gin waste (Shen and Agblevor 2008). The biological processes using microbial strains are the most attractive for the conversion of this waste to bioethanol. The present review explores the various potential microbial strains for conversion processes and different methods for production of bioethanol from cotton gin waste in a cost-effective and environmental-friendly manner.

14.2 Current Status/Availability of Cotton Gin Waste in All Over the World and India

The high level of cotton production is directly related to the high production of wastes and residues. Worldwide, the production of this waste is approximately 3.23 million tonnes per year (Plácido et al. 2013). Cotton is an important natural fiber in these days. Major growth of cotton production was observed since the end of Second World War (WWII). In 2012/2013, developing countries accounted for the most of global cotton mill use (96 %), import (97 %), and production (81 %). But they only account for 52 % of global cotton exports (ICTSD 2013). Cotton has been grown in 90 countries during the year of 2007 especially in countries in which the cotton is being planted on a large scale such as China, India, Brazil, Eastern Pakistan, Europe, Turkey, and Australia (ICTSD 2013). It is also a major crop in some parts of the USA such as Texas, Arizona, New Mexico, California, Kansas, and Missouri. From 2006/2007 to 2014/2015 (Table 14.2), four major cotton-producing countries are India, China, the USA, and Pakistan which consider for approximately three-quarters of world's cotton producer. Cotton is the principal source of natural fibers for textile industries and thus one of the most abundant sources of agro-industrial biomass, with a production of 24 million tonnes in the world in 2006–2007 and an annual increase of 2 % (Sharma-Shivappa and Chen 2008). India has the largest area under cotton production and China is the largest producer of cotton in the world, whereas India is the second largest cotton producer (CCI 2013; Isci and Demirer 2007). Interestingly, China with almost half the area under cotton production as compared to India produces more than 2½ times yield (kg per hectare) of cotton as compared to India (CCI 2013). The area under cotton production in the world is estimated at around 30–31 million hectares (Table 14.1).

Table 14.1 World status on the area and yield of cotton production

Country/ region	Area (million 1,000 HA)			Yield (kg per hectare)		
	2007–2008	2008–2009	2009–2010	2007–2008	2008–2009	2009–2010
World	32.94	30.76	30.44	793	761	740
Australia	0.07	0.16	0.2	2,144	1,991	1,960
Brazil	1.08	0.84	0.82	1,488	1,415	1,443
Turkey	0.52	0.34	0.28	1,298	1,236	1,322
China	6.2	6	5.2	1,299	1,332	1,319
Mexico	0.11	0.1	0.07	1,227	1,240	1,291
Syria	0.19	0.19	0.18	1,252	1,272	1,179
Greece	0.35	0.25	0.2	964	1,002	980
USA	4.25	3.06	3.13	985	911	870
Egypt	0.24	0.13	0.12	880	837	841
Uzbekistan	1.43	1.42	1.3	815	705	737
Pakistan	3	2.9	3	646	676	682
India	9.44	9.41	10.26	554	523	528

Source: USDA, Foreign Agricultural Service

The major cultivators of cotton in the world are America, India, China, Egypt, Brazil, Pakistan, and Eastern Europe (Table 14.2).

14.3 Chemical Composition of Cotton Gin Waste

Cotton gin waste consists of three major structural polymeric components, namely, lignin, cellulose, and hemicellulose (Sindhu et al. 2011; Taherzadeh and Niklasson 2004). These components build up about 90 % of dry matter in lignocelluloses, with the rest consisting of extractive, moisture, and ash (Dehkhoda 2008). In order to exploit cotton gin waste for its fermentable sugars, the chemistry must be understood. Bioethanol yield from biomass is directly related to hemicellulose, cellulose, and sugar concentration in the feedstock (Karunanithy et al. 2008). The lignin cannot be used for bioethanol production (Balat 2011).

14.3.1 Cellulose

Cellulose is having a highly crystallized structure as a result of the existence of hydrogen bonds. In significance to its amorphous region, the crystalline region of cellulose makes it difficult to hydrolyze (Credou and Berthelot 2014). Hydrogen bonds between different layers of the polysaccharides contribute to resistance of crystalline cellulose to degradation. Cellulose is a beta (1–4)-linked chain of glucose molecules which is a polymer of D-glucose units linked by 3-glucoside bonds from the anomeric carbon of one unit to the C-4 hydroxy of the next unit (Fengel and Wegener 1983). The cellulose chains further aggregate into alternating highly ordered regions and amorphous regions in a manner described by the fringed micelle theory proposed by Gerngross et al. in 1932 (Fengel and Wegener 1983). The cellulose fibers are sometimes referred to as the elementary fibrils and/or microfibrils (Sjöström 1993). In the biomass feedstock, cellulose is the main reservoir of glucose, which is the most desired fermentation element (Jeoh 1998). The researchers assumed a cotton gin waste composition of 40 % cellulose, 30 % hemicelluloses, and 25 % of lignin (Agblevor et al. 2003).

14.3.2 Hemicellulose

Hemicelluloses are complex, highly branched polysaccharides that occur in association with cellulose in the cell walls (Klass 1998). The monomers that comprise hemicellulose are hexoses (glucose, galactose, and mannose) and pentoses (arabinose and xylose). Hemicellulose can be classified into three groups, namely, xylans, mannans, and galactans, based on the polymer backbone that is very often homopolymeric with β-1,4 linkages. In softwoods, the primary hemicellulose

Table 14.2 World cotton production from 2005/2006 to 2014/2015 (millions of 480-lb. bales)

Country	2005/2006	2006/2007	2007/2008	2008/2009	2009/2010	2010/2011	2010/2011	2011/2012	2012/2013	2013/2014	2014/2015
China	28.4	35.5	37	36.7	31.5	31.5	30.5	34	35	32	29.5
India	19.1	21.6	24	22.6	24.3	23.8	27.2	29	28.5	30.5	28.5
USA	23.9	21.6	19.2	12.8	12.3	12.6	18.1	15.6	17.3	12.9	15
Pakistan	10.2	9.9	8.9	9	9.4	9.8	8.6	10.6	9.3	9.5	9.5
Brazil	4.7	7	7.4	5.5	5.4	5.4	9	8.7	6	7.5	8.3
Uzbekistan	5.6	5.4	5.4	4.6	4.4	4.4	4.1	4.2	4.5	4.2	4.2
Australia	2.8	1.4	0.6	1.5	1.8	1.8	4.2	5.5	4.6	4.1	3.1
Turkey	3.6	3.8	3.1	1.9	1.7	1.7	2.1	3.4	2.7	2.3	2.9
African franc zone	2.5	2.3	1.6	1.6	1.6	1.6	2.1	3	3.9	4.1	3.9
Turkmenistan	1	1.2	1.3	1.4	1.1	1.1	1.8	1.4	1.6	1.5	1.5
EU-27	2.5	1.6	1.7	1.2	1	1	1.2	1.6	1.5	1.6	1.7
Rest of the world	12.6	10.5	9.8	8.7	8.1	8.1	7.7	9.9	9.3	9	8.7
World	116.7	122	119.9	107.5	102.7	102.7	117	126.6	123	118.1	115.9

Source: USDA, Foreign Agricultural Service

components are galactoglucomannans and arabinoglucuronoxylan, while the principal hemicelluloses in hardwoods are glucomannans and methyl glucuronoxylans (Brigham et al. 1996). Xylan is the most important in terms of the percentage of total hemicellulose found in biomass waste. In the cell wall, the hemicellulose polymers surround and associate with the cellulose core of the microfibrils by means of hydrogen bonds (Terashima et al. 1993).

14.3.3 Lignin

Lignin serves as the bonding element or "cement," between plant fibers, stiffening them and acting as a barrier to degradation of the cell walls (Goldstein 1981). Lignin is a three-dimensional phenyl propane polymer with phenyl propane units held together by ether and carbon–carbon bonds (Sun et al. 2002). It is constructed of three monomers: coniferyl alcohol, sinapyl alcohol, and coumaryl alcohol, each of which has an aromatic ring with different substituents (Brown 2003). The dominant monomeric units in the polymers are benzene rings bearing methoxyl, hydroxyl, and propyl groups that can be attached to other units (Klass 1998). Lignin strengthens the cell structures by stiffening and holding the fibers of polysaccharides together (Fan et al. 1987).

14.4 Different Conversion Processes

In general, the production of bioethanol from lignocellulosic material is based on three principal steps such as pretreatment, saccharification, and fermentation, which is represented in Fig. 14.1 (Carere et al. 2008). The first step aims to reduce the quantity of lignin present in the biomass, thereby making the cellulose and hemicelluloses readily available for the saccharification process. The second step is to extract the monosaccharides present in the cellulose (glucose) and the hemicellulose (xylose, arabinose, galactose, and mannose). Finally, microbial fermentation of the sugars produced during saccharification yields bioethanol. The three-step process can be modified to improve the yield of bioethanol from cotton gin waste (Plácido et al. 2013; Alvira et al. 2010).

14.4.1 Pretreatment Methods for Cotton Gin Waste

Specifically pretreatment is for removing or breaking of lignin, extraction of hemicellulose, decrystallizing cellulose, removing acetyl group from hemicellulose, reducing polymerization of cellulose, and expanding the structure to increase pore value and internal surface area so that hydrolysis of carbohydrate fraction to monomeric sugars can be achieved more rapidly with higher yields

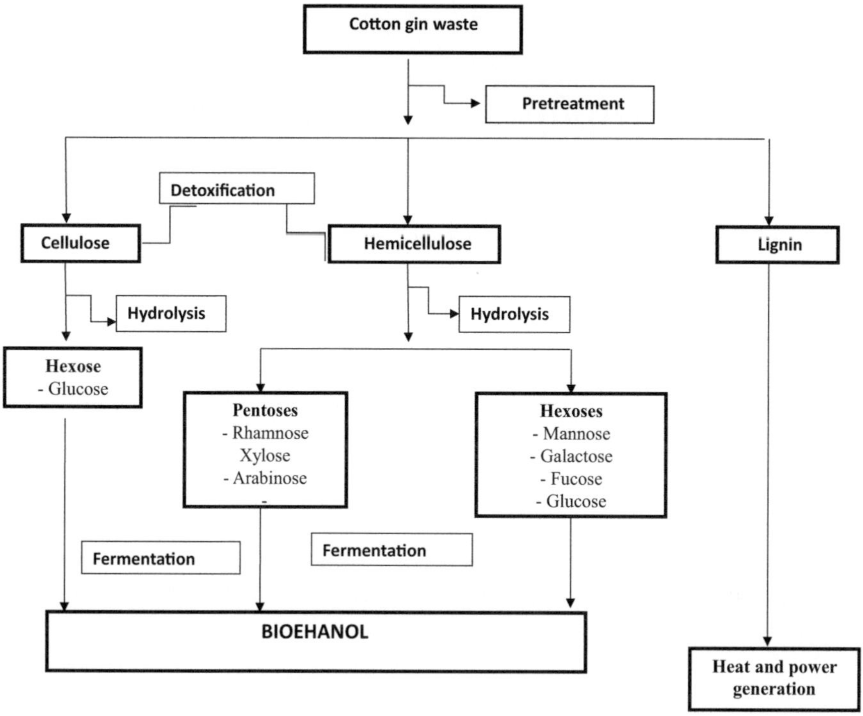

Fig. 14.1 Different conversion processes and the outcome of cotton gin waste to bioethanol (adapted from das Neves and Kimura 2007)

(Aita and Kim 2010). A pretreatment that accomplishes all of these goals is likely to be expensive, so most pretreatment focuses on achieving just a few value-added products. It is clear, however, that different pretreatment methods affect biomass in different ways (Mosier et al. 2005; Wyman et al. 2005). For removal of lignin, a pretreatment is therefore necessary to disrupt or remove lignin from such substrates and thus increase accessibility of cellulose (Kaparaju et al. 2009; Karimi et al. 2006a, b; Zhao et al. 2011). But such pretreatment process is highly expensive and complex the process. Moreover, some of the delignification methods (Safartalab et al. 2014) are found to have influence the biocompatibility of the process. As an advantage, cotton and some viscose wastes fibers due to lack of lignin in their cellulosic structure (Miranda et al. 2007) won't face such problems and thus are suitable substrates for bioethanol production. If the pretreatment is not efficient enough, the resultant residue is not easily hydrolyzable by cellulase enzyme, and if it is more severe, then it produces toxic compounds which inhibit the microbial metabolism (Kodali and Pogaku 2006). The goal of pretreatment on cotton gin waste is shown in Fig. 14.2 (Wyman et al. 2005).

Several methods have been introduced for the pretreatment of lignocellulosic materials prior to enzymatic hydrolysis or digestion. These methods are mainly classified into physicochemical, physical, chemical, and biological pretreatment (Agbor et al. 2011; Berlin et al. 2006; Karimi et al. 2006a; Mtui 2009). In this

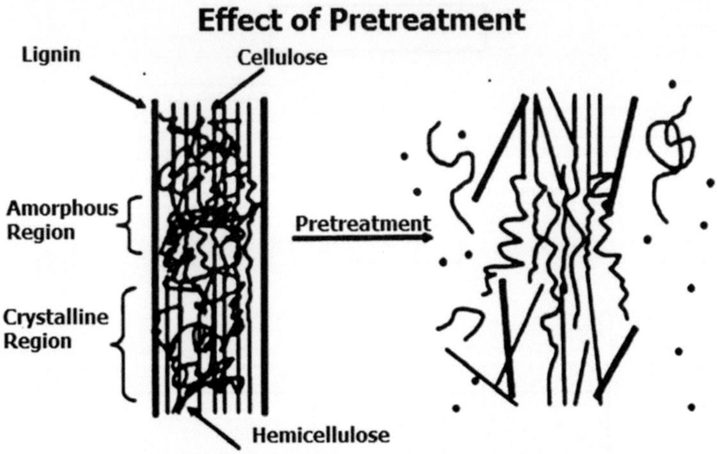

Fig. 14.2 Schematic goals of pretreatment on lignocellulosic material (Wyman et al. 2005)

section, we review these methods, although not all of them have yet been developed enough to be applied for the applications in large-scale processes within the world (Mohammad et al. 2008).

14.4.1.1 Physical Pretreatment

Physical pretreatment develops the accessible surface area and size of pores and reduces the crystallinity and degrees of polymerization of cellulose (Mohammad et al. 2008). Fermentation inhibitors (furfural and phenolic compounds) are generally toxic to cells, which is one of the biggest blockades for lignocellulose biorefinery process. Physical treatment can reduce the production of inhibitor through "fractional conversion" (Wang et al. 1990). Mechanical treatment is one of the physical treatments which reduce biomass size below #20 sieves representing the best mechanical performance (De Sousa et al. 2004). Mechanical pretreatment also increases the digestibility of cellulose and hemicellulose in the lignocellulosic biomass. The use of mechanical chopping (de Sousa et al. 2004), grind milling (Mtui and Nakamura 2005), roll milling (Qi et al. 2005), hammer milling (Iñiguez-Covarrubias et al. 2001; Mani et al. 2004), vibratory milling (Guerra et al. 2006), and ball milling (Inoue et al. 2008) has been already proved successful as a low-cost pretreatment strategy. These mechanical pretreatment techniques are mostly time-consuming and energy expensive to process. The compression milling is apparently the only comminution process that has been tested using a production-scale apparatus (Wyman 1996). Extrusion process is a novel and promising physical pretreatment method for biomass conversion to ethanol production. In extrusion method, the materials are subjected to mixing, heating, and shearing which lead to chemical and physical modifications during the passage through the extruder (Alvira et al. 2010). Screw speed and barrel temperature are generally disrupt the lignocellulose structure causing defibrillation, fibrillation and shortening of the fibers. Accessibility of carbohydrates to enzymatic attack (Karunanithy et al. 2008). Irradiation, for example, gamma rays, electron beam, and microwaves, can improve enzymatic hydrolysis and delignification of lignocelluloses. The combination of radiation and other methods such as acid treatment can further accelerate enzymatic hydrolysis (Ardica et al. 1985; Taherzadeh and Karimi 2008). The liquid hot water method uses compressed hot liquid water (at pressure above saturation point) to hydrolyze the hemicellulose. Xylose recovery is relatively high (88–98 %) and no acid or chemical catalyst is required, which makes it environmentally attractive and economically interesting (das Neves and Kimura 2007; Plácido et al. 2013). Pyrolysis has also been used for pretreated lignocellulosic materials, since biomass can be used as substrate for a fast pyrolysis for thermal conversion of cellulose and hemicellulose into fermentable sugars with good yields (Tomás-Pejó et al. 2008). Hydrothermolysis is one of the conventional approaches, which started as a pretreatment method before hydrolysis (Tian et al. 2011). In this hydrothermal process, water, steam, or both and heat are used (Zhang et al. 2014a).

14.4.1.2 Physicochemical Pretreatment

Both chemical and physical treatment systems mostly dissolve hemicellulose and alteration of lignin structure by providing an improved accessibility of the cellulose for hydrolytic enzymes (Hendriks and Zeeman 2009). Steam explosion is one of the most promising physicochemical methods to make biomass more accessible to cellulase attack (Liu et al. 2014; Szengyel 2000). Basically, this method consists of heating the material using high-pressure steam for a few minutes (Mtui 2009). Steam explosion treatment increases crystallinity of cellulose by promoting crystallization of the amorphous portions. In this process hemicellulose is easily hydrolyzed by steam explosion treatment, and it promotes delignification (Balat 2011). This technique is economically attractive, is an additional potential for energy efficiency, requires less hazardous process chemicals, and offers complete sugar recovery (Tomas-Pejo 2008). The extraction of cellulose from cotton gin waste was studied using a steam explosion technology as a pretreatment process followed by alkali bleaching which produces higher yield of ethanol (El-Zawawy et al. 2011). Addition of H_2SO_4 (or SO_2) or CO_2 in steam explosion of lignocellulosic waste can efficiently improve enzymatic hydrolysis, reduce the production of inhibitory compounds, and lead to more complete liquefaction of hemicellulose, glucan, xylan, mannan, galactan, and arabinan (Jeoh and Agblevor 2001; Sun et al. 2002). Ammonia fiber explosion (AFEX) is one of the alkaline physicochemical pretreatment processes. In this process, the material is subjected to liquid ammonia at high pressure and temperature and a subsequent fast decompression. Similar to the steam explosion, that causes a rapid saccharification of the lignocellulosic material (Abril and Abril 2009). In a typical AFEX process, the dosage of liquid ammonia is 1–2 kg ammonia/kg dry biomass, the temperature is 363 K, and residence time is 30 min (Kumar et al. 2009). The effective parameters in the AFEX treatment are temperature, ammonia loading, time, water loading, blowdown pressure, and number of treatments (Mohammad et al. 2008). AFEX process is more effective for the biomass which has less lignin, and the AFEX pretreatment does not significantly solubilize hemicellulose in comparison to other pretreatment processes such as dilute acid. Furthermore, ammonia must be recycled after the pretreatment to reduce the cost and to protect the environment (Eggeman and Elander 2005; Sun et al. 2002). In CO_2 explosion, 75 % of the theoretical glucose released during 24 h of the enzymatic hydrolysis has been reported (Sun et al. 2002). Carbon dioxide (CO_2) has been considered as an extraction solvent for non-extractive purposes, as a result of many advantages such as availability at relatively low cost, nontoxicity, nonflammability, easy recovery after extraction, and environmental suitability (Mohammad et al. 2008). Ethanol yield up to 83 % of the theoretical value has been achieved for lignocellulosic waste by physico-chemical treatment (Jeoh and Agblevor 2001).

14.4.1.3 Chemical Pretreatment

Common chemical pretreatment methods are dilute acid, alkaline, ammonia, organic solvent, SO_2, CO_2, or other chemicals (das Neves and Kimura 2007). Ozonolysis involves breakdown of lignin and hemicellulose by ozone gas, thereby increasing the biodegradability of the cellulose. The pretreatment is usually carried out at room temperature and is effective for lignin removal without the formation of toxic by-products (Mohammad et al. 2008). Alkali pretreatment refers to the application of alkaline solutions to remove lignin and various uronic acid substitutions on hemicellulose that lowers the accessibility of enzyme to the hemicellulose and cellulose (Silverstein et al. 2007; Han et al. 2009). Alkali pretreatment may be carried out at ambient conditions, but pretreatment time is measured in terms of hours or days rather than minutes or seconds (Mosier et al. 2005). In spite of the advantages, these methods present difficulties from the point of view of the process economy for obtaining fuels (Abril and Abril 2009). Sodium, potassium, calcium ammonium carbonate (Kim et al. 2014), and ammonium hydroxide are appropriate chemicals for pretreatment. Among these, NaOH has been studied the most (Kumar et al. 2009). Alkaline peroxide is an effective method for pretreatment of lignocellulosic biomass. In this process, the lignocelluloses are soaked in pH-adjusted water (e.g., to pH 11–12 using NaOH) containing H_2O_2 at room temperatures for a period of time (e.g., 6–24 h). This method can enhance the enzymatic hydrolysis by delignification (Taherzadeh and Karimi 2008). Organosolvent is used to provide treated cellulose for easier enzymatic hydrolysis, by using an organic or aqueous organic solvent to remove or degrade the complex structure of lignin and hemicellulose (Sindhu et al. 2012). To improve ethanol productivity with few inhibitors generated, a novel process of combined alkaline peroxide pretreatment and semi-simultaneous saccharification and fermentation (SSSF) was developed. Pretreatment with 10 % of H_2O_2 at 160 °C for 2 h followed by SSSF was found to be the optimal combination with remarkably increased ethanol yield around 63.1 % which was about five times more than that of the untreated sample (Zhang et al. 2014b). Pretreatment of lignocellulosic materials can be performed by treatment with ozone, referred to as "ozonolysis" pretreatment. This method can effectively degrade lignin and part of hemicellulose. These pretreatments are usually carried out at room temperature and does not lead to the formation of inhibitory compounds (Travaini et al. 2013). Dilute acid hydrolysis is the oldest technology for converting cellulose biomass to bioethanol. Treatment of cotton gin waste with dilute acid can efficiently improve enzymatic hydrolysis (Mahalakshmi 2011; Kumari and Pramanik 2012). Sulfuric acid is one of the most applied acid, whereas other acids like nitric acid and HCl were also reported (Taherzadeh and Karimi 2007). Most dilute acid convesion are limited to a sugar recovery efficiency of around 50 % (Badger 2002). Dilute acid hydrolysis occurs in two stages to take advantage of the differences between hemicellulose and cellulose. Generally the first stage is performed at low temperature to increase the production of sugar from hemicellulose, and in the second stage, higher-temperature stage is optimized for hydrolysis process of the cellulose portion of the lignocellulosic waste (Farooqi and Sam 2004). The first stage is conducted under mild process conditions (e.g., 0.7 % H_2SO_4, 463 K) to recover five-carbon sugars, while in the second stage, only the remaining solids with more resistant cellulose undergo harsher conditions (488 K, but a milder 0.4 % H_2SO_4) to recover the six-carbon sugars (Hamelinck et al. 2005).

14.4.1.4 Biological Pretreatment

Biological pretreatment generally involves microorganisms such as brown, white, and soft-rot fungi that are used to degrade or decompose complex lignin and hemicellulose. White-rot fungi are the most efficient microbes for biological pretreatment of lignocellulosic biomass (Taherzadeh and Karimi 2007; Kumar et al. 2009; Sun et al. 2002). The most widely studied white-rot organism is *P. chrysosporium*, which is one of the holobasidiomycetes (Sánchez 2009). The influence of fungus treatment on the biochemical composition and degradation of cotton plant by-products (cotton burns and cotton gin trash) by *Pleurotus sajor-caju* was evaluated. Lignin degradation was maximized as the incubation period progressed, whereas the highest losses of cellulose, hemicellulose, acid detergent fiber, and neutral detergent fiber were recorded for treated cotton plant by-product. Biodegradation of cotton stalks and cotton seed hull by the oyster mushroom, *Pleurotus ostreatus*, was studied for higher yield of ethanol (Yang et al. 2013). Nine agro-industrial and forestry by-products were subjected to solid-state fermentation by using *Agrocybe cylindracea* and *Pleurotus ostreatus*, where the process and end products were comparatively evaluated (Koutrotsios et al. 2014). The advantages of biological pretreatment include low energy requirements and mild environmental conditions, while a disadvantage is that the long time period required for lignin degradation (Balat 2011). Lignin biodegradation by white-rot fungi is an oxidative process and phenol oxidases are the key enzymes

(Moreno et al. 2014; Rabinovich et al. 2002). Degradation of lignin by white-rot fungi, which are the most effective microorganisms for biological pretreatment, occurs through the action of lignin-degrading enzymes such as peroxidases and laccases (Kumar et al. 2009; Ibarra et al. 2006). Some of enzymes whose roles have not been fully elucidated include glyoxal oxidase, glucose oxidase, oxidoreductase, and methanol oxidase (Eriksson 2000). Two groups of peroxidases, lignin peroxidases (LiPs) and manganese-dependent peroxidases (MnPs), have been well characterized. Laccase was demonstrated to be present in fungi for the first time by both Bertrand and Laborde in 1896 (Kunamneni et al. 2007). Recently some bacterial laccases have also been characterized from *Azospirillum lipoferum* and *Bacillus subtilis* (Kunamneni et al. 2007). Several white-rot fungi such as *Phanerochaete chrysosporium*, *Ceriporia lacerata*, *Trametes pubescens*, *Cyathus stercoreus*, *Ceriporiopsis subvermispora*, *Pycnoporus cinnabarinus*, and *Pleurotus ostreatus* have been examined on different lignocellulosic biomass showing high delignification efficiency (Marcolongo et al. 2014; Kumar et al. 2009; Shi et al. 2008; Melamane et al. 2007). The biological pretreatment might be used not only for lignin removal but also for biological removal of specific components such as antimicrobial substances and for detoxification to improve its digestion (Taherzadeh and Karimi 2008). *Klebsiella oxytoca* and *Escherichia coli* strains have been genetically engineered to produce microbial biocatalysts that yield bioethanol from lignocellulosic biomass (Jarboe et al. 2007; Peterson and Ingram 2008). Biological delignification processes are being developed for their integration in biomass to ethanol process. Solid-state and submerge state of fermentation are the method of choice for biological delignification. Most of the capital and operating prices for solid-state fermentation can be kept low, and also the lignocellulosic substrate is likely to be the major component of the cost of the delignified product (Saritha et al. 2012). *Pycnoporus cinnabarinus* fungus was compared with commercial enzyme laccases from *Trametes villosa* and *Myceliophthora thermophila* in terms of stability and mediator oxidation rates (Ibarra et al. 2006). *Rigidoporus lignosus*, a white-rot basidiomycete, excreted two oxidative enzymes into the culture medium, laccase and Mn peroxidase, and these two enzymes acted synergistically in solubilizing the lignin (Galliano et al. 1991). Wang et al. (1990) first cloned a lignin peroxidase gene from *Streptomyces viridosporus* T7A into *Streptomyces lividans* and demonstrated that the genetically engineered *S. lividans* expressed significant extracellular 2,4-dichlorophenol peroxidase activity and degraded lignocellulose in solid-state processes. Most ligninolytic microorganisms solubilize or consume not only lignin but also hemicellulose and cellulose (Eggeman and Elander 2005). Cultivation of edible mushrooms such as *Lentinus* spp., *Lentinula* spp., *Leonotis* spp., *Pleurotus* spp., *Agaricus* spp., *Agrocybe* spp., *Volvariella* spp., and *Grifola* spp. is achievable on a wide range of lignocellulosic waste substrates such as wood waste, wheat straw, corncob meal, barley straw, soybean straw, cereal bran, cotton waste, sorghum stalk, hazel nut husks, waste tea leaves, dry weed plants, peanut shells, waste paper, and olive mill wastewater (Morais et al. 2000; Philippoussis et al. 2001; Yildiz et al. 2002; Kalmış and Sargın 2004; Silva et al. 2005; Ozçelik and Pekşen 2007; Peker et al. 2007; Das and Mukherjee 2007; Akyüz and Yildiz 2008; Gaitán-Hernández and Salmones 2008; Rani et al. 2008).

14.4.2 Hydrolysis

Pretreatment is carried out by the breakdown of carbohydrate polymers to free sugar monomers. This method is termed as hydrolysis as the process involves the addition of one water molecule for every glycosidic bond broken. Hydrolysis methods include enzymatic method (fungal or commercial enzyme), steam explosion process, dilute and concentrated acid methods for cotton gin waste have been reported. Among these, acid hydrolysis and enzymatic hydrolysis are mostly used (El-Zawawy et al. 2011; Sindhu et al. 2011). Cellulases, the most commonly used enzymes for depolymerization of cellulose to glucose, consist of three major classes, namely, endoglucanases, exoglucanases, and *b*-glucosidases (Duan and Feng 2010). Biomass processing by enzymatic or microbial hydrolysis commonly involves four biologically mediated transformations: (1) production of saccharolytic enzymes (cellulases, hemicellulases, xylanase, mannanase, and β-glucosidase), (2) hydrolysis of carbohydrate components present in pretreated biomass to sugars, (3) fermentation of hexose sugars (glucose, mannose, and galactose), and (4) fermentation of pentose sugars (xylose and arabinose). Wood et al. (1986) isolated an anaerobic fungus, *Neocallimastix frontalis*, from the rumen of a sheep which produces a highly active extracellular cellulase in solubilizing the hydrogen bond-ordered cellulose in cotton fiber. The cellulose was several-fold more active in solubilizing cotton fiber per unit of endo-1,4-b-glucanase than the cellulase of the aerobic fungus *T. reesei* mutant strain C-30, which is one of the most active cellulases isolated so far (Domingues et al. 2001). Due to the complex structure of hemicelluloses or pentose, several different enzymes are needed for their enzymatic degradation. The two main glycosyl hydrolases depolymerizing the hemicellulose backbone are endo-1,4-β-D-xylanase and endo-1,4-β-D-mannanase (Menon et al. 2010b). Cellulase enzyme extracted by *T. reesei* 3EMS35 mutant hydrolyzed most of cellulose (91 %) in wheat straw within the first 24 h, which is further used for higher ethanol production (Khokhar et al. 2014). The strains like *Trichoderma*, *Aspergillus*, *Penicillium*, and *Alternaria*, isolated from different animal dung manure soils, are reported to have highly lignocellulolytic activity, that is, cellulolytic activity along with lignolytic activity and hemicellulolytic activity (Devi and Kumar 2012).

14.4.3 Fermentation

This review provides an overview of the current status of ethanol fermentation, potential microorganisms, and technology. *Kluyveromyces marxianus* has been considered as a potential species of interest for ethanol production since it can produce ethanol at high temperature from a wide variety of substrates. However, the reason why this yeast can produce ethanol at high temperature is largely unknown (Liang et al. 2014). *H. polymorpha* ferments both glucose and xylose up to 45 °C (Ryabova et al. 2003). This thermotolerant *Debaryomyces* sp. used both pentoses and hexoses to similar extents in sugar mixtures, and a preference for one carbohydrate does not inhibit the consumption of other (Menon et al. 2010a). *Neurospora crassa* is known to produce ethanol directly from the cellulose/hemicellulose, since it produces both the cellulase and xylanase and also has the capacity to ferment the sugars to ethanol anaerobically (Dogaris et al. 2013). Recent studies have employed genetically engineered strains to produce ethanol from sugars with high efficiency by utilizing the *Saccharomyces*, *Pichia stipitis*, and *Zymomonas mobilis* strains tested under RaBIT fermentations to determine their suitability for this platform (Sarks et al. 2014). An efficient conversion of glucose and xylose is an innovative fermentation method that was designed by using coculturing immobilized *Zymomonas mobilis* and *Pichia stipitis* in a modified fermentor for the glucose and xylose fermentation, respectively (Fu et al. 2009). The Gram-negative bacterium *Klebsiella oxytoca* contains native ability to metabolize and transport cellobiose and reduce the need for extracellular cellobiase. Strain P2 is one of the recombinant derivatives in which *Z. mobilis pdc* and *adhB* genes have been integrated into the chromosome for directing the metabolism of pyruvate to ethanol. Maximum rates of ethanol production of this recombinant strain were estimated at temperature 37 °C and pH 5.0 under different stresses such as heat and osmotic stresses (Sootsuwan et al. 2013). Coculturing the strain *K. oxytoca* P2 together with other cellulase-producing microorganisms in direct ethanol fermentation process seems very promising for high-level production of ethanol with minimizing the inhibitory effect of cellobiose (la Grange et al. 2010). *Saccharomyces cerevisiae* was engineered for assembly of minicellulosomes by heterologous expression of a recombinant scaffolding protein from *Clostridium cellulovorans* and a chimeric endoglucanase E from *Clostridium thermocellum* which is helpful for higher yield of ethanol production (Hyeon et al. 2010). Microaeration enhances productivity of bioethanol from LCW using ethanologenic *E. coli* (Okuda et al. 2007); simultaneous saccharification and fermentation (SSF) using recombinant *Saccharomyces cerevisiae* results in high ethanol production of theoretical value (Jingping et al. 2012). The saccharification of the lignocellulosic biomass by the enzymes and the subsequent fermentation of the sugars to ethanol by the yeast *Saccharomyces* and *Zymomonas* mixed with *Kluyveromyces fragilis* has produced improved ethanol production (Szambelan et al. 2004). The main advantage of using SSF for the ethanol bioconversion is the enhanced rate of hydrolysis of lignocellulosic biomass (cellulose and hemicellulose) due to removal of end-product inhibition. SHF (separate hydrolysis and fermentation) is a conventional two-step process where the hemicellulose is hydrolyzed using the enzymes to form the reducing sugars in the first step and sugars, thus formed, are subsequently fermented to ethanol in the second step. The advantage of this process is that each step can be carried out at its optimum conditions. Different fusant strains were obtained by fusing protoplasts of

Saccharomyces cerevisiae and xylose-fermenting yeasts such as *Pachysolen tannophilus*, *Pichia stipitis*, and *Candida shehatae*. These fusants were used for fermentation of glucose–xylose mixture. The highest ethanol-producing fusant was used for further study to ferment hydrolysates produced by acid pretreatment and enzymatic hydrolysis of cotton gin waste (Kumari and Pramanik 2012; Benjaphokee et al. 2012). *Escherichia coli* KO11 was genetically engineered to produce ethanol from pentose and hexose sugars by inserting genes encoding alcohol dehydrogenase and pyruvate decarboxylase from the bacterium *Zymomonas mobilis*. *E. coli* KO11 can efficiently metabolize complex mixtures of sugars (Imamoglu and Sukan 2014). Kim et al. reported that the fermentation of glucose and xylose attained a level of 90 % ethanol production at 12 h using *S. cerevisiae* K35 and *P. stipitis* KCCM 12009 (Kim et al. 2011). Lu et al. found the improvement of robustness and ethanol production of ethanologenic *Saccharomyces cerevisiae* under co-stress of heat and inhibitors (Lu et al. 2012).

14.5 Future Work

Extensive research has been done on the development of advanced pretreatment technologies to produce more digestible biomass to bioethanol. An efficient and cost-effective pretreatment method may need several characteristics (Yang and Wyman 2008; Drapcho et al. 2008) such as (a) maximum fermentable carbohydrate recovery; (b) minimum inhibitors produced from carbohydrate degradation during pretreatment; (c) low environmental impact; (d) low demand of post-pretreatment methods like washing, neutralization, and detoxification; (e) low quantity of water and chemical use; (f) minimal capital cost for reactor; (g) moderately low energy input; and (g) relatively high treatment rate. Therefore, the future research on pretreatment would be focused on the following areas: first, reduction of water and chemical use; second, recovery of carbohydrates and value-added by-products to improve the economic feasibility; third, development of clean delignification with improved economics of pretreatment; fourth, fundamental understanding of pretreatment mechanisms and the relationship between the biomass structure features and enzymatic hydrolysis; and fifth, reduction of the inhibitors such as furfural, 5-HMF and acetic acid which could significantly inhibit enzymatic hydrolysis and fermentation of lignocellulosic materials (Zheng et al. 2009). Because of this, many challenges include development of stable genetically engineered microorganisms, improved gene cloning and sequencing technologies, and enhancement of productions based on economies of scale for more efficient and cost-effective conversions of cotton gin waste into bioethanol. Nano-biotechnology is now being used for LCW nanofibers in plastic composites is concerned (Alemdar and Sain 2008).

14.6 Conclusion

In recent years, much interest has been given to biofuels due to the increasing energy costs and environmental problems. Bioethanol is the most widely used biofuel for transportation worldwide. Cotton gin waste is a potential raw material for biofuel or bioethanol production. Cotton gin waste is one of the cheap and abundant feedstock, which is required to produce fuel bioethanol at reasonable cost. Several pretreatment methods have been presented for lignocelluloses and cotton gin waste materials in order to improve ethanol production. Accumulation of cotton gin waste in large quantities in places where these residues create a disposal problem results not only in deterioration of the environment but also in loss of potential of land fertility. Several novel processes and microorganisms for cotton gin waste have been identified recently. Biological pretreatment systems appear to be the mostly preferred methods, while the use of treated cotton gin waste as bio-based adsorbents of pollutants and the production of degradable biofuels and bio-polymers from cotton gin waste have drawn a lot of research interest. Current and future research trends on cotton gin waste are directed toward the development and applications of engineered organisms to tackle the challenges encountered from using conventional strains.

References

Abril D, Abril A (2009) Ethanol from lignocellulosic biomass. Cien Inv Agr 36(2):177–190

Agblevor FA, Batz S, Trumbo J (2003) Composition and ethanol production potential of cotton gin residues. Appl Biochem Biotechnol 105–108:219–230

Agblevor FA et al (2006) Storage and characterization of cotton gin waste for ethanol production. Resour Conserv Recycl 46(2):198–216

Agbor VB et al (2011) Biomass pretreatment: fundamentals toward application. Biotechnol Adv 29(6):675–685

Aita GM, Kim M (2010) Pretreatment technologies for the conversion of lignocellulosic materials to bioethanol. In: Eggleston G (ed) Sustainability of the sugar and sugar–ethanol industries. American Chemical Society, Washington, DC

Akyüz M, Yildiz A (2008) Evaluation of cellulosic wastes for the cultivation of Pleurotus eryngii (DC. ex Fr.) Quel. Afr J Biotechnol 7(10):1494–1499

Alemdar A, Sain M (2008) Isolation and characterization of nanofibers from agricultural residues: wheat straw and soy hulls. Bioresour Technol 99(6):1664–1671

Alvira P et al (2010) Pretreatment technologies for an efficient bioethanol production process based on enzymatic hydrolysis: a review. Bioresour Technol 101(13):4851–4861

Aquino FL, Capareda SC, Parnell CBJ (2010) Elucidating the solid, liquid, and gaseous products from batch pyrolysis of cotton gin trash. Trans ASABE 53(3):651–658

Ardica S, Calderaro E, Cappadona C (1985) Radiation pretreatments of cellulose materials for the enhancement of enzymatic hydrolysis—II. Wood chips, paper, grain straw, hay, kapok. Radiat Phys Chem (1977) 26(6):701–704

Arthe R et al (2008) Production of bio-ethanol from cellulosic cotton waste through microbial extracellular enzymatic hydrolysis and fermentation. Electron J Environ Agric Food Chem 7(6):2948–2958

Badger P (2002) Ethanol from cellulose: a general review. In: Janick J, Whipkey A (eds) Trends in new crops and new uses. ASHS, Alexandria, VA, pp 17–21
Balat M (2011) Production of bioethanol from lignocellulosic materials via the biochemical pathway: a review. Energy Convers Manage 52(2):858–875
Benjaphokee S et al (2012) Highly efficient bioethanol production by a Saccharomyces cerevisiae strain with multiple stress tolerance to high temperature, acid and ethanol. N Biotechnol 29(3): 379–386
Berlin A et al (2006) Inhibition of cellulase, xylanase and beta-glucosidase activities by softwood lignin preparations. J Biotechnol 125(2):198–209
Brigham J, Adney W, Himmel ME (1996) Hemicellulases: diversity and applications. In: Wyman C (ed) Handbook on bioethanol: production and utilization. Taylor & Francis, Washington, DC, pp 119–141
Brown RC (2003) Biorenewable resources: engineering new products from agriculture. Wiley, New York
Carere CR et al (2008) Third generation biofuels via direct cellulose fermentation. Int J Mol Sci 9(7):1342–1360
CCI (2013) Current cotton scenario. The Cotton Corporation of India Ltd
Credou J, Berthelot T (2014) Cellulose: from biocompatible to bioactive material. J Mater Chem B 2:4767–4788
das Neves M, Kimura T (2007) State of the art and future trends of bioethanol production. Dyn Biochem Process Biotechnol Mol Biol 1(1):1–14
Das N, Mukherjee M (2007) Cultivation of Pleurotus ostreatus on weed plants. Bioresour Technol 98(14):2723–2726
Dehkhoda A (2008) Concentrating lignocellulosic hydrolysate by evaporation and its fermentation by repeated fed-batch using flocculating Saccharomyces cerevisiae. University College of Borås
De Sousa MV, Monteiro SN, d'Almeida JRM (2004) Evaluation of pre-treatment, size and molding pressure on flexural mechanical behavior of chopped bagasse–polyester composites. Polym Test 23(3):253–258
Devi MC, Kumar MS (2012) Production, optimization and partial purification of cellulase by Aspergillus niger fermented with paper and timber sawmill industrial wastes. J Microbiol Biotechnol Res 2(1):120–128
Dogaris I, Mamma D, Kekos D (2013) Biotechnological production of ethanol from renewable resources by Neurospora crassa: an alternative to conventional yeast fermentations? Appl Microbiol Biotechnol 97(4):1457–1473
Domingues FC et al (2001) Production of cellulases in batch culture using a mutant strain of Trichoderma reesei growing on soluble carbon source. Biotechnol Lett 23(10):771–775
Drapcho CM, Nhuan NP, Walker TH (2008) Biofuels engineering process technology. McGraw-Hill, New York
Duan C J, Feng J X (2010) Mining metagenomes for novel cellulase genes. Biotechnol Lett 32(12):1765–1775
Eggeman T, Elander RT (2005) Process and economic analysis of pretreatment technologies. Bioresour Technol 96(18):2019–2025
El-Zawawy WK et al (2011) Acid and enzyme hydrolysis to convert pretreated lignocellulosic materials into glucose for ethanol production. Carbohydr Polym 84(3):865–871
Eriksson K-E (2000) Lignocellulose, lignin, ligninases. In: Schaechter M (ed) Encyclopedia of microbiology, vol 3. Academic, San Diego, CA, pp 39–48
Fan L, Gharpuray MM, Lee Y-H (1987) Cellulose hydrolysis. Springer, Berlin
Farooqi R, Sam AG (2004) Ethanol as a transportation fuel: Canadian policies and challenges in the context of climate change. In Centre for Applied Business Research in Energy and the Environment. University of Alberta, Edmonton, AB
Fengel D, Wegener G (1983) Wood: chemistry, ultrastructure, reactions. Walter de Gruyter, Berlin

Fu N et al (2009) A novel co-culture process with Zymomonas mobilis and Pichia stipitis for efficient ethanol production on glucose/xylose mixtures. Enzyme Microb Technol 45(3): 210–217
Gaitán-Hernández R, Salmones D (2008) Obtaining and characterizing Pleurotus ostreatus strains for commercial cultivation under warm environmental conditions. Sci Hortic 118(2):106–110
Galliano H et al (1991) Lignin degradation by Rigidoporus lignosus involves synergistic action of two oxidizing enzymes: Mn peroxidase and laccase. Enzyme Microb Technol 13(6):478–482
Gavrilescu M (2010) Environmental biotechnology: achievements, opportunities and challenges. Dyn Biochem Process Biotechnol Mol Biol 4(1):1–36
Goldstein IS (1981) Organic chemicals from biomass. CRC, Boca Raton, FL
Guerra A et al (2006) Toward a better understanding of the lignin isolation process from wood. J Agric Food Chem 54(16):5939–5947
Hamelinck CN, van Hooijdonk G, Faaij AP (2005) Ethanol from lignocellulosic biomass: techno-economic performance in short-, middle- and long-term. Biomass Bioenergy 28(4):384–410
Han M et al (2009) Bioethanol production from ammonia percolated wheat straw. Biotechnol Bioprocess Eng 14(5):606–611
Hendriks ATWM, Zeeman G (2009) Pretreatments to enhance the digestibility of lignocellulosic biomass. Bioresour Technol 100(1):10–18
Hyeon J et al (2010) Production of minicellulosomes from Clostridium cellulovorans for the fermentation of cellulosic ethanol using engineered recombinant Saccharomyces cerevisiae. FEMS Microbiol Lett 310(1):39–47
Ibarra D et al (2006) Exploring the enzymatic parameters for optimal delignification of eucalypt pulp by laccase-mediator. Enzyme Microb Technol 39(6):1319–1327
ICTSD (2013) Cotton: trends in global production, trade and policy. p 14
Imamoglu E, Sukan FV (2014) The effects of single and combined cellulosic agrowaste substrates on bioethanol production. Fuel 134:477–484
Iñiguez-Covarrubias G, Lange SE, Rowell RM (2001) Utilization of byproducts from the tequila industry: part 1: agave bagasse as a raw material for animal feeding and fiberboard production. Bioresour Technol 77(1):25–32
Inoue H et al (2008) Combining hot-compressed water and ball milling pretreatments to improve the efficiency of the enzymatic hydrolysis of eucalyptus. Biotechnol Biofuels 1(1):2
Isci A, Demirer GN (2007) Biogas production potential from cotton wastes. Renew Energy 32(5): 750–757
Jarboe LR et al (2007) Development of ethanologenic bacteria. Adv Biochem Eng Biotechnol 108: 237–261
Jeoh T (1998) Steam explosion pretreatment of cotton gin waste for fuel ethanol production. Virginia Polytechnic Institute and State University, Blacksburg, VA
Jeoh T, Agblevor FA (2001) Characterization and fermentation of steam exploded cotton gin waste. Biomass Bioenergy 21(2):109–120
Jingping G et al (2012) A genome shuffling-generated Saccharomyces cerevisiae isolate that ferments xylose and glucose to produce high levels of ethanol. J Ind Microbiol Biotechnol 39(5):777–787
Kalmış E, Sargın S (2004) Cultivation of two Pleurotus species on wheat straw substrates containing olive mill waste water. Int Biodeter Biodegr 53(1):43–47
Kaparaju P et al (2009) Bioethanol, biohydrogen and biogas production from wheat straw in a biorefinery concept. Bioresour Technol 100(9):2562–2568
Karimi K, Emtiazi G, Taherzadeh MJ (2006a) Ethanol production from dilute-acid pretreated rice straw by simultaneous saccharification and fermentation with Mucor indicus, Rhizopus oryzae, and Saccharomyces cerevisiae. Enzyme Microb Technol 40(1):138–144
Karimi K, Kheradmandinia S, Taherzadeh MJ (2006b) Conversion of rice straw to sugars by dilute-acid hydrolysis. Biomass Bioenergy 30(3):247–253
Karunanithy C, Muthukumarappan K, Julson JL (2008) Influence of high shear bioreactor parameters on carbohydrate release from different biomasses. In: 2008 Providence, Rhode Island,

29 June–2 July 2008. American Society of Agricultural and Biological Engineers, St. Joseph, MI

Khokhar Z et al (2014) On-site cellulase production by Trichoderma reesei 3EMS35 mutant and same vessel saccharification and fermentation of acid treated wheat straw for ethanol production. EXCLI J 13:82–97

Kim SB et al (2011) Dilute acid pretreatment of barley straw and its saccharification and fermentation. Biotechnol Bioprocess Eng 16(4):725–732

Kim I et al (2014) Effects of ammonium carbonate pretreatment on the enzymatic digestibility and structural features of rice straw. Bioresour Technol 166:353–357

Klass DL (1998) Biomass for renewable energy, fuels, and chemicals. Elsevier, New York

Kodali B, Pogaku R (2006) Pretreatment studies of rice bran for the effective production of cellulose. Electron J Environ Agric Food Chem 5:1253–1264

Koutrotsios G et al (2014) Bioconversion of lignocellulosic residues by Agrocybe cylindracea and Pleurotus ostreatus mushroom fungi—assessment of their effect on the final product and spent substrate properties. Food Chem 161:127–135

Kumar P et al (2009) Methods for pretreatment of lignocellulosic biomass for efficient hydrolysis and biofuel production. Ind Eng Chem Res 48(8):3713–3729

Kumari R, Pramanik K (2012) Improved bioethanol production using fusants of Saccharomyces cerevisiae and xylose-fermenting yeasts. Appl Biochem Biotechnol 167(4):873–884

Kunamneni A et al (2007) Fungal laccase—a versatile enzyme for biotechnological applications. In: Mendez-Vilas A (ed) Communicating current research and educational topics and trends in applied microbiology. Formatex Research Center, Badajoz, pp 233–245

La Grange DC, den Haan R, van Zyl WH (2010) Engineering cellulolytic ability into bioprocessing organisms. Appl Microbiol Biotechnol 87(4):1195–1208

Lark N et al (1997) Production of ethanol from recycled paper sludge using cellulase and yeast, Kluyveromyces marxianus. Biomass Bioenergy 12(2):135–143

Liang J et al (2014) Alcohol dehydrogenases from Kluyveromyces marxianus: heterologous expression in Escherichia coli and biochemical characterization. BMC Biotechnol 14(1):45

Liu C-G et al (2014) Assessment and regression analysis on instant catapult steam explosion pretreatment of corn stover. Bioresour Technol 166:368–372

Lu Y et al (2012) Improvement of robustness and ethanol production of ethanologenic Saccharomyces cerevisiae under co-stress of heat and inhibitors. J Ind Microbiol Biotechnol 39(1): 73–80

Mahalakshmi M (2011) Bioconversion of cotton waste from textile mills to bioethanol by microbial saccharification and fermentation. Ann Biol Res 2(3):380–388

Mani S, Tabil LG, Sokhansanj S (2004) Grinding performance and physical properties of wheat and barley straws, corn stover and switchgrass. Biomass Bioenergy 27(4):339–352

Marcolongo L et al (2014) The effect of Pleurotus ostreatus arabinofuranosidase and its evolved variant in lignocellulosic biomasses conversion. Fungal Genet Biol 72:162–167

Melamane X, Tandlich R, Burgess J (2007) Anaerobic digestion of fungally pre-treated wine distillery wastewater. Afr J Biotechnol 6(September):1990–1993

Menon V, Prakash G, Prabhune A et al (2010a) Biocatalytic approach for the utilization of hemicellulose for ethanol production from agricultural residue using thermostable xylanase and thermotolerant yeast. Bioresour Technol 101(14):5366–5373

Menon V, Prakash G, Rao M (2010b) Value added products from hemicellulose: biotechnological perspective. Global J Biochem 1(1):36–67

Miranda R et al (2007) Pyrolysis of textile wastes. J Anal Appl Pyrolysis 80(2):489–495

Morais MH et al (2000) Note. Production of shiitake mushroom (Lentinus edodes) on lignocellulosic residues/Nota. Cultivo del hongo shiitake (Lentinus edodes) en residuos lignocelulosicos. Food Sci Technol Int 6(2):123–128

Moreno AD et al (2014) A review of biological delignification and detoxification methods for lignocellulosic bioethanol production. Critical review in biotechnology 0, 1–13

Mosier N et al (2005) Features of promising technologies for pretreatment of lignocellulosic biomass. Bioresour Technol 96(6):673–686
Mtui GYS (2009) Recent advances in pretreatment of lignocellulosic wastes and production of value added products. Afr J Biotechnol 8(8):1398–1415
Mtui G, Nakamura Y (2005) Bioconversion of lignocellulosic waste from selected dumping sites in Dar es Salaam, Tanzania. Biodegradation 16(6):493–499
Okuda N et al (2007) Microaeration enhances productivity of bioethanol from hydrolysate of waste house wood using ethanologenic Escherichia coli KO11. J Biosci Bioeng 103(4):350–357
Ozçelik E, Pekşen A (2007) Hazelnut husk as a substrate for the cultivation of shiitake mushroom (Lentinula edodes). Bioresour Technol 98(14):2652–2658
Peker H et al (2007) Cultivation of Agaricus bisporus on wheat straw and waste tea leaves based compost formulas using wheat chaff as activator material. Afr J Biotechnol 6(4):400–409
Peterson JD, Ingram LO (2008) Anaerobic respiration in engineered Escherichia coli with an internal electron acceptor to produce fuel ethanol. Ann N Y Acad Sci 1125:363–372
Philippoussis A, Zervakis G, Diamantopoulou P (2001) Bioconversion of agricultural lignocellulosic wastes through the cultivation of the edible mushrooms Agrocybe aegerita, Volvariella volvacea and Pleurotus spp. World J Microbiol Biotechnol 17(2):191–200
Plácido J, Imam T, Capareda S (2013) Evaluation of ligninolytic enzymes, ultrasonication and liquid hot water as pretreatments for bioethanol production from cotton gin trash. Bioresour Technol 139:203–208
Qi BC et al (2005) Acidogenic fermentation of lignocellulosic substrate with activated sludge. Chem Eng Commun 192(9):1221–1242
Rabinovich ML, Melnik MS, Bolobova AV (2002) Dedicated to the memory of I.V. Berezin and R.V. Feniksova Microbial Cellulases (Review). Appl Biochem Microbiol 38(4):305–322
Rani P, Kalyani N, Prathiba K (2008) Evaluation of lignocellulosic wastes for production of edible mushrooms. Appl Biochem Biotechnol 151(2-3):151–159
Ryabova OB, Chmil OM, Sibirny AA (2003) Xylose and cellobiose fermentation to ethanol by the thermotolerant methylotrophic yeast Hansenula polymorpha. FEMS Yeast Res 4(2):157–164
Safartalab K, Dadashian F, Vahabzadeh F (2014) Fed batch enzymatic hydrolysis of cotton and viscose waste fibers to produce ethanol. Univ J Chem 2(1):11–15
Sánchez C (2009) Lignocellulosic residues: biodegradation and bioconversion by fungi. Biotechnol Adv 27(2):185–194
Saritha M, Arora A, Lata (2012) Biological pretreatment of lignocellulosic substrates for enhanced delignification and enzymatic digestibility. Ind J Microbiol 52(2):122–130
Sarks C et al (2014) Studying the rapid bioconversion of lignocellulosic sugars into ethanol using high cell density fermentations with cell recycle. Biotechnol Biofuels 7(1):73
Sharma-Shivappa R, Chen Y (2008) Conversion of cotton wastes to bioenergy and value-added products. Trans ASABE 51(6):2239–2246
Shen J, Agblevor FA (2008) Optimization of enzyme loading and hydrolytic time in the hydrolysis of mixtures of cotton gin waste and recycled paper sludge for the maximum profit rate. Biochem Eng J 41(3):241–250
Shi J, Chinn MS, Sharma-Shivappa RR (2008) Microbial pretreatment of cotton stalks by solid state cultivation of Phanerochaete chrysosporium. Bioresour Technol 99(14):6556–6564
Silva EM, Machuca A, Milagres AMF (2005) Effect of cereal brans on Lentinula edodes growth and enzyme activities during cultivation on forestry waste. Lett Appl Microbiol 40(4):283–288
Silverstein RA et al (2007) A comparison of chemical pretreatment methods for improving saccharification of cotton stalks. Bioresour Technol 98(16):3000–3011
Sindhu R et al (2011) Dilute acid pretreatment and enzymatic saccharification of sugarcane tops for bioethanol production. Bioresour Technol 102(23):10915–10921
Sindhu R et al (2012) Organosolvent pretreatment and enzymatic hydrolysis of rice straw for the production of bioethanol. World J Microbiol Biotechnol 28(2):473–483
Sjöström E (1993) Wood chemistry: fundamentals and applications. Gulf, London

Sootsuwan K, Thanonkeo P, Keeratirakha N, Thanonkeo S, Jaisil P, Yamada M (2013) Sorbitol required for cell growth and ethanol production by Zymomonas mobilis under heat, ethanol, and osmotic stresses. Biotechnol Biofuels 6(1):180
Sun Y, Cheng J, Westerman PW (2002) Enzymatic hydrolysis of rye straw and bermudagrass for ethanol production. North Carolina State University, Raleigh, NC
Szambelan K, Nowak J, Czarnecki Z (2004) Use of Zymomonas mobilis and Saccharomyces cerevisiae mixed with Kluyveromyces fragilis for improved ethanol production from Jerusalem artichoke tubers. Biotechnol Lett 26(10):845–848
Szengyel Z (2000) Ethanol from wood—cellulase enzyme production. Lund University
Taherzadeh MJ, Karimi K (2007) Acid-based hydrolysis processes for ethanol from lignocellulosic materials: a review. BioResources 2(3):472–499
Taherzadeh MJ, Karimi K (2008) Pretreatment of lignocellulosic wastes to improve ethanol and biogas production: a review. Int J Mol Sci 9:1621–1651
Taherzadeh M, Niklasson C (2004) Ethanol from lignocellulosic materials: pretreatment, acid and enzymatic hydrolyses, and fermentation. In: Saha B, Hayashi K (eds) Lignocellulose biodegradation. ACS Division of Cellulose and Renewable Materials, American Chemical Society, Washington, DC, pp 49–68
Terashima N et al (1993) Comprehensive model of the lignified plant cell wall. In: Jung HG et al (eds) Forage cell wall structure and digestibility. American Society of Agronomy, Crop Science Society of America, Soil Science Society of America, Madison, WI, pp 247–270
Tian X-F et al (2011) Pretreatment of microcrystalline cellulose in organic electrolyte solutions for enzymatic hydrolysis. Biotechnol Biofuels 4(1):53
Tomas-Pejo E (2008) Realistic approach for full-scale bioethanol production from lignocellulose: a review. J Sci Ind Res 67(November):874–884
Tomás-Pejó E, Oliva JM, Ballesteros M (2008) Realistic approach for full-scale bioethanol production from lignocellulose: a review. J Sci Ind Res 67(11):874–884
Travaini R et al (2013) Sugarcane bagasse ozonolysis pretreatment: effect on enzymatic digestibility and inhibitory compound formation. Bioresour Technol 133:332–339
Wang Z et al (1990) Cloning and expression of a lignin peroxidase gene from Streptomyces viridosporus in Streptomyces lividans. J Biotechnol 13(2–3):131–144
White DH, Coates WE, Wolf D (1996) Conversion of cotton plant and cotton gin residues to fuels by the extruder-feeder liquefaction process. Bioresour Technol 56(1):117–123
Wood TM et al (1986) A highly active extracellular cellulase from the anaerobic rumen fungus Neocallimastix frontalis. FEMS Microbiol Lett 34(1):37–40
Wyman C (1996) Handbook on bioethanol: production and utilization. CRC, New York
Wyman CE et al (2005) Coordinated development of leading biomass pretreatment technologies. Bioresour Technol 96(18):1959–1966
Yang B, Lu Y (2007) The promise of cellulosic ethanol production in China. J Chem Technol Biotechnol 10(August 2006):6–10
Yang B, Wyman CE (2008) Pretreatment: the key to unlocking low-cost cellulosic ethanol. Biofuels Bioprod Biorefining 2(1):26–40
Yang W, Guo F, Wan Z (2013) Yield and size of oyster mushroom grown on rice/wheat straw basal substrate supplemented with cotton seed hull. Saudi J Biol Sci 20(4):333–338
Yildiz S et al (2002) Some lignocellulosic wastes used as raw material in cultivation of the Pleurotus ostreatus culture mushroom. Process Biochem 38(3):301–306
Zabaniotou A, Andreou K (2010) Development of alternative energy sources for GHG emissions reduction in the textile industry by energy recovery from cotton ginning waste. J Cleaner Prod 18(8):784–790
Zhang H et al (2014a) Structure and enzymatic accessibility of leaf and stem from wheat straw before and after hydrothermal pretreatment. Biotechnol Biofuels 7:74
Zhang L et al (2014b) Enhanced fermentability of poplar by combination of alkaline peroxide pretreatment and semi-simultaneous saccharification and fermentation. Bioresour Technol 164: 292–298

Zhao X, Wu R, Liu D (2011) Production of pulp, ethanol and lignin from sugarcane bagasse by alkali-peracetic acid delignification. Biomass Bioenergy 35(7):2874–2882
Zheng Y, Pan Z, Zhang R (2009) Overview of biomass pretreatment for cellulosic ethanol production. Int J Agric Biol Eng 2(3):51–68

Chapter 15
Microalgae: Cultivation and Application

V. Aishvarya, J. Jena, N. Pradhan, P.K. Panda, and L.B. Sukla

15.1 Introduction

Algae are recognised as one of the oldest life form (Falkowski and Raven 1997) and include large group of organisms from different phylogenetic groups, representing many taxonomic divisions. In general they are primitive plants (thallophytes), i.e. lacking roots, stems and leaves, vascular tissues and a sterile covering of cells around the reproductive organ (Khan et al. 2009; Brennan and Owende 2010; Mutanda et al. 2011). Most of the algae contain chlorophyll by virtue of which they get green colour; however, some of the algae are not green but appears brown and red due to the presence of other pigments such as carotenoid (Wang and Chen 2009). Algae are oxygen evolving photosynthetic microorganisms containing chlorophyll 'a' as primary photosynthetic pigments and grow in various aquatic environments (fresh, marine and brackish water streams) including hot springs (Wang and Chen 2009). Some species can grow on rocks, soils, plants, etc., with minimum nutrient requirements (Zhou 2014). Algae are unicellular or colonial. When the cells are arranged end to end, the algae are said to be filamentous that may be unbranched filaments or more intricate branched filaments (Wang and Chen 2009).

Most of the algae are microscopic known as microalgae, whereas some are quite large and known as macroalgae. Microalgae are unicellular or simple multicellular organism with size ranges from 2 to 200 μm, whereas the macroalgae are completely multicellular organism, some growing to over 100 ft in length (Madigan et al. 1997; Wang and Chen 2009; Mutanda et al. 2011). Among the algae,

V. Aishvarya
Hydro and Electrometallurgy Department, CSIR – Institute of Minerals and Materials Technology, Bhubaneswar, 751013 Odisha, India

J. Jena (✉) • N. Pradhan • P.K. Panda • L.B. Sukla
Bioresources Engineering Department, CSIR – Institute of Minerals and Materials Technology, Bhubaneswar, 751013 Odisha, India
e-mail: jayashree123biotech@gmail.com; npradhan@immt.res.in

L.B. Sukla et al. (eds.), *Environmental Microbial Biotechnology*, Soil Biology 45,
DOI 10.1007/978-3-319-19018-1_15

microalgae have been extensively studied for not only due to their ubiquitous nature but also for their simple structure that renders them to grow and proliferate in a wide range of environmental condition (Vonshak 1990; Hu et al. 2007). This simple cellular structure provides a large surface to volume body ratio, which gives them the ability to uptake large amount of nutrients (Brennan and Owende 2010). The mechanism of photosynthesis in microalgae is similar to that of higher plants; however, they are generally considered as the more efficient converters of solar energy than higher plants because of their simple cellular structure (Walter et al. 2005; Spolaore et al. 2006; Khan et al. 2009; Kirrolia et al. 2013).

The present review aims to describe microalgae and its applications, with special emphasis on microalgal biodiesel production and CO_2 sequestration. This work starts by describing microalgae, its chemical composition, cultivation and harvesting system and various applications.

15.2 Microalgae

In applied phycology, the term microalgae covers all unicellular and simple multicellular oxygenic photosynthetic organism with chlorophyll 'a' as common photosynthetic pigment (Richmond 2008). The biodiversity of microalgae is outstanding, and it is estimated that from 200,000 species to several million species exist in nature; however, a very limited number have been studied and analysed (Norton et al. 1996; Mata et al. 2010). Among the microalgae, the most important groups of microalgae in terms of abundance are diatoms, green algae, blue-green algae and golden algae (Table 15.1).

Chlorophyta (green algae) includes a large number of microalgae with great morphological variability like coccoid, unicellular or colonial flagellates and multicellular or multinucleated filaments (Richmond 2008). They contain

Table 15.1 The four important group of microalgae (Khan et al. 2009)

Sl No	Algae	Known species (near about)	Morphology	Storage material	Habitat
1	Diatoms (Bacillariophyceae)	100,000	Unicellular	Chrysolaminarin (polymer of carbohydrates) and TAGs oceans	Fresh and brackish water
2	Green algae (Chlorophyceae)	8,000	Unicellular to leafy	Starch and TAGs	Freshwater, brackish water
3	Blue-green algae (Cyanophyceae)	2,000	Unicellular	Starch and TAGs	Different habitats
4	Golden algae (Chrysophyceae)	1,000	Unicellular	TAGs and carbohydrates	Freshwater

chlorophyll a, b which gives it a complete green colour. Along with chlorophylls, carotenoids are synthesised and accumulated in chloroplast.

15.2.1 Habitats of Microalgae

Microalgae are some of the most robust organisms on earth and are able to grow in almost every habitat in every part of the world. These are able to grow in a wide range of aquatic to terrestrial environment. The aquatic environment includes lacustrine, brackish, fresh water, higher saline water, hot springs, waste water maturation ponds, dams, rivers, marine and coastal areas. Besides all the above places, microalgae grow in rocks (internal and surface), mud, sand and terrestrial plants (tree trunks, branches, shady sides of trees). Interestingly some of the microalgae grow on and in other organism. Main habitats are fresh water, brackish and marine ecosystem. Among all, a number of microalgae are found in brackish water due to the nutritional composition and warmer temperature that results from the mixture of seawater and fresh water (Woelfel et al. 2007). It usually occurs at the river mouth on the coastline, and the microalgae always remain in suspended form due to the rapid water movement (Anandraj et al. 2007; Bhakta et al. 2011).

15.2.2 Microalgae Cell Components

Microalgae are simple eukaryotic plants lacking root, stem and leaves and possess all other common cell components that are present in higher plants, i.e. membrane bound organelles (Golgi body, endoplasmic reticulum, vacuoles, mitochondria, centrioles, plastids) with specialised function. The eukaryotic microalgal cell is surrounded by a thin, rigid cell wall. Some algae have an outer matrix lying outside the cell wall, similar to bacterial capsules. The cell wall provides a barrier between the environment outside the cell and that inside the cell. The cell wall is composed of a network of cellulose fibrils with the addition of polysaccharides such as pectin, xylans, mannans, alginic acids or fucinic acid whereas in some cases only with cellulose fibrils. But in case of diatoms, the cell wall is composed of silica along with protein and polysaccharides which gives a high rigidity to the cell. Cell wall of microalgae is more porous, containing small pores about 3–5 nm, that allows to pass only low molecular weight substances such as water, inorganic ions, gases and nutrients, however impermeable to macromolecules (Wang and Chen 2009).

The nucleus is bound with a nuclear envelope having pores (nuclear pores) and contains nucleolus, chromatin and karyolymph. The nucleus is larger in size, and within the nucleus, the DNA is kept organised. Photosynthesis is carried out in specialised organelles called chloroplast. The chloroplast contains a series of membrane bound sacs called thylakoids that hold the chlorophylls and surrounded by a matrix called stroma. In thylakoid the light reaction of photosynthesis takes place, whereas in stroma the dark reaction of carbon dioxide fixation takes place. In some cases a pyrenoid, a dense proteinaceous area, is present in the chloroplast. This pyrenoid is associated with synthesis and storage of starch material. Mitochondrial structure varies greatly in the algae. Some algae (euglenoids) have discoid cristae (folds in the inner mitochondrial membrane); some have lamellar cristae (green and red algae); and the remaining (golden-brown and yellow-green, brown and diatoms) have tubular cristae (Prescott et al. 2002).

15.2.3 Chemical Composition of Microalgal Cell

The microalgae biomass is mainly composed of proteins, lipids, carbohydrates and nucleic acid. Protein is always the major organic constituent, followed by lipid and then by carbohydrate. Most of the microalgae possess high-protein content and is used as an unconventional source of protein and nutrition supplement (Pulz and Gross 2004; Soletto et al. 2005). The cells are capable of synthesising all essential amino acids which compares favourably with that of other food protein and can be used as an essential source for human and animal nutrition (Priyadarshani and Rath 2012). Microalgal cells contain all essential vitamins such as A, B1, B2, B6, C, E, nicotinate, biotin, folic acid and pantothenic acid (Spolaore et al. 2006; Kirrolia et al. 2013). Carbohydrates in the microalgae are mainly found in the form of starch, glucose, sugars and other polysaccharides (Wang and Chen 2009; Kirrolia et al. 2013). The lipid content of microalgae widely varies from species to species. Generally microalgae possess average lipid content of 5–20 % and can reach up to 80–90 % of dry weight under specific condition (Hu et al. 2008). Algal lipids are mainly composed of glycerol, sugars or bases esterified to saturated and unsaturated fatty acids (short to long chain). Among the fatty acids present in microalgae, polyunsaturated fatty acids such as omega 3, eicosapentaenoic acid (EPA) and docosahexaenoic acid (DHA), and omega 6, γ-linolenic acid (GLA) and arachidonic acid (AA), are of particular interest for nutraceutical application (Spolaore et al. 2006; Bishop and Zubeck 2012). The fatty acid proportion and amount are dependent on the nutritional and environmental factors. Microalgae synthesise and accumulate a high amount of triacylglycerol (TAG) in the range of 20–80 % of dry weight under adverse condition. These TAGs are extracted and converted to biodiesel and used as an alternative renewable fuel (Hu et al. 2008; Chisti 2007).

Microalgae come in various colours due to the presence of different types of photosynthetic pigments (chlorophylls, carotenoids and phycobiliproteins). Different group of algae have different combination of chlorophyll molecules. These molecules have a wide range of commercial application. Among the pigments, carotenoid is a special class of natural fat-soluble pigment having high antioxidant property (Munir et al. 2013). Some of the microalgae (diatoms) contain pigment fucoxanthin that belongs to the group of xanthophyll having several biological properties such as antioxidant, antimicrobial and anticancer activities (Rajauria and Abu-Ghannam 2013).

15.3 Microalgal Growth Requirement

Media used for cultivating microalgae must supply all the necessary components required for growth and maintenance of the organism. Optimal parameters as well as tolerated range are species specific, and each parameter must be determined individually (Lavens and Sorgeloos 1996). All these parameters not only affect photosynthesis and productivity of cell biomass but also influence the pattern, pathway and activity of cellular metabolism and thus change in cell composition (Richmond and Hu 2013). Hence while cultivating microalgae, several factors must be considered, and the most important parameters regulating algal growth include:

- Nutrients quantity and quality
- Light
- Carbon dioxide
- Temperature
- pH
- Turbulence and salinity

15.3.1 Nutrients

Nutrients are inorganic or organic compound other than CO_2 and water, used for growth (Neenan et al. 1986). The media used for culturing the microalgae should have sufficient amounts of nutrients in them in adequate proportion for their proper growth. Minimal nutritional requirements can be estimated using the approximate molecular formula of the microalgal biomass, that is, $CO_{0.48}H_{1.83}N_{0.11}P_{0.01}$. This formula is based on data presented by Grobbelaar (2004). Microalgal culture media that was first introduced by Pringsheim (1950) consists of a biphasic soil-water medium. However, the chemical composition of the media was not defined properly. Hence to overcome the problem, Vonshak (1986) established another microalgal culture media that contain carbon source (organic or inorganic), nitrogen source, trace elements and chelating agents, vitamins, salt content and other ionic components (potassium, magnesium, sodium, sulphate and phosphate) based on cellular composition. A vast number of culture media with various proportions of nutrient have been designed, while some media are derived from analysis of water in the native habitat and ecology of microalgae. Furthermore some media are species specific, while some are very general used for various microalgae. A detailed report on microalgal culture media was available in Culture Collection of

Algae and Protozoa (CCAP) (2013). Some of the commonly used media are described below:

- Bold basal media (Bold 1949; Bischoff and Bold 1963; Stein 1973).
- BG11 media-growth medium for blue-green algae and fresh water green algae Rippka et al. 1979).
- F/2 medium (Guillard and Ryther 1962; Guillard 1975) is a widely used media (generally enriched with sea water) for diatoms and marine microalgae.
- Walne medium (Walne 1970) and ASN-III (Rippka et al. 1979) used for culture of marine cyanobacteria and eukaryotic green algae.

However, due to the complexity and cost of the above culture media, it may not be feasible to use them for large-scale production of microalgal biomass. Alternative growth media comprised of commercially available agricultural-grade fertiliser (NPK, urea, potash, superphosphate) are suitable for large-scale microalgal cultivation. Among the nutrient, nitrogen accounts for about 7–10 % of cell dry weight (DCW) and is an essential constituent of all structural and functional proteins in algal cells. Microalgae have a limited ability to produce nitrogen-stored materials when growing under nitrogen-sufficient conditions. Discoloration of microalgal cell is common in nitrogen-deficient media due to the decrease in chlorophyll content and increase in carotenoids. Further nitrogen limitation shows active and specific degradation of phycobilisomes, and photosynthesis continues at a slow rate. As a result, photosynthetically fixed carbon is diverted from the protein synthesis into the pathways for carbohydrate and lipid synthesis. Further nitrogen deficiency may cause several changes in cell, where nitrogen limitation could activate diacylglycerol acyltransferase that converts acyl-CoA to triglyceride (TAG) (Xin et al. 2010). Nitrogen may be supplied in the form of urea, nitrate and ammonia. Among all the source of nitrogen, nitrate is mostly used; however, urea and ammonia also show similar growth of microalgae. A combination of urea and sodium nitrate for the organism *Scenedesmus* sp. showed highest ash-free dry biomass content with yield of 4.15 ± 0.38 g l^{-1} (Lin and Lin 2011). Further on consumption of urea by the microalgae, the carbon atom present in urea is released as CO_2 and used as carbon source in photosynthesis (Neenan et al. 1986).

Besides nitrogen, phosphorous is another important element to carry out many cellular processes such as energy transfer and biosynthesis of nucleic acid. Other than nitrogen and phosphorous, sulphur, potassium, sodium, iron, magnesium, calcium and trace elements like magnesium, zinc, molybdenum, cobalt and vanadium are also important for microalgal growth (Grobbelaar 2004). Further silicate is used as a major source for some microalgae culture (diatoms) for their cell wall synthesis. Apart from all these, some microalgae need some vitamins such as thiamin (B1), cyanocobalamin (B12) and sometimes biotin.

15.3.2 Light

Like plants, microalgae require light as the main source of energy to carry out fixation of CO_2 into organic matter in the process of photosynthesis. For proper photosynthesis, three variables of light are important such as intensity, spectral quality and photoperiod (light/dark period) (Lavens and Sorgeloos 1996). Usually most of the problem in cultivating microalgae is related to the light intensity as low intensity causes photo-limitation and higher intensity causes photo-inhibition. Light source for microalgae growth may be natural (sunlight) or artificial supplied by fluorescent tubes. The requirement of light varies greatly with culture growth (density) and culture system (depth). Generally, as the microalgae grow and reproduce, biomass density increases. As a result, microalgae distant from the surface are shaded by the microalgal culture present between it and the light source, thus receiving lesser amount of light. In this case the light intensity must be increased to penetrate through the culture (1,000 lux of light for Erlenmeyer flask and 5,000–10,000 is required for larger volume). Further too high light intensity (than needed) may result in photo-inhibition which causes decrease in photosynthetic pigments (Adir et al. 2003). Although the range of solar radiation is very broad, only radiation between 400 and 700 nm can be used by microalgae. This part of the solar spectrum is called "photosynthetic active radiation" (PARS) and accounts for 43 % of the solar radiation (Thimijan and Heins 1983). Microalgae cells cultivated under limited light conditions assimilate carbon towards the synthesis of amino acids and other essential cell constituents, but under saturated light conditions, sugars and starch are formed via the pentose phosphate-reducing pathway, suggesting the dependence of the biomass composition with the light availability (Harun et al. 2014). Fluorescent tubes emitting either in the blue or the red light spectrum may be preferred as these are most active portion of the light spectrum for photosynthesis. Further unlike light intensity, photoperiod (light/dark cycle) plays a great role for microalgae culture. This is because cell division occurs under dark conditions for many unicellular photosynthetic cultures, while for others, cell division occurs both in the dark and the illuminated phase (Harun et al. 2014). Although the photoperiod varies from organism to organism, for industrial applications relating the ratio between the cost of energy and the corresponding biomass production, 12–15 h of illumination is considered as the optimal period (Harun et al. 2014).

15.3.3 Temperature

The optimal temperature for phytoplankton culture is generally between 20 and 24 °C. Further the temperature range varies with culture medium composition and organism cultivated. Most commonly used microalgae tolerate temperatures between 16 and 27 °C. Temperature below the optimal temperature may not kill the microalgae but reduce the growth rate, whereas high temperature will kill most of the microalgae.

15.3.4 Salinity

The total salt concentration mostly depends on the ecological origin of the organism. Salinity changes normally affect microalgae in three ways: osmotic stress, ion stress and changes of cellular ion concentration due to the selective permeability of ion through the membrane. Marine microalgae are extremely tolerant to changes in salinity. Most of the organisms grow best at a salinity that is slightly lower than that of their native habitat. Salinities of 2.0–2.5 % have been found to be optimal for microalgae (Lavens and Sorgeloos 1996).

15.3.5 pH

Most of the microalgae grow in the pH range of 7–9, while the optimum range is 8.2–8.7 (Lavens and Sorgeloos 1996). As the culture grows, pH of the culture medium increases with time as a result of continuous consumption of CO_2. If the pH is not maintained within the optimum pH range, it may result in disruption of many cellular processes, leading to the inhibition of biomass growth. The pH in the growing culture can be maintained either through simple aeration or through addition of extra CO_2.

15.3.6 Aeration and Mixing

Continuous mixing of the culture is essential for successful microalgal biomass production. Mixing provides appropriate distribution of nutrients, light, dissolved CO_2, O_2 elimination, maintenance of pH, temperature gradient and evade algal sediment formation (Lavens and Sorgeloos 1996). Further it improves the gas exchange between the culture medium and the air. Upon algae growth, the dissolved CO_2 in the culture become limited. In this case diffusion of culture with CO_2 as microbubbles may be carried out by enriching aeration (only air containing 0.03 % CO_2) or by mixing pure CO_2 gas with air in case of heavy biomass density (at a rate of 1 % of the volume of air) for proper microalgal growth. This process not only provides CO_2 for better growth but also maintains pH by CO_2/HCO_3^- balance and provides better mixing of the algal culture. Depending upon the culture volume, mixing is achieved by using stirrer in bioreactors or photosynthetic orbital shaker, but for large-scale operations like raceway ponds, the paddle wheels are more suitable. Further the mixing of culture is species specific (Gouveia 2011).

15.4 Biomass Production

Under natural growth conditions, photoautotrophic algae absorb sunlight and assimilate CO_2 from air and nutrients from the aquatic habitats. Hence any artificial production should attempt to replicate and enhance the optimum natural conditions (Brennan and Owende 2009). Microalgae production can be done in a variety of systems either as pure culture or as consortium with minimal sophistication and equipments. Understanding and taking advantage of the biology of algal strains selected for the use in production systems is the foundation for processing feedstock into fuels and products (Chisti 2007).

15.4.1 Growth of Biomass

After selecting the microalgae strain, it is necessary to develop a range of bioprocesses that make its commercialisation viable. Thus, the design and optimisation of bioreactors to cultivate these microorganisms is a major step in transforming scientific findings into a marketable product. From a commercial point of view, a microalgae culture system must have as many of the following characteristics as possible: high area productivity, high volumetric productivity, inexpensiveness (both in terms of investment and maintenance costs), simple control of the culture parameters (temperature, pH, CO_2, turbulence), energy efficiency and reliability (Olaizola 2003). The cultivation methods adopted for microalgae are traditionally either in open ponds, known as high rate ponds (HRP) or raceway ponds (RP), or in enclosed systems known as photobioreactors. Macroalgae have been grown for some time in attached systems; the best examples of these are Algal Turf Scrubber (Craggs and Adey 1995). In order to minimise costs, algae is often grown using sunlight as a free source of light, even though it is variable with daily and seasonal changes in the amount of available light (Molina 1999; Molina and Fernandez 1999). Each system has its own advantages and disadvantages.

15.4.1.1 Open Pond Systems

The classical open-air cultivation systems comprise lakes and natural ponds, circular ponds, raceway ponds and inclined systems. Open-air systems are the most widespread growth systems for microalgae since these systems are easier and less expensive to build, operate more durable and have a larger production capacity than

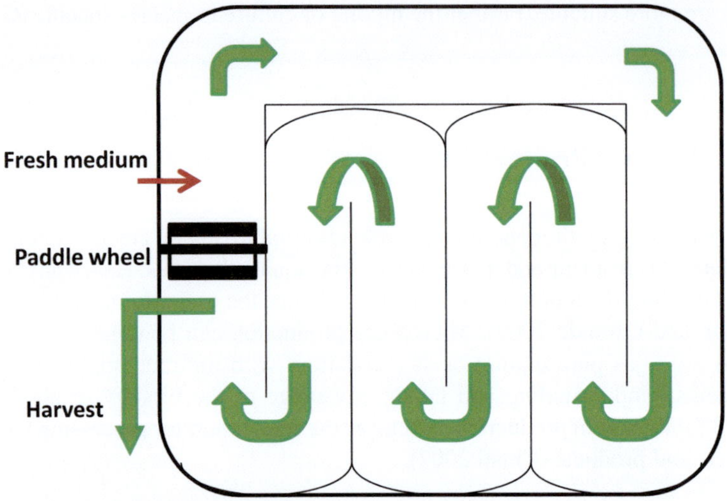

Fig. 15.1 Schematic of a raceway pond

most closed systems. They can utilise sunlight and the nutrients provided through runoff water from nearby land areas or by channelling the water from sewage/water treatment plants thus making it the cheapest method of large-scale algal biomass production (Carlsson et al. 2007).

Raceway ponds are the most commonly used artificial culture system (Fig. 15.1). The pond is usually designed in a 'raceway' or 'track' configuration generally between 0.2 and 0.5 m deep, in which a paddlewheel provides circulation and mixing of the algal cells and nutrients. Raceways are typically made from poured concrete or simply dug into the earth and lined with a plastic liner to prevent the ground from soaking up the liquid. Baffles in the channel guide the flow around bends. Pure carbon dioxide or air-CO_2 mixture is usually supplied through membrane diffusers or pipes along the channels to dispense tiny bubbles. Surface evaporation results in some loss of liquid, but it also helps in regulating the temperature of the medium. In a continuous production cycle, algae broth and nutrients are introduced in front of the paddlewheel and circulated through the loop. Growth period depends upon the species, inoculum density and other physicochemical parameters. Biomass concentrations of up to 1 g l^{-1} and productivities of 10–25 g m^{-2} day^{-1} are possible (Pulz 2001). The largest raceway-based biomass production facility located in Calipatria, CA (USA), occupies an area of 440,000 m^2 to grow *Spirulina* for producing biomass for food (Chisti 2007). Seaweed cultivation (macroalgae) for production of particular compound or product such as carrageenan, liquid fertilisers, agar, etc., has been demonstrated and licensed by CSIR-CSMCRI, Bhavnagar, India. Figure 15.2 shows large-scale microalgae cultivation facility available at CSIR-IMMT comprising of eight raceway ponds with 40,000 l capacity each.

Fig. 15.2 Large-scale microalgae cultivation facility at IMMT (CSIR), Bhubaneswar. Eight raceway ponds fitted with paddle wheels and CO_2 supply system

Although these systems are cost effective and easy to operate, open-air systems present significant technical challenges. Ponds are susceptible to weather conditions, with no control of physical parameters such as water temperature, evaporation and lighting (Molina 1999; Molina and Fernandez 1999). Furthermore, biomass productivity is also limited by contamination with unwanted algal species as well as organisms that feed on algae. Consequently, this strictly limits the species of algae that can be grown in such systems. As a result, very few species with high adaptability to salinity (*Dunaliella*), alkalinity (*Spirulina*) (Carlsson et al. 2007) and nutrient-rich conditions (*Chlorella*) have been successfully cultured so far.

15.4.1.2 Closed Systems: Photobioreactors

Photobioreactors (PBRs) give control on nearly all the biotechnologically important parameters. They present reduced contamination risk and CO_2 losses, reproducible cultivation conditions, controllable hydrodynamics and temperature and flexible technical design (Pulz 2001). Recent advances in microalgae mass culture require closed systems, as many of the new algae and algal high-value products for use in the pharmaceutical and cosmetics industry must be grown free of pollution and contaminants such as heavy metals and microorganisms (Janssen et al. 2003; Tredici 2004).

PBR are generally made of glass/fibre/plastic with sufficient strength against failure. They receive either direct sunlight through the transparent container walls or indirectly via light fibres or tubes that channel it from sunlight collectors. Some PBR have artificial light sources. The ground beneath the solar collector is often painted white or covered with white sheets of plastic to increase the light received by tubes. Mixing is achieved through circulation of culture or sparging gas through culture fluid.

There are several designs of PBR that have been reported, but the basic design in many of them is common. The main categories include (1) tubular (helical, manifold, serpentine and α shaped), (2) flat plate (alveolar panels and glass plates), (3) column (bubble columns and airlift) and (4) stirred tank reactor. A significant amount of work has been carried out in optimisation of different PBR systems for microalgae cultivation (Chaumont 1993; Janssen et al. 2003; Tredici 2004; Carvalho et al. 2006; Li et al. 2007; Aishvarya et al. 2012). A PBR has inlet for fresh feed, outlet for recirculation or harvest and a column for gas removal and settling purpose. Schematic for typical tubular PBR is given in Fig. 15.3. Tubular PBR have horizontal tubes transparent to sunlight enclosing the culture, and the tube diameter is often less than 0.1 m as light penetration is difficult in larger tubes (Chisti 2007). Bubble columns consist of a long column connected to a reservoir,

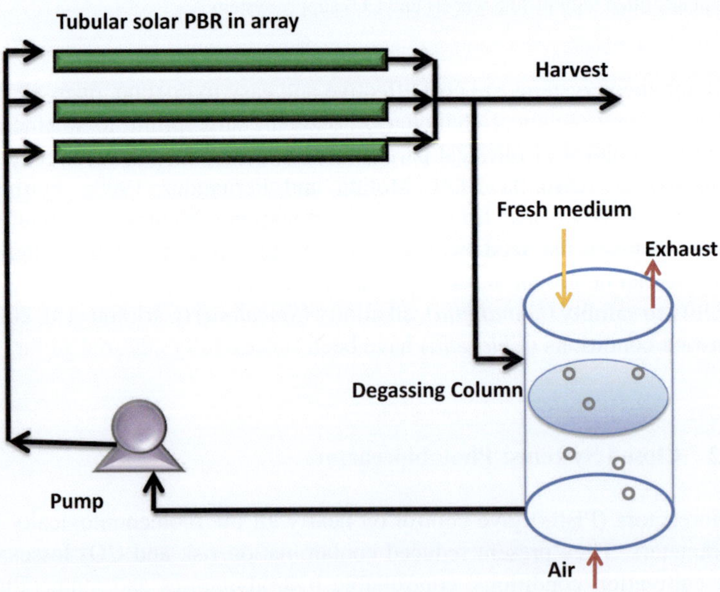

Fig. 15.3 Schematic of tubular PBR using solar light

Table 15.2 Comparison of properties of large-scale cultivation methods

Properties	Paddle wheel RWP	Stirred tank reactor	Tubular reactor	Column reactors
Light utilisation efficiency	Fairly good	Fairly good	Very good	Fairly good
Gas transfer	Moderate	Low-high	Low-high	High
Mixing	Fairly uniform	Almost uniform	Uniform, complete mixing	Almost uniform
Species control and sterility	Poor/none	Very good	Good	Good
Disadvantages	Large area of land required, low productivity	Difficult to scale up, high initial and operational costs	Fouling of tubes especially at bends, difficult to clean	Variable shear stress, high operational costs, difficult to scale up

and the liquid culture and gas are circulated either in same or opposite directions. As the effective surface area exposed to light is less compared to tubular reactors, bubble columns require artificial lighting along their length for good productivity. Due to design limitations, only tubular reactors (serpentine type) have been widely used in scale up yielding reasonable energy and cost efficiency.

A comparison between different modes of cultivation is given in Table 15.2 based on the vital parameters controlling the productivity of biomass (Borowitzka 1999; Brennan and Owende 2010).

15.5 Harvest of Biomass

The first step in the downstream processing of microalgae production is harvesting of cells. This is a major challenge in the microalgae cultivation as enormous amounts of liquid have to be processed to separate microalgae cells from the culture fluid. There are many techniques followed based on the species of algae, target products, the cell size and composition (high or low lipid content) as it decides the economy and efficiency of the process (Brennan and Owende 2010). Microalgae harvesting can be done in two stages involving (1) bulk harvesting and (2) thickening. Some commonly followed procedures are given in Table 15.3 with advantages and disadvantages.

Solid-liquid separation processes can be classified into two kinds. In the first, the liquid is constrained in a vessel and particles can move freely within the liquid (e.g. sedimentation and flotation). In the second kind, the particles are constrained by a permeable medium through which the liquid can flow (e.g. filtration and screening) (Shelef et al. 1984).

Table 15.3 Methods of harvesting microalgae

Stages of harvesting	Methods	Advantages	Disadvantages
Bulk harvesting	Sedimentation	Low cost, useful as first stage in separation to reduce energy input and cost	Settling rate specific to algae species; best for dense non-motile cells; can be slow
	Flotation	Uses air or gas bubbles; rapid than sedimentation	Species specific; oil-laden cells easily separated; air bubbling costs can be high
	Flocculation	Range of techniques available with low to high cost	Removal of flocculants and chemical contamination of harvested biomass
Thickening	Centrifugation	Can handle most algal cell types, efficient harvesting	High capital and operational costs
	Filtration	High concentrations can be achieved, efficient for large cells	Species dependent; clogging or fouling of filters; membrane costs can be high
	Ultrafiltration	Can handle delicate cells	High capital and operational costs

Sedimentation and flotation are gravity separation processes and depend on density of cells. Sedimentation occurs when cells settle due to gravity and can be increased by increasing cell dimension (i.e., coagulation) as sedimentation rate depends on particle size. Flotation is based on the attachment of air or gas bubbles to solid particles, which are then carried to the liquid surface and accumulate as float which can be skimmed off. The density difference between cell and culture medium decides which process is suitable. Oil-laden cells with low cell density can be separated by flotation.

Chemical flocculation is a well-known traditional method used in water treatment. Addition of chemicals to algal cultures to induce flocculation of cells is a routine procedure in various separation technologies. These chemicals can be inorganic such as Al^{3+} in alum, Fe^{3+} in ferric sulphate, $Ca(OH)_2$, etc., or polymeric (cationic, anionic, non-ionic) in nature (natural and synthetic polymers). During flocculation, single cells form larger aggregates that can be easily separated from the medium by simple gravity sedimentation. Flocculation is also performed as a first step before filtration in some cases as the cost and energy demand are significantly reduced if the cells are preconcentrated.

Flocculation can also be achieved by other methods which have been explored in the recent years such as electroflocculation, autoflocculation, flocculation using physical forces, bioflocculation, etc. (Vandamme et al. 2013).

Electroflocculation Flocculation is induced through electrolytic release of metal ions from a sacrificial anode. The efficiency of this method might be improved by changing the polarity of the electrodes. Similar to flocculation by metal salts, electroflocculation results in contamination of the biomass with metals, although to a lesser extent than when metal coagulants are directly used.

Autoflocculation Microalgal cultures spontaneously flocculate when pH increases beyond 9 due to photosynthetic CO_2 depletion. Autoflocculation is known to occur with formation of inorganic precipitates (calcium and magnesium) at high pH, and thus the harvested biomass contains high concentration of minerals.

Physical methods Forces such as ultrasound waves and magnetic nanoparticles have been studied to avoid contamination in the biomass. Magnetite (Fe_3O_4) nanoparticles adsorb directly on microalgae cells which are then separated by applying magnetic field, thus combining flocculation and separation in one step. The nanoparticles can be recovered after harvesting by desorption and reused.

Bioflocculation Certain extracellular polymer substances secreted by some species in the medium result in spontaneous flocculation. Some microalgae species flocculate more readily than others. Bacteria or fungi can also induce bioflocculation with their exopolysaccharides or positively charged hyphae, respectively. The mechanisms behind bioflocculation are poorly understood deserving further research.

15.6 Biotechnological Applications of Microalgae

Microalgae have a number of uses in industries. The algal biomass contains three main components: carbohydrates, protein and lipids/natural oil. High content of these commercially important molecules has seen escalation of research in areas from food industry, nutra- and pharmaceuticals to biofuels and phycoremediation (Fig. 15.4). Microalgal biotechnology has also gained importance due to diverse commercial application, such as value-added products for pharmaceutical purposes, human and animal nutrition, cosmetics, high-value molecules such as fatty acids, pigments and as energy sources (Spolaore et al. 2006; Mata et al. 2010). Recently, various nanoparticles and nanocomposites have also been developed using microalgae for environmental applications (Jena et al. 2013, 2014a). Some of the microalgal species in its raw or semi-decomposed form can be used as an organic biofertiliser. Further algae have seen applications in solving environmental issues as well by means of remediation, biosorption and latest for energy production.

Besides all the above potential application, the use of microalgae for biofuel production and carbon dioxide sequestration on an industrial scale is an important area of research owing to the increasing energy demands, predicted fossil fuels shortage in the near future and environmental concerns such as the production of greenhouse gas carbon dioxide.

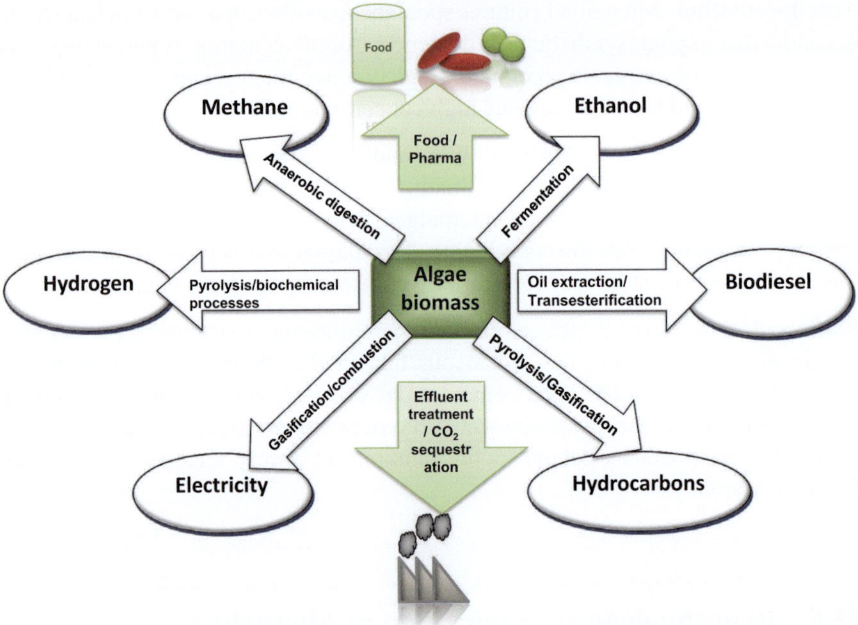

Fig. 15.4 Current industrial uses of algae

15.6.1 Food and Pharmaceuticals

Seaweeds are used as food for both humans and livestock. Due to the high-protein content, *Chlorella* sp. is widely used as health food for human beings and as animal nutritional supplements (Metting 1996). Some microalgae produce valuable by-products like proteins, pigments, biopolymers and fatty acids (docosahexaenoic acid) including antioxidant substances for commercial or pharmaceutical purpose (Priyadarshani and Rath 2012). Carbohydrate exists in several forms in the microalgae, such as starch, glucose, sugars and polysaccharides (Spolaore et al. 2006), and finds its potential applications in food, cosmetic and pharmaceutical industries (Banerjee et al. 2002). Seaweed is rich in many vitamins as A, B1, B2, B6, B12, C and niacin. Algae are also rich in iodine, potassium, iron, magnesium and calcium and are used in production of food supplements (Poulickova et al. 2008). The cyanobacterium *Arthrospira platensis* and *Arthrospira maxima*, more commonly known as '*Spirulina*', are widely consumed by humans as whole food or dietary supplements due to their rich protein and vitamin content. Also, algal hydrocolloids alginate, agar and carrageenan are produced from seaweeds (especially macroalgae) and largely used as viscosity-modifying agents in foods and pharmaceuticals (Mata et al. 2010). Many types of algae are also rich in omega-3 fatty acids and as such are used as diet supplements and as a component of livestock feed. Eicosapentaenoic acid (EPA) and docosahexaenoic acid (DHA) are two relatively rare and valuable fatty acids found in microalgae, and these essential fatty acids are critical structural component of our cell membranes and play a vital role in brain function (Crawford 1990). Blue-green algae contain a wide range of antioxidants in the form of amino acids, vitamins and especially pigments like carotenes (astaxanthin) and potent green-blue pigments (phycocyanins, phycobiliproteins) (Gonzalez et al. 1999) that have opened a new direction in dermatological treatment for radiation damage.

15.6.2 CO_2 Sequestration

Algae being autotrophic in nature can use CO_2 as a nutrient through photosynthetic metabolism, the most highly polluting green house gas. Coupling of CO_2 scrubbing and algae production is widely practiced as it fixes carbon in algal biomass, a renewable energy source (Campbell et al. 2011). With respect to air quality maintenance and improvement, microalgae is being considered as a eminent CO_2 fixation system that can reduce the green house gas emission caused by the burning of fossil fuel for various purpose. As reported, a 1 kg of microalgal biomass can efficiently fix 1.83 kg of CO_2 (Chisti 2007). Flue gases from power plants are responsible for more than 7 % of the total world CO_2 emissions from energy use. Also, industrial exhaust gases contain up to 15 % CO_2, providing a CO_2-rich source for microalgae cultivation and a potentially more efficient route for CO_2 bio-fixation (Mata et al. 2010). Therefore, the use of flue gas emissions from an industrial process unit (e.g. from fuel-fired power plants) as a source of CO_2 for the microalgae growth is envisioned to have a great potential to diminish CO_2 and to provide a very promising alternative to current green house gas emissions mitigation strategies (Danielo 2005). Enhanced sequestration of CO_2 in the form of bicarbonate is possible in mildly alkaline medium (pH-10.0) and can be exploited industrially (Aishvarya et al. 2012). Some species with tolerance to high CO_2 levels have also been identified and studied in these environments. Various microalgal genera including *Chlorella* (Douskova et al. 2009; Ramanan et al. 2010; Borkenstein et al. 2011; Aishvarya et al. 2012), *Scenedesmus* (Morais and Costa 2007), *Chlorococcum*, *Synechococcus*, *Thermosynechococcus* (Hsueh et al. 2009), *Nannochloropsis* (Hsueh et al. 2009), and *Spirulina* (Ramanan et al. 2010) have been reported for CO_2 sequestration ability.

15.6.3 Effluent and Wastewater Treatment

Since most of the nutrients that algae needs are often found in wastes, such as nitrogen, phosphorous and ammonia, utility of algae in wastewater treatment has become a widespread notion. The exponential growth of algae under ideal nutrient loads has led to the idea of algae as a phycoremediation tool (Olguin 2003). Some unicellular green microalgae species are particularly tolerant to sewage effluent conditions, most notably those of the *Chlorella* and *Scenedesmus* genus (Pittman et al. 2011). Growth of microalgae in primary settled sewage water was shown to increase significantly under long photoperiod conditions following addition of CO_2, while increased temperature decreased algal biomass concentration (Ip et al. 1982). Various species of *Chlorella* and *Scenedesmus* can provide very high (>80 %) and in many cases almost complete removal of ammonia, nitrate and phosphate from secondary treated wastewater (Martinez et al. 2000; Zhang et al. 2008; Ruiz-Marin et al. 2010). In small-scale treatments, algae can be used effectively in a pond or tank like setup. Since species control is not a priority in many cases of treatment, consortium can work better than single species. A consortium of algae has been used to treat carpet mill waste streams and simultaneously accumulate lipid for fuel purposes (Chinnasamy et al. 2010). Heavy metal in waste streams from electroplating, ceramics and plastics industries often polluting open water sources are toxic and accumulate in food chain. Biological treatment of such heavy metal containing waters is possible with microalgae by means of biosorption, bioaccumulation, retention and desorption (Chojnacka 2010). Biosorption of heavy metals has been studied on metal ions like Cr^{3+}, Cd^{2+}, Cu^{2+}, etc. (Chojnacka et al. 2005; Jena et al. 2014b).

15.6.4 Microalgae as a Source of Biodiesel

The possibility of using algae as a source of energy received widespread attention after the energy crisis of the 1970s. Algae has several advantages like higher photosynthetic efficiency, higher growth rates and biomass production compared to other energy crops, no competition with food crops for land, less water and nutrient requirements and their ability to accumulate lipids. Microalgal lipids are chemically similar to common vegetable oils (Dunahay et al. 1996; Chisti 2007). Many microalgae can accumulate lipids due to excess photosynthesis, and some species can accumulate high amount of lipids under heterotrophic or environmental stress, such as nutrient deficiency or salt stress (Takagi and Karseno 2006).

Biodiesel is defined as the mono-alkyl esters of fatty acids derived from vegetable oils, animal fats, waste cooking oil and jatropha oil (Felizardo et al. 2006; Chisti 2007). The oil of biodiesel consists of triglycerides in which three fatty acid molecules are esterified with a molecule of glycerol. The nature of fatty acids in biodiesel affects its properties. The percentage saturation and unsaturation of biodiesel determine its fluidity at room temperature, lubricity, viscosity, cetane number and its emission characteristics. This can be important when selecting the biodiesel for a particular application. Traditionally oil from algae is extracted by solvent extraction, Soxhlet extraction and recently by supercritical CO_2 extraction and pyrolysis (Miao and Wu 2004). Solvent extraction is widely done using chloroform and methanol (2:1, v/v) or hexane. The general procedure is to treat the biomass with solvent at 130 °C for 4 h, the extract is then passed through anhydrous sodium sulphate to remove the moisture and the solvent is evaporated under vacuum to get the oil.

Biodiesel from each feedstock is different from the other in that they are made of different proportions of saturated, monounsaturated and polyunsaturated fatty acids. Microalgal oils differ from most vegetable oils in being quite rich in polyunsaturated fatty acids with four or more double bonds like EPA (C20:5n-3; five double bonds) and DHA (C22:6n-3; six double bonds) that occur commonly in algal oils. The stability can bc improved by reducing the extent of unsaturation in microalgal oil with more than four double bonds, by partial catalytic hydrogenation of the oil (Jang et al. 2005), the same technology that is commonly used in making margarine from vegetable oils. Recent research aims to solve this by enhancing the transcription rate of MUFA genes and changing the metabolic pathway to improve the triglyceride conversion.

15.6.5 Energy Source

The concept of biodiesel from algae has accelerated the algae production manifold at different levels. But algal biomass can serve as energy source in more ways than just oil. The biomass left after the extraction of oil is rich with cellular storage products. Ethanol, a valuable fuel, can be produced from the leftover biomass by fermentation. Residual biomass may be used to produce methane by anaerobic digestion, for generating the electrical power necessary for running the microalgal biomass production facility (Spolaore et al. 2006). Although the microalgal biomass can be directly used to produce methane by anaerobic digestion, it cannot compete with the many other low-cost organic substrates that are available (Chisti 2007). After anaerobic treatment, the residue can be used as low-cost fertilisers as they are rich in N and P (Pittman et al. 2011). Research on hydrogen production from algae is another recent venture with algae. Algae biomass can be subjected to pyrolysis to extract hydrocarbons and synthetic gas from the biomass. Gasification of biomass can lead to electricity generation. All these potential utilities have led to integrated biorefinery using algae for multiple end products. A microalgal biorefinery can simultaneously produce biodiesel, animal feed, biogas and electrical power.

15.7 Concluding Remarks

The success of algae cultivation and its application have been one of the highly debated areas in biotechnology. But a strict control of desired conditions and judicious selection of appropriate microalgae species after considerations of its properties can result in predictable outcomes. With the wide range of applications that they offer, microalgae can be considered one of the inevitable options of the future biotech industry.

Acknowledgments The authors acknowledge the financial support from Department of Science and Technology (DST), Govt. of India and are thankful to Director, CSIR-IMMT for the permission to publish.

References

Adir N, Zer H, Shochat S, Ohad I (2003) Photoinhibition a historical perspective. Photosynth Res 76:343–370

Aishvarya V, Pradhan N, Nayak RR, Sukla LB, Mishra BK (2012) Enhanced inorganic carbon uptake by *Chlorella* sp. IMMTCC-2 under autotrophic conditions for lipid production and CO_2 sequestration. J Appl Phycol 24:1455–1463

Anandraj A, Perissinotto R, Nozais C (2007) A comparative study of microalgal production in a marine versus a river-dominated temporarily open/closed estuary, South Africa. Estuar Coast Shelf Sci 73:768–780

Banerjee A, Sharma R, Chisti Y, Banerjee UC (2002) *Botryococcus braunii*: a renewable source of hydrocarbons and other chemicals. Crit Rev Biotechnol 22:245–279

Bhakta S, Das SK, Nayak M, Jena J, Panda PK, Sukla LB (2011) Phyco-diversity assessment of Bahuda river mouth areas of East coast of Odisha, India. Recent Res Sci Technol 2:80–89

Bischoff HW, Bold HC (1963) Phycological studies IV. Some soil algae from Enchanted Rock and related algal species. Univ Tex Publ 6318:1–95

Bishop WM, Zubeck HM (2012) Evaluation of microalgae for use as nutraceuticals and nutritional supplements. J Nutr Food Sci 2:147. doi:10.4172/2155-9600.1000147

Bold HC (1949) The morphology of *Chlamydomonas chlamydogama* sp. nov. Bull Torrey Bot Club 76:101–108

Borkenstein CG, Knoblechner J, Frühwirth H, Schagerl M (2011) Cultivation of *Chlorella emersonii* with flue gas derived from a cement plant. J Appl Phycol 23:131–135

Borowitzka MA (1999) Commercial production of microalgae: ponds, tanks, tubes and fermenters. J Biotechnol 70:313–321

Brennan L, Owende P (2010) Biofuels from microalgae—a review of technologies for production, processing, and extractions of biofuels and co-products. Renew Sustain Energy Rev 14:557–577

Campbell PK, Beer T, Batten D (2011) Life cycle assessment of biodiesel production from microalgae in ponds. Bioresour Technol 102:50–56

Carlsson AS, Van Beilen JB, Moller R, Clayton D (2007) Micro- and macro-algae: utility for industrial applications, 1st edn. CPL, Newbury

Carvalho AP, Meireles LA, Malcata FX (2006) Microalgal reactors: a review of enclosed system designs and performances. Biotechnol Prog 22:1490–1506

Chaumont D (1993) Biotechnology of algal biomass production: a review of systems for outdoor mass culture. J Appl Phycol 5:593–604

Chinnasamy S, Bhatnagar A, Hunt RW, Das KC (2010) Microalgae cultivation in a wastewater dominated by carpet mill effluents for biofuel applications. Bioresour Technol 101:3097–3105

Chisti Y (2007) Biodiesel from microalgae. Biotechnol Adv 25:294–306

Chojnacka K (2010) Biosorption and bioaccumulation—the prospects for practical applications. Environ Int 36:299–307

Chojnacka K, Chojnacki A, Gorecka H (2005) Biosorption of Cr^{3+}, Cd^{2+} and Cu^{2+} ions by blue–green algae *Spirulina* sp.: kinetics, equilibrium and the mechanism of the process. Chemosphere 59:75–84

Craggs RJ, Adey WH (1995) Phosphorus removal from wastewater using an algal turf scrubber. In: 3rd IAWQ international specialist conference on waste stabilization ponds—technology and applications. Pergamon-Elsevier, Joao Pessoa, Brazil

Crawford M (1990) The early development and evolution of the human brain. Ups J Med Sci Suppl 48:43–78

Culture Collection of Algae and Protozoa (CCAP). (2013). Media recipes used by CCAP to maintain microalgal strains. http://www.ccap.ac.uk/media/pdfrecipes.htm

Danielo O (2005) An algae-based fuel. Biofuture 255:1–4

Douskova I, Doucha J, Livansky K, Machat J, Novak P, Umysova D, Zachleder V, Vitova M (2009) Simultaneous flue gas bioremediation and reduction of microalgal biomass production costs. Appl Microbiol Biotechnol 82:179–185

Dunahay TG, Jarvis EE, Dais SS, Roessler PG (1996) Manipulation of microalgae lipid production using genetic engineering. Appl Biochem Biotechnol 578:223–231

Falkowski PG, Raven JA (1997) Aquatic photosynthesis. Blackwater, London, p 375

Felizardo P, Correia MJN, Raposo I, Mendes JF, Berkemeier R, Bordado JM (2006) Production of biodiesel from waste frying oil. Waste Manage 26:487–493

Gonzalez R, Rodriguez S, Romay C, Gonzalez A, Armesto J, Remirez D, Merino N (1999) Anti-inflammatory activity of phycocyanin extract in acetic acid-induced colitis in rats. Pharmacol Res 39:55–59

Gouveia L (2011) Microalgae as a feedstock for biofuels. Springer, Heidelberg

Grobbelaar JU (2004) Algal nutrition. In: Richmond A (ed) Handbook of microalgal culture: biotechnology and applied phycology. Blackwell, Ames, IA, pp 97–115

Guillard RRL (1975) Culture of phytoplankton for feeding marine invertebrates. In: Smith WL, Chanley MH (eds) Culture of marine invertebrate animals. Plenum, New York, pp 26–60

Guillard RR, Ryther JH (1962) Studies of marine planktonic diatoms. *I. Cyclotella nana Hustedt* and *Detonula confervacea* (Cleve) Gran. Can J Microbiol 8:229–239

Harun I, Yahya L, Chik MN, Abdul Kadir NN, Azmir Pang MAM (2014) Effects of natural light dilution on microalgae growth. Int J Chem Eng Appl 5:112–116

Hsueh HT, Li WJ, Chen HH, Chu H (2009) Carbon bio-fixation by photosynthesis of *Thermosynechococcus* sp. CL-1 and *Nannochloropsis oculata*. J Photochem Photobiol 95:33–39

Hu Q, Sommerfeld M, Jarvis E, Ghirardi M, Posewitz M, Seibert M, Darzins A (2008) Microalgal triacylglycerols as feedstocks for biofuel production: perspectives and advances. Plant J 54:621–639

Ip SY, Bridger JS, Chin CT, Martin WRB, Raper WGC (1982) Algal growth in primary settled sewage—the effects of five key variables. Water Res 16:621–632

Jang ES, Jung MY, Min DB (2005) Hydrogenation for low trans and high conjugated fatty acids. Compr Rev Food Sci Food Saf 4:22–30

Janssen M, Tramper J, Mur LR, Wijffels RH (2003) Enclosed outdoor photobioreactors: light regime, photosynthetic efficiency, scaleup, and future prospects. Biotechnol Bioeng 81:193–210

Jena J, Pradhan N, Dash BP, Sukla LB, Panda PK (2013) Biosynthesis and characterization of silver nanoparticles using microalga *Chlorococcum humicola* and its antibacterial activity. Int J Nanomater Biostruct 3:1–8

Jena J, Pradhan N, Nayak R, Dash BP, Sukla LB, Panda PK, Mishra BK (2014a) Microalga *Scenedesmus* sp.: a potential low-cost green machine for silver nanoparticle synthesis. J Microbiol Biotechnol 24:522–533

Jena J, Pradhan N, Aishvarya V, Nayak RR, Dash BP, Sukla LB, Panda PK, Mishra BK (2014b) Biological sequestration and retention of cadmium as CdS nanoparticles by the microalga *Scenedesmus-24*. J Appl Phycol. doi:10.1007/s10811-014-0499-8

Khan SA, Rashmi Hussain MZ, Prasad S, Banerjee UC (2009) Prospects of biodiesel production from microalgae in India. Renew Sustain Energy Rev 13:2361–2372

Kirrolia A, Bishnoi NR, Singh R (2013) Microalgae as a boon for sustainable energy production and its future research & development aspects. Renew Sustain Energy Rev 20:642–656

Lavens P, Sorgeloos P (1996) Manual on the production and use of live food for aquaculture. FAO fisheries technical paper 361

Li M, Hu C et al (2007) Outdoor mass culture of the marine microalga *Pavlova viridis* (Prymnesiophyceae) for production of eicosapentaenoic acid (EPA). Cryptogamie Algologie 28:397–410

Lin Q, Lin J (2011) Effects of nitrogen source and concentration on biomass and oil production of a *Scenedesmus rubescens* like microalga. Bioresour Technol 102:1615–1621

Madigan MT, Martinko JM, Parker J (1997) Brock biology of microorganisms, 8th edn. Prentice Hall, Upper Saddle River, NJ

Martinez ME, Sanchez S, Jimenez JM, Yousfi FE, Munoz L (2000) Nitrogen and phosphorus removal from urban wastewater by the microalga *Scenedesmus obliquus*. Bioresour Technol 73:263–272

Mata TM, Martins AA, Caetano NS (2010) Microalgae for biodiesel production and other applications: a review. Renew Sustain Energy Rev 14:217–232

Metting FB (1996) Biodiversity and application of microalgae. J Ind Microbiol Biotechnol 17:477–489

Miao X, Wu Q (2004) High yield bio-oil production from fast pyrolysis by metabolic controlling of *Chlorella protothecoides*. J Biotechnol 110:85–93

Molina EG (1999) Microalgae mass culture methods. In: Flickinger MC, Drew SW (eds) Encyclopedia of bioprocess technology: fermentation biocatalysis and bioseparation. Wiley, New York

Molina EM, Fernandez FG (1999) Scale-up of tubular photobioreactors. In: 8th Tri annual international conference on applied algology (8th ICAA). Kluwer, Montecatini Terme, Italy

Morais MG, Costa JAV (2007) Carbon dioxide fixation by *Chlorella kessleri*, *C. vulgaris*, *Scenedesmus obliquus* and *Spirulina* sp. cultivated in flasks and vertical tubular photobioreactors. Biotechnol Lett 29:1349–1352

Munir N, Sharif N, Naz S, Manzoor F (2013) Algae: a potent antioxidant source. Sky J Microbiol Res 1:22–31

Mutanda T, Ramesh D, Karthikeyan S, Kumari S, Anandraj A, Bux F (2011) Bioprospecting for hyper-lipid producing microalgal strains for sustainable biofuel production. Bioresour Technol 102:57–70

Neenan B, Feinberg D, Hill A, McIntosh R, Terry K (1986) Fuels from microalgae: technology status, potential, and research requirements. Report no. SERI/SP-231-2550, Solar Energy Research Institute, Golden, CO

Norton TA, Melkonian M, Andersen RA (1996) Algal biodiversity. Phycologia 35:308–326

Olaizola M (2003) Commercial development of microalgal biotechnology: from the test tube to the marketplace. Biomol Eng 20:459–466

Olguin EJ (2003) Phycoremediation: key issues for cost-effective nutrient removal processes. Biotechnol Adv 22:81–91

Pittman JK, Dean AP, Olumayowa O (2011) The potential of sustainable algal biofuel production using wastewater resources. Bioresour Technol 102:17–25

Poulickova A, Hasler P, Lysakova M, Spears B (2008) The ecology of freshwater epipelic algae: an update. Phycologia 47:437–450

Prescott LM, Harley JP, Klein DA (2002) Microbiology, 5th edn. McGraw-Hill, New York

Pringsheim E (1950) The soil-water culture technique for growing algae. In: Culturing of algae. Charles F. Kettering Foundation, Dayton, OH, pp 19–26

Priyadarshani I, Rath B (2012) Commercial and industrial applications of micro algae—a review. J Algal Biomass Util 3:89–100

Pulz O (2001) Photobioreactors: production systems for phototrophic microorganisms. Appl Microbiol Biotechnol 57:287–293

Rajauria G, Abu-Ghannam N (2013) Isolation and partial characterization of bioactive fucoxanthin from *Himanthalia elongata* brown seaweed: a TLC-based approach Hindawi Publishing Corporation. Int J Anal Chem. doi:10.1155/2013/802573, Article ID 802573, 6 pages

Ramanan R, Kannan K, Deshkar A, Yadav R, Chakrabarti T (2010) Enhanced algal CO_2 sequestration through calcite deposition by *Chlorella* sp. and *Spirulina* platensis in a mini-raceway pond. Bioresource Technol 101:2616–2622

Richmond A (2008) Handbook of microalgal culture: biotechnology and applied phycology. Wiley-Blackwell, New York

Richmond A, Hu Q (2013) Handbook of microalgal culture: applied phycology and biotechnology. Wiley, Oxford, 736 page

Rippka R, Deruelles J, Waterbury JB, Herdman M, Stainer RY (1979) Generic assignment, strain histories and properties of pure cultures of cyanobacteria. J Gen Microbiol 111:1–61

Ruiz-Marin A, Mendoza Espinosa LG, Stephenson T (2010) Growth and nutrient removal in free and immobilized green algae in batch and semi-continuous cultures treating real wastewater. Bioresour Technol 101:58–64

Shelef G, Sukenik A, Green M (1984) Microalgae harvesting and processing: a literature review. A subcontract report prepared by SERI for U.S Department of energy

Soletto D, Binaghi L, Lodi A, Carvalho JCM, Converti A (2005) Batch and fed-batch cultivations of *Spirulina* platensis using ammonium sulphate and urea as nitrogen sources. Aquaculture 243:217–224

Spolaore P, Joannis-cassan C, Duran E, Isambert A (2006) Commercial applications of microalgae. J Biosci Bioeng 101:87–96

Stein JR (ed) (1973) Handbook of phycological methods: culture methods and growth measurements. Cambridge University Press, Cambridge, 448 pp

Takagi M, Karseno YT (2006) Effect of salt concentration on intracellular accumulation of lipids and triacylglyceride in marine microalgae *Dunaliella* cells. J Biosci Bioeng 101:223–226

Thimijan RW, Heins RD (1983) Photometric, radiometric and quantum light units of measure: a review of procedures for interconversion. Hortscience 18:818–822

Tredici MR (2004) Mass production of microalgae: photobioreactors. In: Richmond A (ed) Handbook of microalgal culture: biotechnology and applied phycology. Blackwell, Oxford, pp 178–214

Vandamme D, Foubert I, Muylaert K (2013) Flocculation as a low-cost method for harvesting microalgae for bulk biomass production. Trends Biotechol 31:233–239

Vonshak A (1986) Laboratory techniques for the cultivation of microalgae. In: Handbook of microalgal mass culture. CRC, Boca Raton, FL, pp 117–145

Vonshak A (1990) Recent advances in microalgal biotechnology. Biotechnol Adv 8:709–727

Walne PR (1970) Studies on the food value of nineteen genera of algae to juvenile bivalves of the genera Ostrea, Crassostrea, Mercenaria and Mytilus. Fish Invest 26:162

Walter TL, Purton S, Becker DK, Collet C (2005) Microalgae as bioreactor. Plant Cell Rep 24:629–641

Wang J, Chen C (2009) Biosorbents for heavy metals removal and their future. Biotechnol Adv 27:195–226

Woelfel J, Schumann R, Adler S, Hubener T, Karsten U (2007) Diatoms inhabiting a wind flat of the Baltic Sea: species diversity and seasonal variations. Estuar Coast Shelf Sci 75:296–307

Xin L, Hu HY, Ke G, Sun YX (2010) Effects of different nitrogen and phosphorus concentrations on the growth, nutrient uptake and lipid accumulation of a freshwater microalga *Scenedesmus* sp. Bioresour Technol 101:5494–5500

Zhang ED, Wang B, Wang QH, Zhang SB, Zhao BD (2008) Ammonia-nitrogen and orthophosphate removal by immobilized *Scenedesmus sp* isolated from municipal wastewater for potential use in tertiary treatment. Bioresour Technol 99:3787–3793

Zhou W (2014) Recent advances in microalgal biotechnology. OMICS, Foster, CA

Chapter 16
Advances in Manganese Pollution and Its Bioremediation

A.P. Das, S. Ghosh, S. Mohanty, and L.B. Sukla

16.1 Introduction

Manganese (Mn) is the twelfth most plentiful element present in the world (Wei et al. 2012; Das et al. 2011; Pakarinen 2011), and the majority of the deposits are in the form of different oxides. Environmentally, Mn is found as a major and minor component in more than 100 naturally occurring minerals (Gadd 2010). The major components are birnessite (δMnO_2), psilomelane [Ba, $Mn^{2+}(Mn^{4+})_8O_{16}(OH)_4$], manganite (MnOOH), pyrolusite, vernadite (MnO_2), hausmannite (Mn_3O_4), and braunite (Mn, Si_2O_3) (Ehrlich 2002), while minor amount includes micas such as biotite and ferromagnesian minerals such as pyroxenes and amphiboles (Trost 1958). Manganese rarely exists in its pure elemental state, and environmental existence of Mn oxidation state ranges from +2 to +3 to +4 (Ehrlich and Newman 2009).

Manganese nodules are mineral contents found at the bottom of the sea and are made up of manganese and iron oxides. Based on their geological position, sediment, and mineral deposition of the region, the nodule size varies from 12 to 20 cm in diameter. Mn nodules cover about 30 % of the ocean belts in the form of solid precipitate from the surrounding water which are formed in three ways: (1) hydrogenetic nodules gathered by means of precipitation directly from wintry ambient seawater, (2) diagenetic nodules gathered from ocean floor sediment waters and modified by chemical interaction with different sediments present in the ocean (Kenneth 2010), and (3) specific microbes formed by the interaction of

A.P. Das (✉) • S. Ghosh • S. Mohanty
Centre of Biotechnology, Bioengineering Laboratory, Siksha O Anusandhan University, Bhubaneswar 751003, Odisha, India
e-mail: alok1503@gmail.com

L.B. Sukla
Bioresource Engineering Department, CSIR – Institute of Minerals and Materials Technology, Bhubaneswar, 751013 Odisha, India

L.B. Sukla et al. (eds.), *Environmental Microbial Biotechnology*, Soil Biology 45,
DOI 10.1007/978-3-319-19018-1_16

different communities of organism with their physical surroundings exist in the ocean (Li et al. 2008). Apart from this, other aspects that are supposed to be significant to Mn nodule creation are the supply of metals to growing surface, presence of nucleus, corrosive and erosive forces caused by benthic currents of the Antarctic Bottom water, occurrence of semiliquid surface layer on the seafloor, bioturbation, and internal stratigraphy of individual nodules. Mn nodules take place on the ocean floors broadly on the vast sediment-covered rolling plains of the depths of about 4,000–6,500 m (Elaine 2011). The MnO_2 ores are produced commonly due to earthly weathering progression and are established in a diversity of surroundings throughout the globe with varied outer structural features, organic chemistry, and physical characteristics (Ostwald 1988). At the center of the ocean, spreading concentrations of Mn have a greater proportion in deep-sea hydrothermal vents where biological invention shows confirmation for special biogeochemical processes in which microbes play a crucial role.

Mn is found as a foremost reserve in countries like Russia, Australia, Gabon, Brazil, South Africa, and India (Das et al. 2012a). India has a huge deposit of Mn ore and produces around two million tons of high-grade ores annually. In India, manganese ores are categorized into different grades depending on the deposits with 7 % of high-grade, 8 % of medium-grade, 34 % of furnace-grade, and the left 51 % of low-grade ores (Acharya et al. 2004). In the general mineral map of India, the state of Odisha presents an important position in comprising the manganese ore content (Mishra et al. 2009). Present estimation of manganese deposition includes low-grade ore that attains several billion tons of about 22 % Mn content but high-grade ores having more than 44 % Mn content with 680 million tons (Das et al. 2012a). Enormous quantity of manganese wastes is generated due to widespread mining and mineral processing and anthropological activities. During the dispensation and mining processes, a huge amount of rejected Mn is produced as waste and does not have a market value for which handing out and consumption route are not available (Venugopal 2010). The insistence for manganese has increased significantly due to the boost in the manufacturing of steel and the mounting shortage of natural resources (Acharya et al. 2003). The unprocessed liberation and unorganized administration of these metallic wastes have resulted in the enlistment of Mn to the atmosphere contributing to contamination of soil substrates, groundwater pollution, shortage of nutrients, decrease in biological diversity, and lots of severe ecological tribulations (Liu et al. 2008). The enormous loss of untreated metals released results in undue loss of valuable element like manganese for its high concentration of Mn content. More than 15 % in general of unprocessed mining effluents are released into the environment (Ge et al. 2004).

The extraction of industrial valuable metals and minerals in different possible methods has been an important concern due to environmental pollution and imposed health effects. For instance, several proposed physical and electrochemical recovery processes have been established to obtain valuable metals from the mining sites, but these techniques do not have much implication in remediating the effluents. So to overcome from these difficulties, diverse reductive and oxidative leaching processes have been implemented for the processing of low-grade Mn ores including leaching with organic reductants, bioleaching, and electroreductions. Among these processes, bioleaching with Mn-solubilizing microorganisms is the most promising and foremost operated technique applied in a large scale which makes the processes less economically viable. By improving the efficiency of microorganisms interconnected to metal-solubilizing activities, the expansion of bioremediation technology focuses on predominant remediation of manganese. The microbial species with Mn-solubilizing ability are a key step in evaluating the significance of microbial ecological role in manganese biomining.

This chapter endeavors a whole development in the research on manganese environmental pollution, its ecotoxicity, and recent approaches for Mn bioremediation. It is concluded that this chapter offers an effective technique for the recovery of manganese through Mn-solubilizing microorganism.

16.1.1 Occurrence and Resources

Deposition of manganese ore is generally of sedimentary source (Bekker et al. 2010). Deposits of Mn are found in the Paleoproterozoic Hotazel formation in South Africa conserving more than four billion heaps of Mn ore (13.5 billion metric tons (mt) of Mn), creating the world's largest land-based economic reserve at Kalahari manganese field (Kirschvink et al. 2000). Cyanobacterium provides a clear-cut explanation of the manganese formations, one with possible intense insinuation for the progression of life (Kirschvink 1992). MnO_2 is the main widespread mineral and is distributed all over the globe (Zhu et al. 2010). Juarez-Santillan (2010) reports the existence of Mn^{2+} in the environment, and sediments are fairly alkaline plus reducing circumstances with usual electrical conductivity and medium-to-high cation replace ability. In order of predominance, mainly the cost-effective rating, Mn ores are resulted from hydrothermal vents connected through transitional volcanic rocks and classified into sedimentary rock hosted, volcanic rock hosted, and karst hosted (Maynard 2010).

The average size of the nodules appears to be around 30 cm with a growth rate that differs starting about 1–200 million years with a regular assortment of 3–4 mm/my (Achurra et al. 2009). Pertaining to the spatial allocation of manganese nodules on the seabed, the researchers reveal that the Mn nodules seem to have a structure which can be as small as golf balls or as big as large potatoes. The majority of the Mn nodules are found on the top with a distance of 1 m from the ocean layer sediment, while some of the nodules are seen at a depth of 5.5–6 m.

The entire global production of Mn alloys achieved 17.7 million mt in 2011 growing to assure demand from steel mills. China remained to be the world's major Mn alloy-producing nation. The International Manganese Institute calculates that its alloy production exceeded ten million mt in 2011 accounting for almost 58 % of global output. In India, Manganese Ore India Limited (MOIL) is the major Mn ore manufacturer, and it reports for 40 % of India's entire manufacture. MOIL reported whole resources of Mn ore in the nation as 240 million tons, with reserves category of 138 million tons accounting a total of 378 million tons. State wise, Odisha secures top position with 44 % followed by Karnataka 22 %, Madhya Pradesh 16 %, Maharashtra 8 %, Andhra Pradesh 4 %, and Jharkhand and Goa 3 % each. Rajasthan, Gujarat, and West Bengal together shared the remaining about 3 % resources of the total 138 million tons of resources.

16.2 Production and Usage

Worldwide Mn is the fourth majorly used metal in stipulations of tonnage following iron, aluminum, and copper. The outlook of steel industry depends directly on the universal demand for manganese (Zhang 2007). More than 95 % of the globally produced manganese is used in desulfurization and strengthening of steels (Hariprasad et al. 2009), and the other 5 % is utilized in battery and chemical industries (Pakarinen 2006) (Fig. 16.1). Manganese improves strength, toughness, hardness, workability, and finally quality of steels and is used in steel-alloying applications (Acharya et al. 2004). Universal requirement for Mn depends straightly on the position of steel manufacturing. The global Mn ore manufacturing for the year 2010 is 47.21 million tons, as compared to 35.41 million tons during the year 2009 indicating a strong 33.32 % expansion in accessibility. It is expected that in the approaching time, the manufacture of Mn ore shall further enhance. The global annual consumption of Mn is above 1,500,000 tons, and it is anticipated to raise (Zhang 2007). According to the International Manganese Institute, the resources of Mn minerals are limited. Universal deposits of high-grade Mn ores are reducing at a disquieting speed due to the hasty enhancement in the requirement for metals. Demand for Mn will be concentrated in two countries, namely, China and India. For example, India's national steel policy projects a production of 110 million mt of steel by 2020 from a current production level of approximately 60 million mt. Several key economic factors suggest that these countries will continue leading the demand for steel in the medium term.

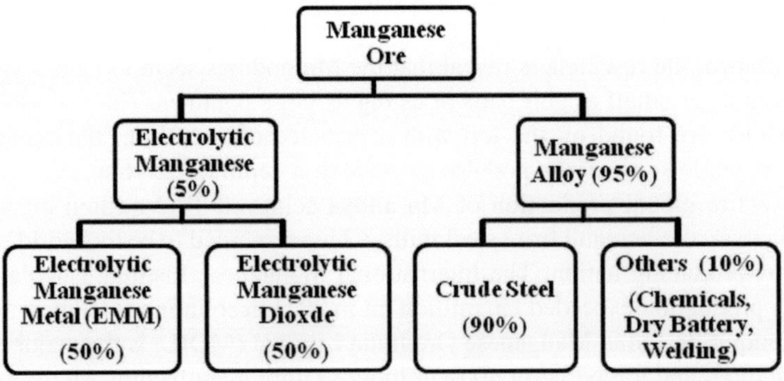

Fig. 16.1 Key applications of manganese ore

Universal requirement for manganese depends directly on the requirements of the steel industry. There are many grades of steel and each necessitates a different quantity of manganese. Consumption is calculated by the average requirement of manganese per ton of steel. Manganese has been expansively used in dry-cell batteries, metallurgy, glass materials, ceramic objects, pigments, dyes, glass, fireworks, soil, food, and medicine. In alloy industries, manganese is also used with Cu or Al to reduce corrosion, in batteries as MnO_2, for coloring ceramics, and in glasses, and in chemical industries, it is used as permanganate for oxidation reactions. More than 95 % of the total mined manganese is used in steel and metal industries, thus increasing steel toughness, potency, and rigidity by oxidizing sulfur and phosphorus impurities and others that include electrolytic Mn, electrolytic Mn ore metal (EMM), and electrolytic Mn dioxide (EMD). Everyday metallic objects have need of some percentage of manganese in their manufacture. Pyrolusite (MnO) has been used in glassmaking. To prevent metal corrosion, about 1 % Mn is used in thin aluminum beverage cans. Manganese in dry-cell batteries helps to reduce production of waste matter. At present, there is no appropriate alternate for Mn in steel manufacturing. Hence, steel production has an encouraging effect on Mn demand. It is anticipated that the manufacture of manganese ore shall further rise in the upcoming years. India has vast deposits of manganese ore which is among the most important minerals of the country and produces more than two million tons of high-grade ores every year (Das et al. 2011). India is one of the principal manufacturers of Mn ore in the earth. Concentration less than 35 % of Mn ore is currently ignored because extraction methods applied to low-grade Mn ore are not economically viable. Vastly mounting requirement for Mn has made it increasingly important to extend methods for cost-effective resurgence of Mn from low-grade ores and other secondary sources. The requirement for Mn ore has enhanced greater than before the rising shortage of natural resources due to boost in the production of steel (Xin et al. 2011, 2012). Apart from the proficient worldwide production of manganese, scarcity of high-grade manganese ore in global markets still exists for manufacturing steels in industries that leads to the addition of prices for both Mn ore and alloy (Doshi 2007). Key applications of manganese ores are explained in Fig. 16.1.

Besides an efficient Mn manufacturing globally, a scarcity of high-grade Mn ore in global markets still exists for carbon steels which have seen a cost increase for both Mn ores and alloys (Doshi 2007). There will be a massive requirement gap involving the accessibility and requirement of Mn. In view to attain the anticipated requirement of ferro-alloys, there will be enormous prerequisite of Mn, and till date, no satisfactory substitute for manganese in steelmaking has been identified.

16.3 Mn Pollution and Health Issues

Exposure of enormous amounts of mine effluents has resulted due to widespread mining and untreated anthropogenic and industrial activity. The present acceptable discarding system for manganese solid waste (MSW) is landfill, which causes environmental pollution, soil and groundwater contamination, and other severe ecological contamination. Thus, for environmental protection and sustainable development, utilization of EMSW has become an issue. The foremost cause of Mn exposure is water and air pollution. Mn-contaminated area and occupational exposure include mining industries, steel construction, battery manufacturing, welding, fossil fuel combustion, and in chemical engineering (Santamaria 2008). In the manufacturing industries of ferromanganese, iron, dry-cell battery, and steel industries, Mn exposure primarily occurs by breathing of manganese fumes or dusts. Electrolytic Mn slag is extremely formed from the electrolytic manganese metal (EMM) manufacturing as a major effluent. Approximately 8–9 tons of the waste is released into the surroundings per ton of EMM (Duan et al. 2011). These hold several dangerous metal pollutants, and the unprocessed release causes severe contamination of the surrounding (Ge et al. 2004). The utilization of batteries has enlarged in the previous years due to their adaptability, reduced price, and necessities of the electronic production (Gallegos et al. 2013). Batteries find their applications in radios, remotes, calculators, recorders, cameras, watches, and in several additional electronic items (Sayilgan et al. 2009). Usually, batteries are consumed rapidly and are thrown away; this fact makes it a major environmental and health threat (Kierkegaard 2007). The waste batteries cause a serious concern due to their toxicity, abundance, and permanence in the environment. Inhabitants living close to mining extraction sites (Riojas-Rodríguez et al. 2010) and alloy industries (Menezes-Filho et al. 2009) are at threat of extreme Mn exposure. Research investigations on ecological exposure to Mn have effectively applied scalp hair Mn and air Mn as natural and ecological indicator of exposure (Viana et al. 2014).

Manganese is a recognized neurotoxic representative since the last 175 years; till date, numerous investigations were conceded recounting indications of Mn intoxication in humans, rodents, and cellular models (Michalke and Fernsebner 2014). Chronic and acute exposure during mining, welding, and smelting, nutritional exposure, and further untreated ecological and manufacturing anthropogenic effluents of this metal contaminant lead to dangerous health effect and are classified medically by the numerous disorders of neurotoxicity including Parkinson's disease, psychiatric symptoms, manganism (Zoni et al. 2012), cognitive and neurodegenerative diseases, and other disturbances in the motor system (Sriram et al. 2012). A list of disease caused by Mn exposure is provided in Table 16.1. Overexposure to the metal can also lead to neurodegenerative damage and neurological dysfunction (Cordova et al. 2012; Yokel 2009). Mn toxicity can gain the right to enter the body by the central nervous system (CNS) and transport by means of the pulmonary epithelium into the lymph or blood and mucociliary clearance of the heavy metal in

Table 16.1 Diseases caused by manganese exposure

Source		Disease reference
Air	Memory loss, sleep disturbance, irritability, anxiety disorders, gait disturbance, and Parkinson's disease	Zoni et al. (2012), Sadek et al. (2003)
Nutrition	Deposition in blood and transport to other regions like the brain and neuron causing neural disorder	Ordonez-Librado et al. (2010), Roth (2006)
Water	Adverse health effect, neurotoxins	Sankar et al. (2014), McArthur et al. (2012)

the gastrointestinal tract from the lung (Roth 2006). Mn cytotoxicity is described in a cascade of signaling pathways which activates Mn-induced apoptosis leading to cell death (Roth 2006).

The indecent management effects in contributing soil contamination into the surrounding environment resulting in the destruction of soil texture, deficiency of nutrients, devastation of biological landscape, and pollution of groundwater and shows a high decline in Biodiversity, etc. (Azizi et al. 2012; Behera and Sukla 2012). The untreated release of heavy metals to the surrounding results in an enormous loss of valuable element like manganese, due to elevated Mn content in it, more or less 4 % in w/w in general, occasionally still as high as 15 % (Ge et al. 2004). The generation of a massive quantity of mine effluents resulted due to the widespread mining, mineral processing, and other human activities.

Mn is highly necessary for redox reactions, photosynthesis in cyanobacteria (Ogawa et al. 2002), defense from UV light, scavenging of micronutrient trace metals, upholding of an electron acceptor reservoir for the employment in anaerobic respiration, and safety against oxidative stress in bacteria (Spiro et al. 2010). It is a crucial trace compound which plays an important role in antioxidant resistance and forms a component of a superoxide dismutase (MnSOD) which generally characterizes the modified cancer cells (Gerber et al. 2002). The minimum necessary content of manganese in human body supports for maintaining the proper metabolism of cholesterol, proteins, carbohydrates, and amino acids which considerably have a great effectiveness on the activity of different enzymes concerned in metabolic and redox processes. Mn in elevated content interferes DNA replication and repair in bacteria and causes mutations in organism and mammalian cells. A large need of industrial application of manganese, huge manufacturing, and mining has made Mn a potential environmental pollutant which causes various diseases affecting human activity (Duka et al. 2011). Exposure to toxic levels of Mn is known to cause reproductive problems, immunological dysfunction, and other lethal consequences inside the human body. Occupational Mn exposure of miners and welders (Flynn and Susi 2009) leads to neurological dysfunction (Dobson et al. 2004) associated with dystonic movements, cortical structures, and basal ganglia (Perl and Olanow 2007; Roth 2009).

16.4 Recovery of Mn

The current approaches offered in hydrometallurgical industries for the processing of high-grade Mn consist of precipitation, solvent extraction, sulfide precipitation, persulfate addition, the sulfacid process, etc. (Das et al. 2012b). A range of hydrometallurgical techniques from low-grade ores are investigated and developed for the recovery of Mn (Youcai et al. 2014). Manganese recovery is carried out by acid leaching, as reducing agents are essential to change manganese dioxide into low-valence manganese (II) compounds. The reductants used generally comprise gaseous sulfur dioxide, corncob and glucose, and sulfur (Zhang et al. 2013). Mn recovery directly by reductive acid uses various reducing agents such as pyrite, hydrogen peroxide, sulfate, organic acids, cane molasses, and Mn-reducing microorganisms (Liu et al. 2004). Regrettably, every technique has its own restrictions and is not cost-efficiently reasonable for metal recovery from inferior-grade ores (Fisher 2010). Chemical methods could be used to remove manganese, but it will produce potential hazardous chemicals such as ammonia and soda ash into the environment (Han et al. 2013). The chemical techniques applied for Mn removal from electrolytic manganese slag offer some drawback such as huge utilization of strong acid, small removal efficiency, stringent necessity of equipments, and being quite unsafe (Gallegos et al. 2013). The reward of reprocessing Mn from a profitable ecological and industrial point of view depends on several features, counting transportation, recycling techniques, and resources to be utilized. In the mining industry, the bioleaching process has been successfully substituting hydrometallurgical techniques due to their ecological and financial benefits over these conventional methods (Das et al. 2011; Han et al. 2013). A few of the advantages of bioleaching technique are low energy utilization, less environmental discharge, harmless, resourcefulness and cost-effectiveness of the technique, low-grade resource, and option of on-site handling (Hasan et al. 2012). Hence, appropriate method for optimum utilization of low-grade ores requires to be evolved. Currently, restoration of Mn from electrolytic Mn effluents draws emergent considerations due to the rising shortage of natural resources. Reduction, roasting, and acid leaching methods are used for treating low-grade Mn ore with reacting acids like sulfuric acid and oxalic acid, respectively, with a highest efficiency of 93.44 % (Zhang et al. 2013). Mn wastes and bioleaching technology involved for their beneficiation are provided in Table 16.2.

16.4.1 Bioleaching of Mn

The capability of diverse microorganisms to solubilize metals has been exploited in the advance and appliance of biotechnology for recovering metals from ores and wastes (Johnson 2014). Bioleaching is a novel technique connecting microbiology and metallurgy for extracting precious metals from waste ores (Li et al. 2009)

Table 16.2 Mn wastes and bioleaching technology involved for their beneficiation

Waste	Method	Microorganism	Efficiency (%)	Reference
Spent Zn–Mn batteries	Bioleaching	Sulfur-oxidizing bacteria	94	Xin et al. (2012)
Multi-heavy metals	Biogenic Mn oxide production	*Brachybacterium* sp. strain Mn32	22.7	Wang et al. (2009)
Mn-electrolyzed slag	Bioleaching	*Fusarium* sp. *Serratia* sp.	82.5 76.9	Cao et al. (2012)
Low-grade ores	Bioleaching	*Staphylococcus epidermidis*	80	Das et al. (2012b)
Low-grade ores	Bioleaching	*Penicillium citrinum*	64.6	Acharya et al. (2004)
Electrolytic Mn residue	Bioleaching	Pyrite-/sulfur-leaching bacteria	81/93	Xin et al. (2011)
Indian Ocean polymetallic nodules	Bioleaching	*Aspergillus niger*	91	Mehta et al. (2010)
Spent Zn–Mn batteries	Bioleaching	Bioleaching bacteria species	89	Xin et al. (2012)
Black shale ore	Bioleaching	Fungal and bacterial species	88	Anjum et al. (2012)

involving much lower temperatures, reduced energy costs, and smaller carbon footprints. However, current commercial bioleaching processes depend on the blasting, crushing, grinding, and sieving of ores, which is assumed to utilize 5 % of the entire universal power production (Johnson 2013). Microbes solubilize metals into their water-soluble forms and are extracted through this procedure (Ilyas et al. 2012). The "manganese-oxidizing group of microorganisms" is a miscellaneous group of microorganisms illustrated phylogenetically and classified by the ability to catalyze the oxidation soluble Mn (II) to insoluble manganese oxides (Nealson 2006).

The ability to suitably transform Mn(II) to either Mn(III) or Mn(IV) has been found in a various isolated groups of bacteria including *Proteobacteria*, *Actinobacteria*, and *Firmicutes* (Tebo et al. 2005) and from diverse environments (Templeton et al. 2005; Dick et al. 2006; Anderson et al. 2009). Although questions regarding the oxidation of manganese still remain unclear and unanswered, it is assumed that different species use different mechanisms and probably have different reasons for oxidizing manganese. Reviews entailing potential mechanisms and various purposes for bacterial manganese oxidation have been published by Tebo et al. (2005) and Das et al. (2011).

A diversity of microbe has been described to reduce manganese both enzymatically and nonenzymatically. The microbes capable of reducing Mn enzymatically convey this method as a form of respiration. Mn^{2+}-oxidizing microbes are suggested to reduce the Mn under poor oxygen situation. Because anaerobic growth of these microbes with Mn oxide has not been recognized, the practical implication of this progression in these microbes is not determined. A motivating opportunity for the utility of microbial Mn^{2+} oxidation revealed that Mn oxides oxidize complex humic materials, liberating low molecular mass organic compounds for microbial development. This entail that Mn^{2+}-oxidizing microbes might stay alive in a low nutrient situation by generating a strong oxidizing agent able to degrade biologically recalcitrant carbon pools. Fungal Mn peroxidases commence the formation of Mn (III) complexes which can consequently oxidize phenolic complexes.

Although abundant research point to precise roles for enzymatic Mn^{2+} oxidation by bacteria, explicit confirmation for diverse efficient models is still missing. Simply, clarification of the mechanisms of the process in specific bacterial species and recognition of the cellular components concerned will ultimately provide insight in the purpose(s) of Mn^{2+} oxidation. Reduction of Mn (IV) to Mn (II) is mainly vital when attempting biorecovery of Mn from low-grade ores to liquid aqueous phase. Diverse atmospheres from where plentiful manganese oxidizers have been recognized consist of deep-sea manganese nodules, rice fields, low-grade ores, manganese-rich surface, and lake sediments. Additionally, Mn^{2+}-oxidizing microbes are isolated from ecological samples polluted with xenobiotics, such as aromatic hydrocarbons or other heavy metals (Raynal and Pruden 2008). Information about the manganese-oxidizing bacteria and fungus isolated and identified from different habitats is shown in Tables 16.3 and 16.4.

The recovery of manganese from manganiferous ores by bioleaching with different kinds of microorganisms has been extensively investigated by many workers, and microbiological processes also have been proposed to be less hazardous. In recent years, the recovery of Mn from electrolytic manganese slag draws increasing attentions due to the growing scarcity of natural resources. Manganese-based industries produced a huge amount of electrolytic manganese slag each year and from the electrolytic manganese metal (EMM) industry as a main solid waste (Cao et al. 2012). The recovery of Mn from manganiferous ores and electrolytic manganese slag using bioleaching with different microorganisms has been reported (Li et al. 2009). As regards microbial extraction of metals from low-grade oxide

Table 16.3 Manganese-oxidizing bacteria isolated and identified from different habitats

Organism	Habitat	Reference
Bacillus sp., *Caulobacter* sp., *Chromobacterium* sp., *Cytophaga* sp.	Washington Lake	Eileen and James (1982)
Arthrobacter sp., *Micrococcus* sp., *Bacillus* sp.	Desert varnish	Palmer et al. (1986)
Vibrio sp., *Enterobacteriaceae* sp., *Micrococcus* sp., *Bacillus* sp.	Indian Ocean	Chandramohan et al. (1987)
Staphylococcus sp., *Vibrio* sp., *Synechococcus* sp.	Tuticorin harbor	Maruthamuthu et al. (2002)
Acidianus brierleyi, *Thiobacillus ferrooxidans*, *Thiobacillus thiooxidans*	Marine manganese nodules	Konishii and Asai (1995)
Bacillus mesentericus, *Bacillus mycoides*, *Bacillus cereus*, *Brevibacillus centrosporus*	Karelian lakes (former USSR)	Troshanov (1969)

Table 16.4 Manganese-oxidizing fungus isolated and identified from different habitats

Organism	Habitat	Reference
Aspergillus foetidus	Lateritic ores	Tang et al. (2006)
Aspergillus niger	Polymetallic nodules of Indian Ocean	Mehta et al. (2010)
Penicillium funiculosum, Aspergillus foetidus, Penicillium simplicissimum	Australian ores	Valix et al. 2001
Penicillium citrinum	Manganese mine India	Acharya et al. (2004)
Aspergillus awamori, Aspergillus ochraceus, Cladosporium resinae, Penicillium chrysogenum, Rhizopus sp.	Bauxite ores	Ogurtsova et al. (1989)

ores, leaching of valuable metals from ocean nodules by a marine isolate, leaching of manganese from manganese ore by *Penicillium citrinum* (Acharya et al. 2001), and leaching of nickel form lateritic nickel ore by heterotrophic microorganism are reported (Sukla et al. 1993). Metal dissolution characteristic of fungal activity using *A. niger* to leach copper and zinc from an oxidized ore and treatment of domestic low-grade lateritic ore of limonitic and hematitic type by *Aspergillus* sp. and *Penicillium* were studied and are considered as a better alternative as they hold the promise (Couillard et al. 1991).

16.5 Conclusion

Manganese-associated research is heading in diverse approaches with an assorted outlook. High levels of manganese exposure in all forms have led to widespread health problems. The elevated intake of Mn is reported to cause detrimental health consequence, and in extreme amount, it is a neurotoxin where the newborn are principally at risk (Sankar et al. 2014). It is well established that maternal exposure to Mn is linked with cellular abnormalities and Mn consumption through water is connected with neurotoxic possessions in kids. Hence, new techniques with innovative advances for rapid and sensitive detection of manganese are necessary. Rapid increasing requirement for Mn and overexploitation of natural resources are resulting in a sharp exhaustion of mineral reserves. The low- to medium-grade manganese ores have not been utilized properly due to lack of efficient technology. Recently, with an improved perceptive of the mineral recovery procedure and the beginning of novel expertise, exploitation of lean-grade ores has established escalating consideration.

The current status of bioleaching in developing countries is encouraging. It is expected that in the coming years, several new commercial-scale bioleaching plants will be installed. It is likely that heap leaching will continue to be the choice for low-grade ores and tailings, while tank bioleaching technology will probably increase its application for gold, copper, nickel, and other base metal concentrates. The use of thermophilic bacteria and archaea will be a major contribution, increasing the leaching rates and metal recoveries and allowing for the treatment of recalcitrant ores. One area for research is a comprehensive study of the microbial composition of both bioheap systems and stirred-tank reactors. A heterogeneous and complex microflora, composed of acidophilic, heterotrophic, and autotrophic microorganisms, exists with commercial bioprocessing systems. The dynamics of the microbial population also change with time and conditions in the bioprocessing system. There is a need to both define and understand the potential interactions among the components of the microflora. This has the potential for improving bioleaching and mineral biooxidation through definition of how the components of the system interact to bring about bioleach processes.

This chapter describes the current information on manganese-associated ecological contamination, Mn composite-stimulated manganese toxicity and neurotoxicity, current chemical and biological techniques employed for Mn recovery, the significance of these procedure with respect to bioleaching technique, and bioleaching microorganisms. Knowledge concerning the detailed Mn species must be employed to the queries concerning the relation of Mn to the cellular understanding of Mn-induced neurotoxicity. In the current scenario, as power and ecological restriction turns into extra demand, there is a superior need to recycle mining wastes. With enhanced insightful techniques and the beginning of novel expertise, exploitation of low-grade ores has established escalating consideration. Bioremediation of huge amounts of mine tailings has achieved improved consideration since it is pioneering and cost-effective. We believe that cellular level perceptive of microorganisms concerned in manganese biorecovery will assist in improving the process greatly preferred in bioprocessing of waste ores.

Acknowledgment The authors would like to acknowledge the Department of Biotechnology (DBT), Government of India, for providing financial support [BT/PR7454/BCE/8/949/2012] for carrying out related studies.

References

Acharya C, Kar RN, Shukla LB (2001) Microbial extraction of manganese from low grade Mn-ore. Trans Indian Inst Met 54(3):99–103

Acharya C, Kar RN, Sukla LB (2003) Studies on reaction mechanism of bioleaching of manganese ore. Miner Eng 16:1027–1030

Acharya C, Kar RN, Sukla LB, Misra VN (2004) Fungal leaching of manganese ore. Trans Indian Inst Met 57(5):501–508

Achurra LE, Lacassie JP, Le Roux JP et al (2009) Manganese nodules in the Miocene Bahia Inglesa formation, north-central Chile: petrography, geochemistry, genesis and palaeoceanographic significance. Sediment Geol 217:128–139

Anderson CR, Dick GJ, Chu ML et al (2009) Aurantimonas manganoxydans, sp. nov. and Aurantimonas litoralis, sp. nov.: Mn(II) oxidizing representatives of a globally distributed clade of alpha-Proteobacteria from the order Rhizobiales. Geomicrobiol J 26:189–198

Anjum F, Shahid M, Akcil A (2012) Biohydrometallurgy techniques of low grade ores: a review on black shale. Hydrometallurgy 117–118:1–12

Azizi D, Shafaei SZ, Noaparast M, Abdollahi H (2012) Modeling and optimization of low-grade Mn bearing ore leaching using response surface methodology and central composite rotatable design. Trans Nonferrous Met Soc 22(9):2295–2305

Behera SK, Sukla LB (2012) Kinetics study for lateritic Chromite overburden leaching by organic (oxalic) acid. Elixir Pollut 53:11890–11893

Bekker A, Slack JF, Planavsky N et al (2010) Iron formation: the sedimentary product of a complex interplay among mantle, tectonic, oceanic, and biospheric processes. Econ Geol 105:467–508

Cao J-b, Li X-m, Ouyang Y-z et al (2012) Manganese-electrolysed slag treatment: bioleaching of manganese by *Fusarium* sp. Environ Technol 33(11):1307–1312

Chandramohan D, Loka Bharathi PA, Nair S, Matondkar SGP (1987) Bacteriology of ferromanganese nodules from the Indian Ocean. Geomicrobiol J 5:17–31

Cordova FM, Aguiar AS, Peres TV, Lopes MW, Gonçalves FM (2012) In vivo manganese exposure modulates Erk, Akt and Darpp-32 in the striatum of developing rats, and impairs their motor function. PLoS One 7(3), e33057

Couillard D, Chartier M, Mercier G (1991) Bacterial leaching of heavy metals from aerobic sludge. Bioresour Technol 36:293–302

Das AP, Sukla LB, Pradhan N, Nayak S (2011) Manganese biomining: a review. Bioresour Technol 102:7381–7387

Das AP, Swain S, Panda S et al (2012a) Reductive acid leaching of low grade manganese ores. Geomaterials 2:70–72

Das AP, Shukla LB, Pradhan N (2012b) Microbial recovery of manganese using staphylococcus epidermidis. Int J Nonferrous Metall 1:9–12

Dick GJ, Lee YE, Tebo BM (2006) Manganese (II)-oxidizing *Bacillus* spores in Guaymas basin hydrothermal sediments and plumes. Appl Environ Microbiol 72:3184–3190

Dobson AW, Erikson KM, Aschner M (2004) Manganese neurotoxicity. Ann N Y Acad Sci 1012:115–129

Doshi J (2007) Bioleaching of lateritic nickel ore using chemolithotrophic micro organisms (*Acidithiobacillus ferrooxidans*)

Duan N, Xin B, Chen B et al (2011) Extraction of manganese from electrolytic manganese residue by bioleaching. Bioresour Technol 102(2):1683–1687

Duka YD, Svetlana I, Ilchenko MM et al (2011) Impact of open manganese mines on the health of children dwelling in the surrounding Area. Emerg Health Threats J 4:7110. doi:10.3402/ehtj.v4i0.7110

Ehrlich HL (2002) Geomicrobiology, 4th edn. Dekker, New York

Ehrlich HL, Newman DK (2009) Geomicrobiology, 5th edn. CRC/Taylor and Francis, Boca Raton, FL

Eileen G, James TS (1982) Widespread distribution of ability to oxidize manganese among freshwater bacteria. Appl Environ Microbiol 4:509–511

Elaine B (2011) Secretariat of the Pacific Community Deep Sea Minerals: Manganese nodules, a physical, biological, environmental, and technical review, vol. 1B. Secretariat of the Pacific Community (SPC)

Fisher KG (2010) Manganese removal in base metal hydrometallurgical processes

Flynn MR, Susi P (2009) Neurological risks associated with manganese exposure from welding operations—a literature review. Int J Hyg Environ Health 212(5):459–469

Gadd GM (2010) Metals, minerals and microbes: geomicrobiology and bioremediation. Microbiology 156:609–643

Gallegos MV, Falco LR, Peluso MA et al (2013) Recovery of manganese oxides from spent alkaline and zinc–carbon batteries. An application as catalysts for VOCs elimination. Waste Manage 33:1483–1490

Ge XX, Cai GP, Zeng GM (2004) Study on harmless disposal and comprehensive utilization of manganese sulfate waste residue. Chinas Manganese Ind 22:11–14

Gerber GB, Leonard A, Hantson P (2002) Carcinogenicity, mutagenicity and teratogenicity of manganese compounds. Crit Rev Oncol Hematol 42:25–34

Han M, Zhao Z, Gao W et al (2013) Study on the factors affecting simultaneous removal of ammonia and manganese by pilot-scale biological aerated filter (BAF) for drinking water pre-treatment. Bioresour Technol 145:17–24

Hariprasad D, Dash B, Ghosh MK et al (2009) Mn recovery from medium grade ore using a waste cellulosic reductant. Indian J Chem Technol 16:322–327

Hasan HA, Abdullah SRS, Kofli NT et al (2012) Effective microbes for simultaneous bio-oxidation of ammonia and manganese in biological aerated filter system. Bioresour Technol 124:355–363

Ilyas S, Chi R, Bhatti HN et al (2012) Column bioleaching of low-grade mining ore containing high level of smithsonite, talc, sphaerocobaltite and azurite. Bioprocess Biosyst Eng 35:433–440

Johnson DB (2013) Development and application of biotechnologies in the metal mining industry. Environ Sci Pollut Res 11:7768–7776

Johnson DB (2014) Biomining-biotechnologies for extracting and recovering metals from ores and waste materials. Curr Opin Biotechnol 30:24–31

Juarez-Santillan LF, Lucho-Constantino CA, Vazquez-Rodríguez GA, Ceron-Ubilla NM, Beltran-Hernandez RI (2010) Manganese accumulation in plants of the mining zone of Hidalgo. Mexico Bioresour Technol 101:5836–5841

Kenneth P (2010) Manganese nodules and the age of the ocean floor. J Creation 24(3):82–86

Kierkegaard S (2007) EU battery directive, charging up the batteries: squeezing more capacity and power into the new EU battery directive. Comput Law Secur Rep 23:357–364

Kirschvink JL (1992) The Proterozoic biosphere: a multidisciplinary study. Cambridge University Press, Cambridge, pp 51–52. ISBN 0521366151

Kirschvink JL, Gaidos EJL, Bertani E et al (2000) Paleoproterozoic snowball earth: extreme climatic and geochemical global change and its biological consequences. Proc Natl Acad Sci USA 97:1400–1405

Konishii K, Asai S (1995) Bioleaching of marine manganese nodules by acidophilic sulfur-oxidizing bacteria. In: Proceedings of the ISOPE – Ocean mining symposium, Tsukuba, Japan, pp 21–22

Li H, Zhang Z, Tang S et al (2008) Ultrasonically assisted acid extraction of manganese from slag. Ultrason Sonochem 15:339–343

Li HL, Li AM, Cao CM et al (2009) Study on bioleaching of manganese from electrolytic manganese residue. Chin J Environ Eng 3:1667–1672

Liu H, Teng C, Cheng Y (2004) A semi-empirical model for bacterial growth and bioleaching of *Acidithiobacillus* spp. Chem Eng J 99:77–87

Liu YG, Zhou M, Zeng GM et al (2008) Bioleaching of heavy metals from mine tailings by indigenous sulfur-oxidizing bacteria: effects of substrate concentration. Bioresour Technol 99:4124–4129

Maruthamuthu S, Palanichamy S, Manickam ST, Rajendran A (2002) Microfouling of manganese-oxidizing bacteria in Tuticorin harbour waters. Curr Sci 82:7–10

Maynard JB (2010) The chemistry of manganese ores through time: a signal of increasing diversity of earth surface environments. Econ Geol 105:535–552

McArthur JM, Sikdar PK, Nath B et al (2012) Sedimentological control on Mn and other trace elements, in groundwater of the Bengal Delta. Environ Sci Technol 46:669–676

Mehta KD, Das C, Pandey BD (2010) Reductive leaching of valuable metals from Indian Ocean nodules by bacillus circulans. In: Proceedings of the XI international seminar on mineral processing technology, NML Jamshedpur, India

Menezes-Filho JA et al (2009) High levels of hair manganese in children living in the vicinity of a ferro-manganese alloy production plant. Neurotoxicology 30(6):1207–1213

Michalke B, Fernsebner K (2014) New insights into manganese toxicity and speciation. J Trace Elem Med Biol 28:106–116

Mishra PP, Mohapatra BK, Mahanta K (2009) Upgradation of low-grade siliceous manganese ore from Bonai-Keonjhar Belt, Orissa, India. J Miner Mater Char Eng 8:47–56

Nealson KH (2006) National Research Council (NRC). Recommended dietary allowances, 10th. Washington DC: National Academic Press, 1989. Manganese-oxidizing bacteria. Prokaryotes 5:222–231

Ogawa T, Bao DH, Katoh H et al (2002) A two-component signal transduction pathway regulates manganese homeostasis in Synechocystis 6803, a photosynthetic organism. J Biol Chem 277:28981–28986

Ogurtsova LV, Karavaiko GI, Avakyan ZA, Korenevsii AA (1989) Activity of various microorganisms in extracting elements from bauxite. Microbiology 58:774–780

Ordonez-Librado JL, Anaya-Martínez V, Gutierrez-Valdez AL et al (2010) Manganese inhalation as a Parkinson disease model. Parkinsons Dis 2011:612989

Ostwald J (1988) Mineralogy of the Groote Eylandt manganese oxides: a review. Ore Geol Rev 4:3–45

Pakarinen J (2006) Production of Manganese by Electrolysis. Thermo kinet pheno hydrome pro Espoo: TKK

Pakarinen J (2011) Recovery and refining of manganese as by-product from hydrometallurgical processes. ISBN 978-952-265-135-8

Perl DP, Olanow CW (2007) The neuropathology of manganese-induced Parkinsonism. J Neuropathol Exp Neurol 66:675–682

Palmer FE, Staleyj T, Murray R, Counseltl GE, Adamsj B (1986) Identification of manganese oxidizing bacteria from desert varnish. Geomicrobiol J 4:343–360

Raynal M, Pruden A (2008) Aerobic MTBE biodegradation in the presence of BTEX by two consortia under batch and semi-batch conditions. Biodegradation 19:269–282

Riojas-Rodríguez H et al (2010) Intellectual function in Mexican children living in a mining area and environmentally exposed to manganese. Environ Health Perspect 118(10):1465–1470

Roth JA (2006) Homeostatic and toxic mechanisms regulating manganese uptake, retention, and elimination. Biol Res 39:45–57

Roth JA (2009) Are there common biochemical and molecular mechanisms controlling manganism and Parkinsonism. Neuromolecular Med 11:281–296

Sadek AH, Rauch R, Schultz PE (2003) Parkinsonism due to manganism in a Welder. Int J Toxicol 22:393–401

Sankar MS, Vega MA, Defoe PP et al (2014) Elevated arsenic and manganese in groundwaters of Murshidabad, West Bengal, India. Sci Environ 488–489:570–579

Santamaria AB (2008) Manganese exposure, essentiality & toxicity. Indian J Med Res 28:484–500

Sayilgan E, Kukrer T, Civelekoglu G et al (2009) A review of technologies for the recovery of metals from spent alkaline and zinc–carbon batteries. Hydrometallurgy 97:158–166

Spiro TG, Bargar JR, Sposito G, Tebo BM (2010) Bacteriogenic manganese oxides. Acc Chem Res 43:2–9

Sriram K, Lina GX, Jeffersona AM (2012) Manganese accumulation in nail clippings as a biomarker of welding fume exposure and neurotoxicity. Toxicology 291:73–82

Sukla LB, Panchanadikar VV, Kar RN (1993) Microbial leaching of lateritic nickel ore. World J Microbiol Biotechnol 9:255–257

Tang L, Li X, Li L, Mu G, Liu G (2006) The effect of 1-(2-pyridylazo)-2-naphthol on the corrosion of cold rolled steel in acid media: Part 2: Inhibitive action in 0.5 M sulfuric acid. Mater Chem Phys 97:301

Tebo BM, Johnson HA, McCarthy JK et al (2005) Geomicrobiology of manganese (II) oxidation. Trends Microbiol 13:421–428

Templeton AS, Staudigel H, Tebo BM (2005) Diverse Mn (II)-oxidizing bacteria isolated from submarine basalts at Loihi Seamount. Geomicrobiol J 22:127–139

Troshanov EP (1969) Conditions affecting the reduction of iron and manganese by bacteria in the ore-bearing lakes of the Karelian Isthmus. Microbiology 38:528–535

Trost WR (1958) The chemistry of manganese deposits. Mines Branch Res Rep R8. Department of Mines and Technical Surveys, Ottawa, Canada

Venugopal R (2010) Solid waste processing for industrial utilization—a few case studies. In: Proceedings of the XI international seminar on mineral processing technology. http://eprints.nmlindia.org/2666

Viana GF, de Carvalho CF et al (2014) Noninvasive biomarkers of manganese exposure and neuropsychological effects in environmentally exposed adults in Brazil. Toxicol Lett 231 (2):169–178, PMID: 24992226

Valix M, Usai F, Malik R (2001) Fungal bioleaching of low grade laterite ores. Miner Eng 14 (2):197–203

Wang W, Shao Z, Liu Y et al (2009) Removal of multi-heavy metals using biogenic manganese oxides generated by a deep sea sedimentary bacterium – Brachybacterium sp. Strain Mn32. Microbiology 155(6):1986–1996

Wei Z, Hiller S, Gadd GM (2012) Biotransformation of manganese oxides by fungi: solubilization and production of manganese oxalate biominerals. Environ Microbiol 14(7):1744–1753

Xin B, Chen B, Duan N et al (2011) Extraction of manganese from electrolytic manganese residue by bioleaching. Bioresour Technol 102(2):1683–1687

Xin B, Jiang W, Li X et al (2012) Analysis of reasons for decline of bioleaching efficiency of spent Zn–Mn batteries at high pulp densities and exploration measure for improving performance. Bioresour Technol 112:186–192

Yokel RA (2009) Manganese flux across the blood-brain barrier. Neuromolecular Med 11:297–310

Youcai L, Qingquan L, Lifeng L et al (2014) Study on hydrometallurgical process and kinetics of manganese extraction from low-grade manganese carbonate ores. Int J Mining Sci Technol 24:567–571

Zhang JS (2007) Current challenge and chance in China's Mn-industry. China Manganese Ind 1:6–9

Zhang YB, You ZX, Li GH et al (2013) Manganese extraction by sulfur-based reduction roasting-acid leaching from low-grade manganese oxide ores. Hydrometallurgy 133(1):126–132

Zhu G, Zhao Y, Cheng Z (2010) Thermal analysis and kinetic modeling of manganese oxide ore reduction using biomass straw as reductant. Hydrometallurgy 105:96–102

Zoni S, Bonettia G, Lucchinia R (2012) Olfactory functions at the intersection between environmental exposure to manganese and Parkinsonism. J Trace Elem Med Biol 26:179–182

Index

L.B. Sukla et al. (eds.), *Environmental Microbial Biotechnology*, Soil Biology 45,
DOI 10.1007/978-3-319-19018-1

L

M

N

MIX
Papier aus verantwortungsvollen Quellen
Paper from responsible sources
FSC® C105338

If you have any concerns about our products,
you can contact us on
ProductSafety@springernature.com

In case Publisher is established outside the EU,
the EU authorized representative is:
Springer Nature Customer Service Center GmbH
Europaplatz 3, 69115 Heidelberg, Germany

Printed by Libri Plureos GmbH
in Hamburg, Germany